MW01492170

Global Perspectives on Health Geography

Series Editor

Valorie Crooks, Department of Geography, Simon Fraser University,
Burnaby, Canada

Global Perspectives on Health Geography showcases cutting-edge health geography research that addresses pressing, contemporary aspects of the health-place interface. The bi-directional influence between health and place has been acknowledged for centuries, and understanding traditional and contemporary aspects of this connection is at the core of the discipline of health geography. Health geographers, for example, have: shown the complex ways in which places influence and directly impact our health; documented how and why we seek specific spaces to improve our wellbeing; and revealed how policies and practices across multiple scales affect health care delivery and receipt.

The series publishes a comprehensive portfolio of monographs and edited volumes that document the latest research in this important discipline. Proposals are accepted across a broad and ever-developing swath of topics as diverse as the discipline of health geography itself, including transnational health mobilities, experiential accounts of health and wellbeing, global-local health policies and practices, mHealth, environmental health (in)equity, theoretical approaches, and emerging spatial technologies as they relate to health and health services. Volumes in this series draw forth new methods, ways of thinking, and approaches to examining spatial and place-based aspects of health and health care across scales. They also weave together connections between health geography and other health and social science disciplines, and in doing so highlight the importance of spatial thinking.

Dr. Valorie Crooks (Simon Fraser University, crooks@sfu.ca) is the Series Editor of Global Perspectives on Health Geography. An author/editor questionnaire and book proposal form can be obtained from Publishing Editor Zachary Romano (zachary.romano@springer.com).

Marynia A. Kolak • Imelda K. Moise

Editors

Place and the Social-Spatial Determinants of Health

Editors
Marynia A. Kolak
Department of Geography & Geographic
Information Science
University of Illinois Urbana-Champaign
Urbana, IL, USA

Imelda K. Moise
Department of Medical Education
Dr. Kiran C. Patel College of Allopathic
Medicine
Nova Southeastern University
Fort Lauderdale, FL, USA

ISSN 2522-8005   ISSN 2522-8013   (electronic)
Global Perspectives on Health Geography
ISBN 978-3-031-88462-7   ISBN 978-3-031-88463-4   (eBook)
https://doi.org/10.1007/978-3-031-88463-4

This work was supported by University of Illinois Urbana-Champaign, University of Miami, University of British Columbia, Gothenburg University and Brown University.

Cover Photograph: Paulina Arias Caballero. Mural by Héctor Duarte of the first building acquired in 2020 by the Pilsen Housing Cooperative (PIHCO), January 3, 2025.

This Springer imprint is published by the registered company Springer Nature Switzerland AG
The registered company address is: Gewerbestrasse 11, 6330 Cham, Switzerland

If disposing of this product, please recycle the paper.

Preface

This book addresses the urgent need to consider the social determinants of health (SDoH) to improve health outcomes, reduce costs, and promote health equity. It explores how incorporating a spatial perspective enriches the concept of SDoH, lending to the notion of social-spatial determinants of health (SSDoH). Given the multidimensional and intersecting nature of health determinants, a spatial perspective is crucial for defining, measuring, understanding, and operationalizing SDoH. Structural factors that influence population health patterns often leave spatial footprints, such as the legacies of redlining or the placement of vulnerable groups near polluted locales. Without an explicit spatial approach, analysis can be biased, misinterpreted, or even impossible. Adopting a geographic view is essential for SDoH studies theoretically, analytically, and technologically. *Place and the Social-Spatial Determinants of Health* brings together geographers, sociologists, clinicians, public health researchers, architects, data scientists, and thinkers to provide rich discussions on the current state of the field and visions for the future.

Urbana, IL, USA   Marynia A. Kolak
Fort Lauderdale, FL, USA   Imelda K. Moise

Acknowledgements

We, the editors, express our heartfelt appreciation to all contributors for their invaluable time and expertise as well as their enduring commitment to the successful completion of this book. A special thanks to the various anonymous reviewers whose critical insights on the book's content have been indispensable. To these individuals, we offer our sincere gratitude and the hope that the finished work meets or even exceeds their expectations. We are also grateful to Kamaria Barronville, the lead editorial assistant for this book project. Kamaria is grateful to Drs. Kolak and Moise for extending the offer to join them on this journey. She is also thankful to her husband and son who inspire her daily.

Furthermore, we are grateful for constructive comments shared by Dr. Valorie Crooks, especially in early stages of the process, as well as the health geography community who shared feedback across multiple stages of this book generation, especially at the International Medical Geography Symposium in Edinburgh in 2021 and the American Association of Geographers at Honolulu in 2024.

Dr. Marynia A. Kolak was supported in part by the SDOH & Place Project by the Robert Wood Johnson Foundation for this work. Dr. Kolak is grateful to her family for their support during the long editing days, especially son Dante Kolak and his grandparents James and Czeslawa Kolak, as well as Kicia Churchill Kolak for her lap support. They are also thankful for support from colleagues at the Healthy Regions & Policies Lab for ongoing discussions, engagement, and work in support of this book project. We are also grateful to students in Dr. Imelda K. Moise's Geography and Inequalities Lab (GaIL) at the University of Miami who helped with editing the different book chapters; Abigail Lin Adera and Lola Whittingham-Ortiz.

To make this text accessible to all communities, we are also grateful to the support by Dr. Avery Everhart (via the University of British Columbia in Vancouver, Canada), Dr. Qinyun Lin (via the School of Public Health and Community Medicine, University of Gothenburg in Gothenburg, Sweden), Dr. Diana S. Grigsby-Toussaint, as well as Dean Leonidas Bachas, College of Arts & Sciences at the University of Miami, and the Robert Wood Johnson Foundation.

Finally, we extend our deepest gratitude to Zachary Romano from Springer Nature Publishers for his pivotal role in accelerating the publication process of this book and for his collaborative spirit. His contribution was essential to the realization of this work. Additionally, we are immensely thankful to the committed and seasoned team at Springer Nature Publishers, whose efforts ensured the prompt production of this book. Our thanks go to each staff member for maintaining an exceptionally high level of professionalism throughout the book's production.

Contents

About the Authors

Elizabeth Ackert, Ph.D. is an Associate Professor of Geography at the University of California, Santa Barbara. She is a social demographer with expertise in population geography, immigration, urban geography, and health geography. Her research focuses on racial/ethnic and immigration-related variation in exposure to different types of contexts (schools, neighborhoods, and communities) and the consequences of that differential exposure for outcomes in domains such as education, health, development, and well-being. Dr. Ackert's research has been funded by NIH and NSF, and her work has been published in journals including Health Affairs, Demography, Journal of Marriage and Family, and Social Science Research.

Eileen E. Avery, Ph.D. is an Associate Professor of Sociology at the University of Missouri. Her research focuses on residential mobility, neighborhood effects, health and well-being, and social control in urban and rural contexts. Recent and ongoing projects examine rural gentrification, perceptions of social cohesion as it relates to victimization distress, fear, experiences with police, and health and income inequalities.

Madison Avila completed their BA in Sociology and an MA in Demographic and Social Analysis at UC Irvine. Their research interests pertain to gender and work, focusing on women in traditionally male-dominated occupations. More specifically, they research women employed in STEM professions. In the future, Avila wants to research women's retention within STEM occupations and how organizational culture, division of household labor, and parenthood may impact women's persistence in these professions.

Kamaria Barronville, Ed.D., M.F.A. is a Senior Research Assistant at the Healthy Regions and Policies Lab at the University of Illinois Urbana-Champaign. Kamaria is an educator and researcher with a rich background in urban educational leadership and social justice advocacy. She received her EdD in Urban Educational Leadership at Morgan State University, where her research focused on historical trauma. Kamaria's academic journey includes an MFA in Creative Writing from Full Sail University and a BA in English from Florida Atlantic University.

Susan Cassels, Ph.D.   is a Professor in the Department of Geography and Director of the Broom Center for Demography at the University of California, Santa Barbara. She studies and teaches topics broadly related to health geography, demography, and social epidemiology. The central focus of her current research is on geographic mobility, sexual health, and HIV prevention among sexual and gender minorities living in Los Angeles. Specifically, she is interested in differential exposures to social and physical environments and how those exposures align with HIV and substance use risk.

Brianna Chan   is interested in using geospatial methods to analyze the role of the natural and built environment in driving health and social disparities. Ultimately, she aims to help heal the relationships between humans and their environments and further promote equitable access to better health in our communities. Brianna received her bachelor's degree in environmental health from the Gillings School of Global Public Health at UNC-Chapel Hill and is currently a Ph.D. student in Geography at the University of California, Santa Barbara, where she is working on research related to population health, neighborhood effects, and spatial equity.

Anjali Choudhury   is a Master's student at the University of Washington in Seattle. She is passionate about improving access to safe water and sanitation in underserved communities. She also worked as a research assistant in Dr. Moise's Geography and Inequalities Lab (GaIL) on an ongoing research project investigating the impact of onsite sewer failure and SDOH on ecosystem services. Because of her work on this project, she was awarded the summer of 2023 Beyond the Book fellowship, an award for research-based learning through the College of Arts and Sciences.

Donna Rooney, Ed.D.   is a Senior Lecturer at the University of Technology Sydney (UTS) where the primary focus of her research revolves around adult learning. She draws from a range of conceptual resources, including socio-material, public pedagogy, ecological, and practice-based theories. Donna's research has involved studies of adult learning in various spheres, spanning from public spaces, community settings, and workplaces to public institutions and higher education. She has been the recipient of several research awards including a university medal and an industry award for novel methodologies. Donna's research interests inform her teaching practices where she is a passionate educator involved in curriculum design and teaching postgraduate coursework in a Master of Education (Learning and Leadership).

Avery R. Everhart, Ph.D.   is a socio-spatial scientist and theorist whose work has been published in a diverse array of fields that range from biomedical informatics to feminist philosophy. Her latest research considers how transgender and other marginalized communities navigate the complex and overlapping health and legal systems. She uses geographic information systems and spatial analysis to quantify both patterns of access to care and structural and system inequities that shape where care is available. Dr. Everhart is an Assistant Professor of Geography at the University of British Columbia, which is situated on the unceded territories of the xwməθkwəy̓əm (Musqueam) people. She is also Co-founder and Distinguished Fellow of the Center for Applied Transgender Studies.

Praveena K. Fernes is a 2020 Marshall Scholar in the UK, where she is a Ph.D. candidate at the London School of Hygiene & Tropical Medicine. Her current research explores place-based experiences of people who are homeless and seek drug and alcohol services in London, with a special focus on relations of access. Her work strives to advance health equity through transdisciplinary research and using narrative as a tool for change.

Lizandra Garcia Lupi Vergara, M.Sc. is an architect, urban planner, and occupational safety engineer in the area of ergonomics by production engineering at the Federal University of Santa Catarina and Professor at UFSC in Undergraduate and Graduate Studies in Production Engineering and Architecture (PosARQ). She is Vice President of the Brazilian Ergonomics Association, Supervisor of the Ergonomics Laboratory, and leader of the research group *GMETTA* (Multidisciplinary Group of Work Ergonomics and Applied Technologies). She is a Senior Ergonomist Certified by ABERGO, Coordinator of the Technical Group—GT of Ergonomics of the Built Environment and Accessibility—and Member of the Editorial Board of IJIE Magazine (Iberoamerican Journal of Industrial Engineering). Her research interests are human factors in health and safety, accessibility, lean ergonomics, user experience, assistive technology, and healthy aging.

Daniel Grafton is a geographer (Ph.D. candidate) whose research focuses on Arctic tourism and the part climate change plays in shaping tourist experiences. He is learning and investigating the dynamics of stakeholder perceptions of moral responsibility for climate change and role of last-chance tourism in rural areas. Other research interests include scientific tourism, historical geography, cinematic cartographies, feminist theory, and gentle geographies.

Gregory Nick Gibson is a medical student at the Warren Alpert Medical School of Brown University. He earned dual BAs in Sociology and Science, Technology, and Society (STS) from Brown University. His research interests center on racial, spatial, and socioeconomic health disparities, with particular focus on women's health and HIV/AIDS prevention.

Diana S. Grigsby-Toussaint, Ph.D. is a tenured Associate Professor of Behavioral and Social Sciences and Epidemiology at the Brown University School of Public Health. As a social epidemiologist, her work is grounded primarily in theoretical approaches from epidemiology, nutrition, and geography. She is particularly interested in working with vulnerable and racial/ethnic populations across various stages of human development. In addition to the NIH, Dr. Grigsby-Toussaint's research has been supported by the Robert Wood Johnson Foundation, the United States Department of Agriculture, and the National Science Foundation, and her work has been featured in the Huffington Post, the Dallas Morning News, and the Chicago Tribune.

Annemarie G. Hirsch, Ph.D., M.P.H. is an epidemiologist and Professor in the Department of Population Health Sciences at Geisinger, an integrated health system in Pennsylvania. With more than 15 years of experience working with electronic health record (EHR) data, Dr. Hirsch has studied novel applications of health system data for epidemiologic and health service research. As the director of the Geisinger-Johns Hopkins Bloomberg School of Public Health Center for Community Environment and Health, her research focuses on how the social, built, and natural environments impact health. Dr. Hirsch's research is funded by the National Institutes of Health and the Centers for Disease Control and Prevention.

Nick Hopwood, Ph.D., M.D. (Honoris Causa) is Professor of Professional Learning at the University of Technology Sydney (UTS). Nick's research is grounded in the discipline of education, applied in diverse contexts including schools, families, and health services. He teaches in postgraduate coursework Master of Education (Learning and Leadership) as well as contributes to research education and researcher development across UTS. Nick is an active mentor of early and mid-career colleagues and is on the Executive of the International Society for Cultural-historical Activity Research (ISCAR). He co-edits Studies in Continuing Education and is on the editorial board for Mind, Culture, and Activity and the African Journal of Education Studies. Nick was awarded an Honorary Doctor of Medicine from the University of Linköping (Sweden) in 2019 and appointed as Extraordinary Professor at the University of Stellenbosch since 2016

Gabrielle Husted is a Ph.D. student in the Geography Department at UCSB. She earned a Bachelor of Science in nursing from the University of Portland and a master's in Geography from UCSB in 2023. In between her time at the University of Portland and beginning her studies at UCSB, she worked in oncology and public health. Now, Gabrielle investigates the interactivity of people, places, and environment (built and natural) as they relate to public health challenges in our society. Of particular interest are social determinants of health, environmental hazards, and health outcomes. She is working with the Population Health in Geography (PHiG) lab, and her current projects include exploring environmental exposures and geosocial determinants of health as they impact development and prevalence of Alzheimer's disease and related dementias (ADRD).

Ayodeji Iyanda, Ph.D. is a community health geographer specializing in health disparities from a geographic perspective. With a doctorate in Geographic Information Science (GIS) and a multidisciplinary background, Dr. Iyanda has contributed significantly to the field of health and medical geography. His research on spatial social determinants of health showcases his expertise in geospatial health analysis, emphasizing a commitment to interdisciplinary collaboration and advancing our understanding of health challenges through innovative spatial approaches. This work reflects Dr. Iyanda's dedication to revealing the impact of historical place-based discrimination on health and his passion for addressing pressing societal issues.

Sofia Kaloper is a first-year geography MA/Ph.D. student at the University of California, Santa Barbara. Her broad research interests are in population and health geography, social determinants of health, spatial analysis, spatial statistics, and location-based health interventions. Her current work focuses on quantifying uncertainty in HIV-risk hotspots.

Trace Kershaw, Ph.D. focuses on the social and structural determinants of health among adolescents and emerging adults. His current focus is using innovative technological methods to understand how social and geographic context influences their behaviors and health. He is the Chair of the Department of Social and Behavioral Sciences, the Director of two HIV training grants (Yale AIDS Prevention Training, Research Education Institute for Diverse Scholars), and the Director of the Center for Interdisciplinary Research on AIDS (CIRA).

Danielle C. Kuhl, Ph.D. is an Associate Professor of Sociology at Bowling Green State University. Her research areas focus on neighborhood stratification, life course sociology and victimology, health, and social cohesion. Recent scholarship examines geographic disparities in social processes that influence well-being, place-based influences on substance use and mortality, and developmental mechanisms that link victimization to adult relationships and violence.

Katherine A. Lester, Ph.D. is a medical geographer and spatial statistician with a concentration on mental and behavioral health. Her recent research investigates the spatial relationship between suicide, mental illness, and race and ethnicity. Lester also explores GIS applications to spatial patterns of disease and health. At the University of Southern California, Lester teaches several courses within the Spatial Sciences Institute including maps and spatial reasoning, statistics for the spatial sciences, and introduction to spatial analytics.

Qinyun Lin, Ph.D. is a Senior Lecturer in the School of Public Health and Community Medicine, University of Gothenburg. Most broadly, she is a methodologist invested in humanizing quantitative research to better understand mechanisms underlying inequities, particularly health and educational disparities, and, ultimately, informing policymaking. Towards this, she seeks to advance and expand existing quantitative methods and the interpretation of inferences by incorporating human interactions, social and spatial contexts, and mediating factors within these analyses.

Jailos Lubinda, Ph.D. is a Senior Research Officer at Malaria Atlas Project, Telethon Kids Institute, Western Australia. His research focuses on socio-ecological, eco-epidemiological, and geospatial disease modeling of high-resolution global malaria morbidity and mortality estimates for the World Malaria Report and the Global Burden of Disease Study. He supports fine-scale disease risk mapping for operational and public health decision-making, aiding efficient resource use through targeted interventions. Other interests include modeling arboviruses, malaria, COVID-19, and health system strengthening. His collaborations cut across applied natural, social, and global health approaches to inform intervention coverage, uptake/use, and other socioeconomic and demographic factors driving disease.

Hannah Malak is an MA/Ph.D. student in the Geography Department at the University of California, Santa Barbara. She is broadly interested in the spatial variation of health and healthcare access among vulnerable populations, especially among immigrants, people of color, children, and the elderly. In her future research, she hopes to investigate how sociodemographic and environmental factors impact climate vulnerability and resilience in the context of health outcomes and mortality. Hannah is also interested in open science and science communication. She is particularly passionate about improving the accessibility of scientific knowledge and knowledge production in the discipline of geography.

Ben Moscona is interested in environmental and climate solutions designed with culture and people in mind. Their recent research work is focused on experimental economics in agricultural settings using remote sensing. Moscona currently works on monitoring, reporting, and verification (MRV) of methane emissions.

Oliver Mweemba, Ph.D. is a Lecturer, Researcher, and Head of the Department of Health Promotion and Education at the University of Zambia's School of Public Health. He has over 10 years of research experience in applying socio-behavioral theories and methods to complex public health, health promotion, and healthcare interventions in Africa. He has been the Principal Investigator and Co-investigator on US NIH, Canadian IDRC, and Wellcome Trust-funded projects conducted in Zambia. He has published over 60 peer-reviewed journal articles in international journals and has served on the editorial boards of *Health Promotion International Journal, Critical Public Health Journal, and BMC Women's Health Journal.*

Bruna Luísa Poffo Nobre is a civil production engineer at the Federal University of Santa Catarina. She is a member of the Multidisciplinary Group of Work Ergonomics and Applied Technologies and the Ergonomics Laboratory. Her research focuses on promoting the well-being of individuals in their work environments through the analysis of processes, data, and human factors.

Ijeoma Opara, Ph.D., L.M.S.W., M.P.H. is an Associate Professor of Public Health in the Department of Social and Behavioral Sciences at Yale School of Public Health. She is also the Founder and Director of the Substances and Sexual Health

(SASH) Lab and a Co-director of the Yale AIDS Prevention Training Program (Y-APT). Her research interests focus on HIV/AIDS, STI, and substance use prevention for urban youth, racial and gender-specific prevention interventions for Black girls, and community-based participatory research with urban youth.

Joseph R. Oppong, Ph.D. is a medical geographer with a geographic focus on Africa and North America. His research interests include social vulnerability, neighborhood characteristics and HIV/AIDS, emerging diseases and health challenges in Africa, and applications of GIS to understanding spatial patterns of disease and health. At the University of North Texas, Oppong teaches a variety of courses including quantitative methods in geography, medical geography, and contemporary sub-Saharan Africa. Oppong is also a Fulbright Scholar, mentoring graduate faculty and establishing student support-focused graduate programs in Ghana.

Widya A. Ramadhani, Ph.D., M.Arch. is a design researcher at Perkins Eastman. She holds a BArch from Universitas Indonesia and an MArch and Ph.D. in architecture from the University of Illinois Urbana-Champaign. Her research, teaching, and creative practices are centered on person–environment transactions to navigate the changes across the life course. She is particularly interested in understanding the physical, social, and cultural factors to inform the design of the built environment to support the health and well-being of older adults in achieving successful aging in place.

Sean C. Reid is a Ph.D. student in the Geography Department at the University of California, Santa Barbara. Sean's research interests are broadly in urban and population dynamics, health geography, and demography. His current research is focused on migration of sexual and gender minorities in the United States and contextual factors that influence their health outcomes. Sean received his BS in Geography from the University of Utah focusing on GIS and remote sensing. He received his MA in Geography at the University of California with a focus on migration and social determinants of health among sexual and gender minorities.

Tim Rhodes, Ph.D. is Professor of Public Health Sociology with expertise in qualitative research at the London School of Hygiene and Tropical Medicine (UK) and University of New South Wales (Australia). He uses qualitative research methods and analyses of narrative to study evidence-making and intervention in the field of health, linked to epidemics, drug use, and addictions.

Wendy A. Rogers, Ph.D. is Shahid and Ann Carlson Khan Professor of Applied Health Sciences at the University of Illinois Urbana-Champaign. Her primary appointment is in the Department of Kinesiology and Community Health. She has an appointment in Educational Psychology and is an affiliate of the Beckman Institute, Illinois Informatics Institute, Center for Social and Behavioral Science, and Discovery Partners Institute. She received her BA from the University of Massachusetts, Dartmouth, and her MS (1989) and Ph.D. (1991) from the Georgia

Institute of Technology. She is a Certified Human Factors Professional (BCPE Certificate #1539). Her research interests include design for aging, technology acceptance, human-automation interaction, aging in place, human-robot interaction, aging with disabilities, cognitive aging, skill acquisition, and training. She is the Director of the McKechnie Family LIFE Home and the Health Technology Education Program, Program Director of CHART (Collaborations in Health, Aging, Research, and Technology), and Director of the Human Factors and Aging Laboratory.

Bryce Puesta Takenaka, M.A., M.P.H.   is a PhD Candidate in the Department of Social and Behavioral Sciences at Yale School of Public Health, and a research fellow in the Yale School of Medicine, Center for Clinical Investigation. His research interests grapple with the issues of settler colonialism, racial capitalism, militarization, carceral infrastructures, and environmental justice-related impacts on health outcomes. He leans into community-driven approaches and transnational epistemologies to inform participatory and radical spatial practices for alternatives to state-sanctioned violence. Bryce earned a Master of Arts in History of Science and Medicine from Yale University, a Master of Public Health in Epidemiology from Saint Louis University, and a Bachelor of Science in Public Health from Lindenwood University.

Juliana Tasca Tissot, Ph.D.   is a Professor in the Faculty of Architecture and Urbanism at the Federal University of Pelotas, Brazil. She holds a BArch from Paranaense University, Brazil, MS and Ph.D. in architecture from the Federal University of Santa Catarina, Brazil. Her research, teaching, and practices are centered on the person–environment. She is interested in human factors and the design of physical environments that promote healthy aging and support the well-being of older adults. She is a specialist in aging in place certified by the National Association of Home Builders.

Brooke Ury, MPH   is a recent graduate of Brown Univeristy, where she received her Bachelor's in Computer Science and Masters in Public Health. As a Fulbright Open Research Fellow and researcher, her work focuses on the impact of the social and built environment on health, including green space and meteorological exposures. In the future, she plans to utilize her technical skills and experience in computer science to advance public health research in environmental health and health equity.

Sigrid Van Den Abbeele   is a Ph.D. student in the Geography Department at the University of California, Santa Barbara. She is interested in applying statistical methods to study how systemic inequality impacts health and healthcare access at various spatial scales. Recently, she investigated how factors like residential segregation, population composition, and immigrant destination typology relate to access to safety net primary care facilities. Moving forward, she hopes to examine how changes in policy and population composition impact healthcare access. Sigrid is also interested in the scholarship of teaching and learning, specifically teaching practices that create more inclusive learning environments.

Esaú Casimiro Vieyra   is a second-year geography Ph.D. student at the University of California, Santa Barbara. His research interests lie at the intersection of Latino/a/x immigration, health, policy, and spatial data science. Esau's current work looks at spatial patterns and demographic characteristic variations in origin-destination dyads in Mexico-US migration flows. Prior to joining UC Santa Barbara, Esau received a BA in political science from CSU Bakersfield and a master's in public policy from UC Riverside.

Ran Xu, Ph.D.   is an applied statistician and Associate Professor in the Department of Allied Health Sciences, University of Connecticut. His methodological research interests include applied data science and systems science. He has developed new statistical and simulation tools for longitudinal analysis, contagion effects in social networks, and implementation science within the organizational context. He also has a broad substantive health-related research interest in disease prevention and health promotion. His works have appeared in high-impact peer-review journals such as The Lancet Planetary Health, Nature Communications, Health Affairs, and Organizational Research Methods.

Yvonne Young, Ph.D., M.Hum.   recently completed their doctoral studies at the University of Technology, Sydney, Australia. Her professional experience as an educator has led to a sustained interest in supporting families with young children. In 2018, she was awarded the Vice Chancellor's Research Scholarship. Her research focuses on families with young children living in adverse circumstances who need to access multiple services and how that can be accomplished in a way which meets the needs of these families.

About the Editors

Marynia A. Kolak, Ph.D., M.F.A., M.S. is a health geographer and spatial epidemiologist integrating a socio-ecological view of health, spatial data science, and a human-centered design approach to investigate regional and neighborhood health equity. Their research centers on how "place" impacts health outcomes in different ways, for different people, from opioid-risk environments to chronic disease clusters. Kolak is interested in defining the mechanisms and methodological definitions of social-spatial determinants of health and building communities of practice to move the science and policy on SDoH forward. Building on their experience as the Principal Investigator of the US Covid Atlas and policy work in decision support development, Kolak's current research seeks to define and understand multidimensional access to health resources over time as well as develop the next generation of conceptually grounded, participatory web applications for community health. Kolak is an Assistant Professor at the Department of Geography and GIScience at the University of Illinois at Urbana-Champaign; Academic Director of the *Healthy Regions and Policies Lab*; Visiting Fellow at the Center for Spatial Data Science at the University of Chicago; Associate Editor at *Preventing Chronic Disease, Journal of Maps, and Cartography and Geographic Information Science*; and a member of the DEI Committee at the Society for Epidemiologic Research. She is also the Vice Chair of the *Health and Medical Geography* Specialty Group at the American Association of Geographers (AAG) and a Lincoln Excellence Scholar from 2024 through 2026, receiving the *Emerging Scholar Award* in Health Geography at AAG in 2022.

Imelda K. Moise, Ph.D., M.P.H. is an applied health geographer and monitoring and evaluation (M&E) expert whose scholarly contributions span health geography, population health, well-being and applied spatial statistical methods. Her current research focuses on generating evidence for precision public policy for families and communities at risk, understanding human behavior in infectious and vector-borne disease transmission, and leveraging to reduce health inequities at the individual, community, and health system scales. She is a recipient of a diverse array of honors for her research and mentorship, such as the College of Arts & Sciences' Scholarly and Creative Recognition Award, a series of Provost Awards from 2020 through 2023, and the University of Miami College of Arts & Sciences' Gabelli Senior Scholar Award for her cross-disciplinary contributions. Her mentorship excellence has been acknowledged with the Luis Glaser Mentorship Award and the James W. McLamore Social Science Award. Her expertise in culturally responsive evaluation theory and practice has been recognized with a fellowship from the American Evaluation Association. As a Fulbright Specialist, her contributions to academia have been further celebrated, and she has been a contender for the University of Miami Graduate School Faculty Mentor Award in both 2019 and 2020. Prior to joining the faculty at the University of Miami, Dr. Moise supported USAID-funded health programs in low-middle-income countries on various scopes of work, spent 5 years in Illinois coordinating federally funded research projects and program evaluation for state agency initiatives and ongoing programs, and spent 6 years as a Peace Corps technical trainer in Zambia. Moise is the Editorial Board Member of the *Annals of the American Association of Geographers (the Annals)* where she assists with the Geographic Methods section, an Academic Editor for *PLOS One*, and winner of the Best Paper Award from the *Journal of Map and Geography Libraries*.

List of Figures

List of Tables

Part I
Introduction

Chapter 1
Introducing the Social-Spatial Determinants of Health

Imelda K. Moise and Marynia A. Kolak

The Challenge of Social Determinants of Health

This book addresses the growing call to action on the social determinants of health (SDoH) to improve health outcomes, reduce costs, and promote health equity. It explores how a spatial perspective influences, expands, and enriches the concept of SDoH, potentially redefining it as the social-spatial determinants of health (SSDoH). Given the multidimensional and intersecting nature of health determinants, a spatial perspective is crucial for defining, measuring, understanding, and operationalizing SDoH (Kolak et al., 2020; Moise, 2020). Structural factors that influence population health often leave spatial footprints, such as the legacies of redlining or the placement of vulnerable groups near polluted areas. Without an explicit spatial approach, analyses can be biased, misinterpreted, or even impossible without spatial infrastructure. Therefore, adopting a geographic view is essential for SDoH studies in theoretical, analytical, and technological contexts. The scale of this challenge demands an extraordinary response. This prompted us to assemble a leading interdisciplinary group of scholars to explore these issues and answer key questions:

I. K. Moise (✉)
Department of Medical Education, Dr. Kiran C. Patel College of Allopathic Medicine (NSU MD), Nova Southeastern University, Fort Lauderdale, FL, USA
e-mail: imoise@nova.edu

M. A. Kolak
Department of Geography & Geographic Information Science, University of Illinois Urbana-Champaign, Urbana, IL, USA
e-mail: mkolak@illinois.edu

© The Author(s) 2026
M. A. Kolak, I. K. Moise (eds.), *Place and the Social-Spatial Determinants of Health*, Global Perspectives on Health Geography,
https://doi.org/10.1007/978-3-031-88463-4_1

What is the spatial perspective of SDoH? How have social scientists like geographers, epidemiologists, and sociologists measured, defined, understood, and used SDoH to advance health and health equity? Are the spatial structural factors that drive, reinforce, and perpetuate health outcomes significant enough to consider the SSDoH, rather than just the social?

To effectively improve health equity and provide more patient-centered care for the world's growing population of 8 billion, the health community, scholars, funders, and policymakers are addressing challenges in capturing and using SDoH data in both clinical and social contexts. They are also tackling issues that hinder the formation of strong cross-sector data-sharing partnerships. A spatial perspective rooted in place considers the influence of structural determinants on place-based policies, resource allocation, and spatially heterogeneous public health effects. This topic requires immediate attention due to the increasing burden and worsening inequities in some health outcomes, despite decades of efforts to change individual behaviors. It is crucial to use existing data effectively (Harrison & Dean, 2011; Frieden, 2011; Keppel et al., 2002).

The need to address SDoH and health disparities in public health research, policy, and practice cannot be overstated. *First*, the link between social, political, economic, and environmental conditions, context, and health inequities is increasingly evident (McKinlay, 1979; Göran & Whitehead, 1991; Marmot et al., 2008). *Second*, policy initiatives, programs, and investments aimed at addressing SDoH and improving the health of vulnerable populations are proliferating (Braveman & Gottlieb, 2014; National Academies of Sciences, 2017; Bradley et al., 2016; Thornton et al., 2016; Williams & Mohammed, 2013). Evidence shows that concentrated poverty disproportionately affects racial and ethnic minorities across all SDoH (Noonan et al., 2016). For example, extensive research has shown that systemic racism seeks to ensure that Black people continue to live in poverty, impacting their health and ability to access health care.

Recent studies have shown that SDoH disproportionately affect Black people, contributing to racial health disparities during the COVID-19 pandemic (Dalsania et al., 2021). Data from the Centers for Disease Control and Prevention (CDC) and other research indicate that Black communities experience higher disease burden and complications, resulting in increased rates of morbidity and mortality compared to White (Walker et al., 2016). The CDC Health Disparities and Equities Report consistently highlights that Black and other communities of color face greater health disparities across all areas compared to their White counterparts. Despite the growing number of publications in this area, which show how training and clinical practice tools are evolving to better support patients, engage communities, and advocate for social change, there are multiple conceptual models, lists, and frameworks. This can lead to dilution and confusion (Lucyk & McLaren, 2017).

While there is a common understanding of SDoH measures and existing frameworks, including community and mapping tools like the Health Resources and Services Administration (HRSA)'s Area Deprivation Index (ADI), the National Equity Atlas, the Opportunity Nation, and Child Trends' Opportunity Index, available SDoH screening tools often focus on limited domains and lack metric alignment and validity. Additionally, SDoH data components are often collected in disparate points across multiple domains, and there is no universal agreement on a single definitive set of SDoH concepts (measure composition and structure), which can lead to elusive SDoH outcomes.

Overview of the Text

This book explores and provides case studies on how SDoH concepts and measures have been defined and operationalized with a spatial perspective. It carefully discusses the barriers and facilitators posed by a lack of standardization, metric alignment, and inconsistencies in SDoH data and measures. Various viewpoints challenge existing approaches and theories of SDoH, with some extending into new directions, others drawing historical inspiration, and some spanning multiple disciplines. These perspectives redefine SDoH and/or SSDoH theories with robust new case studies and illustrative empirical applications. New perspectives expand how we can integrate SSDoH thinking into daily practice, from building environments that support healthy aging to teaching and discussing SDoH in classrooms. The authors of the chapters in this book employ, probe, redefine, create, and expand SDoH frameworks to examine SDoH from a spatial perspective. They present a unique view of the current understanding of "Place and the Social-Spatial Determinants of Health." Although this theme runs through all the chapters, the contributions represent different perspectives and approach the issues from various angles, thanks to an interdisciplinary and transdisciplinary team of experts.

Guided by SDoH frameworks, the main body of the book is structured into five parts: Part I introduces the topic; Part II conceptualizes SSDoH through five chapters, outlining theoretical approaches, identifying literature gaps, and advocating for theoretical integrity. Part III integrates these determinants into practice, Part IV discusses methodological approaches, and Part V provides empirical illustrations of SSDoH. Kolak and Barronville, in Chap. 2, revisit geographic thinking in health, reviewing social theories like the socio-ecological view of health and salutogenesis from a spatial perspective, and endorse rigorous theoretical frameworks and a trauma-informed approach to advance SDoH science and policy. Chapter 3, by Takenaka et al., describes the conceptualizations and challenges of current SDoH thinking, exploring social-spatial determinants through transformative critical praxis, ecosocial theory, and intersectionality lenses to move scholars beyond static notions of place to other social positions and power relations, centering on power and justice. This chapter concludes by proposing an integrated analytical framework to advance the definition of SSDoH.

In Chap. 4, Iyanda examines the health impacts of historical place-based discrimination (H-PBD) on minority populations from a critical geography perspective, drawing evidence from various countries. The author highlights the adverse effects of spatial segregation on health and life expectancy, advocating for comprehensive approaches to address the structural and institutional factors perpetuating health inequalities. Case studies from Philadelphia, Baltimore, and Trenton illustrate how redlining contributed to racial and residential segregation and poverty in these cities. The chapter concludes by emphasizing the importance of recognizing intersectionality in addressing H-PBD's role in SSDoH.

Chapter 5 applies SSDoH constructs, such as Latinx destination type measured at the county level and Latinx-White residential segregation measured at the county and census tract levels, to examine the interplay of Latinx community attributes and the availability of federally qualified health centers (FQHCs). The findings indicate that FQHC supply is lower in new versus established Latinx destination counties, but there are more FQHCs in counties and tracts with higher Latinx segregation.

Chapter 6 extends the SSDoH framework by using methods from medical geography and spatial epidemiology, grounding in health and human rights law, and empirical approaches to intersectionality to offer a transdisciplinary theory of access to health and healthcare. The chapter begins with a synthesis of literature from each of these fields, highlighting their overlapping concerns with defining, measuring, and promoting access to healthcare and accessibility more broadly. This synthesis underscores the need to expand upon these theories of access by combining insights from each field. The chapter concludes with potential avenues for applying this theory in empirical analyses across different socio-legal and geopolitical contexts.

In the third part of the book, Part III, a set of five chapters provide case studies on integrating SSDoH in practice to improve child and family connections, elderly safety, health system fortification, and college teaching. In Chap. 7, Hirsch delves into the use of SSDoH measures within the U.S. health system, particularly focusing on electronic health records (EHR), risk prediction, resource allocation, reimbursement models, and surrogates for individual-level SDoH. The chapter provides an overview of how these measures are being utilized, along with specific examples, highlighting the indices used, the rationale behind their selection, data sources, and the impact of their application. In Chap. 8, Young et al. apply spatial theory to conceptualize how spaces within place-based Child and Family Learning Centers foster connections that initiate integrated service delivery (ISD) for families with children. The authors identify three practices that create spaces that are safe and rich in intersecting trajectories, producing connections with depth in the moment and reshaping early childhood experiences. The power of informal places proves instrumental in building meaningful and engaging human experiences that support healthy families. Chapters 9 and 10 shift focus to the architectural aspects of built environments and interiors, including objects that promote healthy aging. Tissot and colleagues propose a protocol with design guidelines to create safe environments for older adults, enhancing safety and promoting aging in place. Their findings categorize design and safety recommendations while considering human factors and design attributes.

In Chap. 10, Vergara et al. provide a comprehensive literature review of studies examining the application of assistive technologies for older adults in their homes. The chapter highlights two main categories of technologies: leading and support. Among the technologies discussed are robotic assistance, games and virtual reality, mobile applications, and technological integration. The authors also address challenges such as variability among older adults, limited sample sizes, and technology design. The conclusion emphasizes the potential of these technologies to improve the quality of life for older adults, promote social interactions, and support healthy aging. Part Three concludes with Chap. 11 by Sigrid et al., which focuses on teaching and learning about the geospatial determinants of health. This chapter, written by students of the same course, discusses core concepts from the course and applies the SDoH framework to three case studies: food insecurity, heat exposure over the life course, and environmental racism. The students reflect on how their own perspectives influenced their learning throughout the course. The chapter concludes with lessons learned, reflections on equity, and suggestions for optimizing approaches to teaching and learning.

In Part IV of the book, five chapters present methodological approaches and techniques to assess SSDoH. Chapter 12, authored by Moise and Choudhury, explores methods for operationalizing SDoH indicators. They provide a comprehensive guide for researchers, policymakers, and practitioners, addressing key questions and presenting future directions and emerging trends in the field. The chapter emphasizes how these methods can advance health equity and social justice, concluding with a discussion on the importance and challenges of operationalizing SDoH measures, along with best practices and recommendations for rigorous and ethical implementation. Chapter 13 by Avery and Kuhl discusses various types of gentrification, the mechanisms through which they are hypothesized to impact health, and their operationalization and components within an SDoH framework. In Chap. 14, Cassels and Reid introduce the concept of relative vulnerable time of exposures, arguing that this can lead to differential health outcomes by altering the quality or magnitude of exposure or by affecting the pathway from exposure to amplify adverse health outcomes. This concept is illustrated through qualitative research case studies on geographic mobility and health among Black and Hispanic sexual minority men. Chapter 15, authored by Lin and Xu, proposes a social-spatial network approach for investigating health inequity in syndemics research, linking innovative perspectives across methodological domains to better measure and model complex relationships. To conclude this section, Chap. 16 by Fernes and Rhodes provides a thoughtful discussion on walking as a qualitative methodology for SSDoH research. They explore how space materialized in care relations and practices, using ethnographic walking interviews from a participatory research project with people who use drugs and experience homelessness in London.

The final part of the book, Part V, a suite of chapters provide empirical illustrations and modeling motivations of SSDoH. In Chap. 17, Oppong and Lester delve into the promise and potential of GIS for the spatial analysis of disease and health care, highlighting common methodological challenges such as autocorrelation, spatial support, and small data. They advocate for the availability of robust and comprehensive surveillance data on the geography of environmental conditions, disease agents, and health outcomes over time to fully realize the potential of GIS. Chapter 18 features Lubinda and Mweemba's case study on the application of GIS and the SSDoH framework in Zambia, focusing on delineating malaria and understanding the complexities of its transmission and control, particularly the effectiveness of interventions in Zambia's high-burden areas. Finally, Chap. 19 by Ury et al. explores the history of the Interstate Highway System and its impact on minority communities, proposing potential policy changes and considerations to mitigate and reduce the detrimental effects of highways on these communities, offering important insights for current and future projects.

Concluding Remarks

The papers in this volume significantly enhance our understanding of the current SSDoH frameworks used across various specialties and disciplines. They delve into concepts, integration in practice (e.g., EHR), impacts on different subgroups, and methodological approaches and techniques. The book presents well-conceived case studies that are empirically, methodologically, and theoretically grounded. By focusing on theoretical grounding and fostering discussions across multiple disciplinary perspectives, it challenges researchers, scientists, and scholars to deepen their understanding of the mechanisms through which health and well-being are produced across space and time. This effort advances the conceptual work of understanding place and the SSDoH. Moreover, the book holds significant value for policy circles and practice. It serves as a vital resource for community-based organizations, international organizations, government workers at various levels, students, and foundations, aiding in the monitoring of community health and the improvement of equity and disparities globally. The case studies are particularly informative for policymakers and practitioners as they address health inequalities and the dual burden of community risk and inadequate health system infrastructure. This work aims to minimize negative health outcomes and enhance community-level population health, well-being, and equity. Additionally, the book is a valuable complementary reader for courses in health and medical geography, medical sociology, spatial epidemiology, GIS and spatial analysis, public health practice, and global public health.

References

Bradley, E. H., Canavan, M., Rogan, E., Talbert-Slagle, K., Ndumele, C., Taylor, L., & Curry, L. A. (2016). Variation in health outcomes: The role of spending on social services, public health, and health care, 2000–09. *Health Affairs, 35*(5), 760–768.
Braveman, P., & Gottlieb, L. (2014). The social determinants of health: It's time to consider the causes of the causes. *Public Health Reports, 129*(Suppl. 2), 19–31.
Dalsania, A. K., Fastiggi, M. J., Kahlam, A., Shah, R., Patel, K., Shiau, S., Rokicki, S., & DallaPiazza, M. (2021). The relationship between social determinants of health and racial disparities in COVID-19 mortality. *Journal of Racial and Ethnic Health Disparities*, 1–8.
Frieden, T. R. (2011). Forward: CDC health disparities and inequalities report—United States, 2011. *MMWR Supplements, 60*(1), 1–2.
Göran, D., & Whitehead, M. (1991). Policies and strategies to promote social equity in health. *Stockholm: Institute for Future Studies, 27*(1), 4–41.
Harrison, K. M., & Dean, H. D. (2011). Use of data systems to address social determinants of health: A need to do more. *Public Health Reports, 126*(Suppl. 3), 1–5.

Keppel, K. G., Pearcy, J. N., & Wagener, D. K. (2002). Trends in racial and ethnic-specific rates for the health status indicators: United States, 1990–98. *Healthy People 2000 Stat Notes, (23)*, 1–16.

Kolak, M., Bhatt, J., Park, Y. H., Padrón, N. A., & Molefe, A. (2020). Quantification of neighborhood-level social determinants of health in the continental United States. *JAMA Network Open, 3*(1), e1919928.

Lucyk, K., & McLaren, L. (2017). Taking stock of the social determinants of health: A scoping review. *PLoS ONE, 12*(5), e0177306.

Marmot, M., Friel, S., Bell, R., Houweling, T. A. J., & Taylor, S. (2008). Closing the gap in a generation: Health equity through action on the social determinants of health. *The Lancet, 372*(9650), 1661–1669.

McKinlay, J. B. (1979). A case for refocusing upstream: The political economy of illness. In *Patients, physicians and illness: A sourcebook in behavioral science and health* (pp. 9–25).

Moise, I. (2020). Variation in risk of COVID-19 infection and predictors of social determinants of health in Miami–Dade County, Florida. *Preventing Chronic Diseases, 17*, E124.

National Academies of Sciences, Engineering, and Medicine. (2017). *Communities in action: Pathways to health equity.*

Noonan, A. S., Velasco-Mondragon, H. E., & Wagner, F. A. (2016). Improving the health of African Americans in the USA: An overdue opportunity for social justice. *Public Health Reviews, 37*, 1–20.

Thornton, R. L. J., Glover, C. M., Cené, C. W., Glik, D. C., Henderson, J. A., & Williams, D. R. (2016). Evaluating strategies for reducing health disparities by addressing the social determinants of health. *Health Affairs, 35*(8), 1416–1423.

Walker, R. J., Williams, J. S., & Egede, L. E. (2016). Influence of race, ethnicity and social determinants of health on diabetes outcomes. *The American Journal of the Medical Sciences, 351*(4), 366–373.

Williams, D. R., & Mohammed, S. A. (2013). Racism and health II: A needed research agenda for effective interventions. *American Behavioral Scientist, 57*(8), 1200–1226.

Part II
Conceptualizing Social-Spatial Determinants of Health

Chapter 2
Re-Rooting the Social Determinants of Health within a Trauma-Informed, Integrated Health Geography

Marynia A. Kolak and Kamaria Barronville

Introduction

The social determinants of health (SDoH) have been broadly described by the World Health Organization (WHO) as the nonmedical factors influencing health, or the conditions in which people live, work, and play (World Health Organization, 2010). As the public health community shifted attention toward the "causes of the causes" in the emergence of the twenty-first century (Braveman & Gottlieb, 2014), more research expanded to consider how factors like income and education influence health outcomes, including hypothesizing pathways and mechanisms of action. Articles referencing the "social determinants of health" over time on the *PubMed database*,[1] the golden standard for indexing biomedical and life sciences literature, saw an increase in 2020 (see Fig. 2.1); this rise builds on prior momentum since 2000 of SDoH topics. As the COVID-19 pandemic progressed, identifying SDoH factors that shaped emerging disparities in case infections, deaths, and vaccination trends proved crucial (Abrams & Szefler, 2020; Paul et al., 2020; Rollston & Galea, 2020). Researchers were not only documenting disparities of SDoH and health outcomes, but also applying SDoH frameworks to multiple disease outcomes, distilling intersections of SDoH and related topics, and demanding

---

[1] Concepts were searched on *PubMed*. Resulting citations over time were then downloaded. Citations were prepped for visualization with *tidyverse* and *lubridate*, and plotted using *ggplot* and *ggthemes* in R. Code, data, and documentation can be accessed at https://github.com/Makosak/rootingHealthGeog.

M. A. Kolak (✉) · K. Barronville
Department of Geography & Geographic Information Science, University of Illinois
Urbana-Champaign, Urbana, IL, USA
e-mail: mkolak@illinois.edu

© The Author(s) 2026
M. A. Kolak, I. K. Moise (eds.), *Place and the Social-Spatial Determinants of Health*, Global Perspectives on Health Geography,
https://doi.org/10.1007/978-3-031-88463-4_2

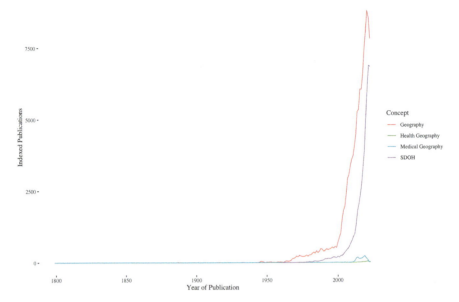

Fig. 2.1 Publications are indexed in *PubMed* by concept, by year of publication. The years shown are from 1799 to the end of 2023 to ensure a full and comparable range. Data from *PubMed*

more nuance in methodological approaches (Lin et al., 2022a, b; Obinna, 2021; Webb Hooper et al., 2022).

At the same time, for over two hundred years the discipline of geography has served as an integral discipline to connect social, political, economic, and environmental components of health outcomes. As public health approaches have shifted (back) toward a wider view of the factors that drive health and produce health inequities, there is a critical need for integrated health geography to connect disparate, often disconnected, conceptual threads. Recent approaches build on the historic, conceptual underpinnings of disease ecology, political ecological interpretations, research on therapeutic geographies, and the socioecological view of health. A spatial perspective can support conceptual, technological, and analytical approaches in the measurement of the social or structural determinants of health, and further reorganize complex, nuanced frameworks in a meaningful way. To move the science forward, we need to specify SDoH frameworks with more nuance and care: in other words, linking to, extending, and eventually surpassing the existing models of health and place by working across (instead of against) intellectual domains of research and knowledge production.

In this chapter, we provide an abridged overview of geography and its history in health and SDoH research. Then, we review major conceptual frameworks of SDoH study, highlighting how key constructs are organized systematically across individual, community, and environmental positionalities. Connecting conceptual frameworks, by explicitly modeling spatial dependence and spatial heterogeneity, provides analytical opportunities to identify multiple forms of health inequity within a system with precision. Finally, we advocate for a trauma-informed perspective to bring new understanding to the growing field of SDoH research. This approach ensures the *study* of health and place does not reproduce the inequality or health disparities but rather contributes to positively impacting individuals and communities. We conclude by arguing that an integrated view of health geography may serve as a holistic framework to connect, decompose, and understand critical topics in public health and SDoH study.

The Roots of Geography in Health and SDoH Thinking

Research on SDoH considers the context of family, community, built environment constructions, economic opportunities, and policy at multiple scales of influence. Several of these measures have long been a feature of geographic study, especially within medical and health geography subspecialties when applied to models of health and well-being. Geographer Tom Koch has traced these histories over the decades, from his classic text on "Social epidemiology as medical geography: Back to the future" published in 2009 to "Back to the future: Covid-19 and the recurring debate over social determinants of disease, and health" in 2022 (Koch, 2009, 2022). Koch's archival study illustrates how the history of epidemiology and public health is *spatial* and *geographic*, with the spatiality of disease (and ultimately, well-being) serving as both critical groundwork in the field of disease studies since the late eighteenth century and as part of its maturation over two hundred years (Koch, 2009). Indeed, a review of journals indexed in *PubMed* shows how geography (and medical geography, separately) is cited throughout the 1800s (especially in the last 50 years) at a similar pacing as the first half of the twentieth century (see Fig. 2.2). Geographic thinking at the earliest points of indexed health studies is more conceptual, used for hypothesis generation, as well as communication. Cartographic products—maps—are used for "analysis and argument," organizing data collection, but not statistical investigation until the nineteenth century (Koch, 2009).

As industrialization and technological advances reordered and rearranged entire populations, driving new forms of societal organization, complexities, and human adaptations, the study of disease kept pace to facilitate new understandings of emerging infectious disease and public health patterns. Statistical health sciences emerged in the early 1800s, profoundly transforming data collection approaches to link spatially located population data with social, economic, environmental, political, and health outcomes as done by Guerry in 1830 or Farr in 1852 (as cited by Koch, 2009; Krieger, 2000). Nancy Krieger, a social epidemiologist well known for integrating spatial perspectives and a focus on disparities, similarly traces these roots of place-based approaches to health. Ecological approaches to disease exploring the interactions between multiple factors (what today may be termed "determinants") and place are exemplified in work by John Snow on the cholera epidemic in 1850s London. However, this approach ebbs and flows for over a century (Koch, 2009; Krieger, 2000) for complex reasons, including: change of focus after two

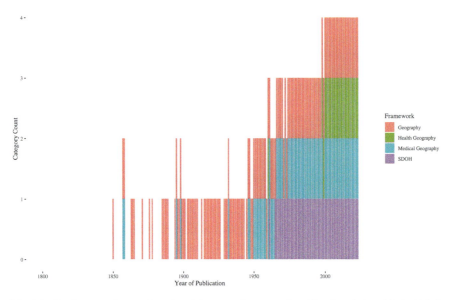

Fig. 2.2 Total count of categories of each framework across all publications indexed in *PubMed*. The years shown are from 1799 to the end of 2023. Data from *PubMed*

world wars; and, more importantly, the lack of systematic census data collection at small area scales across all populations. Koch notes that the first Cancer Atlas in the United States was not published until the 1970s, and only for "Caucasian" or White residents (Koch, 2009).

In this context, we may better understand the emergence and "flat line" of geography in medical literature (as indexed on *PubMed)* over the first century and a half. Citations grew after 1960, with more use of the term "medical geography" separately from geography, as well as the beginnings of formalizing the "social determinants of health" as a term (see Figs. 2.1 and 2.2). Geographic factors bring new insight into health service delivery in Chicago (Morrill & Earickson, 1968). Medical geography is referred to as a "peripheral field" (Mullins, 1966, p. 230) engaging both multidisciplinary and interdisciplinary perspectives; at the same time, dozens of texts globally reference intersections of a specific disease and its geography across a multitude of places (e.g., Canadian Medical Association, 1965; Fuchs, 1959; Janiszewska-Zygier, 1966; Phillips, 1959). In 1961, anthropologist Gordon Macgregor wrote about "Social Determinants of Health Practices" and reported on a Public Health Service project investigating local public health practice, with a focus on the rural Great Plains region in the United States (Macgregor, 1961; Smithsonian Institution, 2024). The project summarized social determinants related to individual health behavior, community-level organizations and procedures impacting health, and additional social structures influencing planning and care across communities. Its multiscalar approach to health, embedded within geographic context, recommends integrated regional organization and cooperation to support new and continuing medical services. Ecological and multifactor dimensions influencing health may also be seen in the work of medical geographer Jacques May and others during this time (see discussion by Koch, 2009), though some distance persists between geographic literature and mainstream health and epidemiological work.

By the 1970s and 1980s, incorporating regional perspectives on social inequalities that consider drivers of social injustices and geographic health disparities continues to persist (albeit slowly) in the medical literature. Black's report on *Inequalities in Health* in 1980 hones in on the relationship with socioeconomic status and health in Great Britain, while the impact of socioeconomic disadvantage is increasingly considered in new epidemics like HIV/AIDS and emerging infectious diseases (Koch, 2022). New computational approaches and spatial statistical techniques enable early Geographic Information Science (GIScience) applications like Openshaw's study on spatial clusters of leukemia (Openshaw et al., 1988), building on Waldo Tobler's case for computer-assisted mapping in behavioral science (Tobler, 1967). Health applications exploded with advances in computer hardware and software services by the 1990s, supporting health data management, health services modeling, new spatial statistical techniques to support understanding disease ecology, and further research translation efforts.

Figure 2.1 illustrates a skyrocketing incorporation of articles citing "geography," and a few years later, the "social determinants of health" after 2000. The term "health geography" also begins to regularly appear in *PubMed* citations (Fig. 2.2). These trends reflect multiple strands of knowledge building co-occurring, sometimes with siloed disciplinary fervor. Medical geography sees a "cultural turn" in the 1990s, shifting away from mapping disease or cool discussion of service provisioning towards critical discussions of power, representation, meaning, and the mechanisms of health (Kearns, 1993). At the same time, GIScience is defined as a new scientific discipline (Goodchild, 1991) with GIS applications continuing to transform statistical techniques, health department protocol for disease surveillance, and mapping techniques. This explosion is referred to as the "spatial turn in health" in *Science* by Richardson et al. (2013), building on the new types of data and technologies continuing to emerge. During and after 2020, these literature strands across geography and health disciplines continue to fortify understandings on the spread of the COVID-19 Pandemic through modeling and statistics, as well as its unequal impact on communities, using multidisciplinary perspectives.

Need for Robust, Conceptual SDoH Frameworks

While there exists a long and complex history of studying the role of place, social context, and the built environment within health, current and common frameworks of the "social determinants of health" tend to be simplistic in application. For example, the *Healthy People 2030* initiative by the US Health and Human Services office groups SDoH into five main domains: economic stability, education access and quality, healthcare access and quality, neighborhood and built environment, and social & community context (U.S. Dept of Health and Human Services, n.d.). The WHO set up a Commission on Social Determinants of Health and developed a framework in 2010, distilling structural and intermediary determinants of SDoH with a focus on conceptualizing the health system itself as a social determinant of health (World Health Organization, 2010). Yet, specifics on how SDoH mechanisms work are not often seen in broad, SDoH domain-focused frameworks to inform practice (Thimm-Kaiser et al., 2023). There is also a lack of standardization in how SDoH factors are modeled, making scientific innovation in this study more difficult. A survey in 2019 by Elias et al. explored the lack of consensus across factors of the SDoH, surveying the literature and grouping similar indicators across twelve sub-categories (built from the original five main groups of Healthy People initiatives): demographics, economic stability, employment, education, food environment, health and healthcare, housing, neighborhood and built environment, physical activity and lifestyle, safety, social and community context, and transportation and infrastructure. They found an overall lack of consensus across tools seeking to measure SDoH categories, recommending more collaboration across and within disciplines. While determining distinct factors and indicators of SDoH is useful for public health thinking and brainstorming policy interventions, the lack of applications detangling the nuance across and between measures should be troubling. SDoH indicators may be associated with health outcomes, but they are also likely to be associated with *each other* (Fuchs, 2017), suggesting we are missing the core phenomena driving change in communities and health. While innovative new methods can be used to expand our understanding of SDoH measures (Kolak et al., 2020; Moise, 2020; Moise et al., 2024), we still need robust conceptual frameworks to frame learning gained in the human-environment mechanisms influencing health.

There is no singular model of the SDoH. We thus need to specify a conceptual approach when discussing SDoH for variable selection, consideration of mechanism and pathways, and overall research into what supports a healthy place/life. We explore several models below that could serve as potential SDoH frameworks for moving discussion and the science forward, and with a geographic perspective extend that to include views of the social-spatial determinants of health. This review is not meant to be exhaustive, but rather highlight flavors of the deep knowledge available to build upon when engaging discussions on the social-spatial determinants of health for research, modeling, and/or intervention planning. We highlight *salutogenesis, socioecological theory,* and *fundamental causes theory* as case studies. Each view may bring a specific focus on place, region, or (spatial) scales, enabling connection to hundreds of years of research in geography and health. Beyond these, we also recommend review of *risk environments* (Collins et al., 2019; Cooper et al., 2016; Rhodes, 2002), *the exposome* (Deguen et al., 2022; Gudi-Mindermann et al., 2023; Hajat et al., 2013; Kruize et al., 2014; Wild, 2005), *political ecology* (Hanchette, 2008; Harper, 2004; Neumann, 2009; Richmond et al., 2005), *intersectionality* (Bowleg, 2012; Crenshaw, 1989; D'Ignazio & Klein, 2020; Harari & Lee, 2021), and the *Reading and Wien tree model* (Reading & Wien, 2009) when determining the most appropriate social-spatial determinants of health model

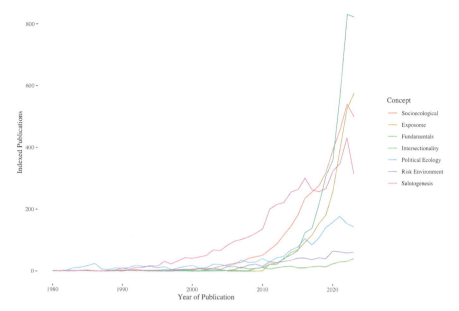

Fig. 2.3 Publications indexed in *PubMed* by social theory concept, by year of publication. The years shown are from 1980 to the end of 2023. Data from *PubMed*

for a specific study or application. Figure 2.3 displays citations and emergence of these theoretical approaches in the biomedical literature.

Salutogenesis: Defining Healthy Places

The salutogenic model of health emerged from Aaron Antonovsky in the late twentieth century was refined over decades in health promotion literature, and is reemerging in the context of SDoH research (Vinje et al., 2017). Salutogenesis encourages a focus on increasing the factors that improve health and resilience, rather than focusing solely on avoiding factors that harm health (i.e., a *pathogenic approach* common to Western medical thinking). From this perspective, the study of health and well-being may be distinctly different from that of disease, differentiating upstream (e.g., disease prevention) and downstream (e.g., curative medicine) efforts. A strong sense of coherence emerges at the individual level as a way to manage health, in response to life stressors. Mittelmark and Bull (2013) have argued that the salutogenic model could be useful in health promotion in further understanding the nature of well-being as something more than simply the absence of pain, suffering, or need for medical care. Identifying resources is crucial to building a sense of coherence across a lifetime, and may be further honed as informed interventions (Bauer et al., 2020) or theorized within a synergistic, salutogenic/asset-modeling approach (Pérez-Wilson et al., 2021).

Contemporary studies continue to emphasize the individual-focused nature of salutogenic research, while also exploring how it can be supported by health organizations, social networks, policy influences, and conducive or supportive living environments. The idea that a strong sense of coherence can serve as a protective factor in a person's health network was further supported by Wind (2021), who cautioned against seeing it merely as a factor within the broader framework of SDoH. The incorporation of an SDoH perspective within a positive, asset-based salutogenic approach has been advocated by researchers like (Gallego-Osorio et al., 2021), who argue for its potential to address issues of risk, exposure, and vulnerability. This integration is evident in the application of salutogenic principles in urban planning and the design of therapeutic landscapes and green spaces, which are increasingly recognized for their health-promoting benefits.

This model's principles are also applied in GIS to map and analyze health-promoting environments, demonstrating how digital tools can support the salutogenic approach in creating healthier living conditions. Such applications highlight the model's relevance and adaptability to various fields, emphasizing its role in promoting a holistic and proactive approach to health and well-being. The integration of salutogenic principles into environmental design extends further into the development of health-promoting urban spaces. This approach aligns with the broader goal of creating sustainable, healthy cities through the strategic inclusion of green spaces, which are known to offer significant mental and physical health benefits (Gallego-Osorio et al., 2021).

The concept of therapeutic geographies also plays a crucial role in this context. Therapeutic geographies explore how particular environments can contribute to health and well-being. For example, the creation of healing gardens and the utilization of urban green spaces as places for rest and rejuvenation are practical applications of this concept, blending the salutogenic model with geographic and urban planning methodologies to enhance public health outcomes (Pérez-Wilson et al., 2021). Furthermore, the use of GIS technologies allows for the spatial analysis of health data in relation to environmental conditions, facilitating targeted interventions that are informed by salutogenic principles. This can include the mapping of green spaces relative to urban populations, assessing accessibility, and ensuring that these health-promoting environments are available to diverse communities, thereby addressing social inequalities in health accessibility (Wind, 2021).

By applying a salutogenic lens, planners and health professionals can more effectively design urban environments that not only mitigate health risks but also actively contribute to the well-being of the population. This holistic approach is increasingly important as cities continue to grow, and the need for sustainable, health-supportive urban environments becomes more critical (Mittelmark & Bull, 2013). The integration of salutogenic principles into urban planning and GIS exemplifies a shift toward more dynamic and health-focused approaches to managing urban development and public health. It represents a proactive strategy that emphasizes resilience and well-being, potentially transforming the way cities are planned and managed for the health of their inhabitants. This strategic application of the salutogenic model highlights its versatility and relevance in addressing modern public health challenges within urban settings.

Socioecological Theory: A View Across Spatial Scales

Socioecological theory, also referred to as social ecological or ecosocial theory, provides a comprehensive framework for understanding health outcomes by examining the dynamic interplay between individuals and their social and physical environments. Rooted in the health promotion work of Bronfenbrenner (1977), socioecological theory posits that health is influenced by multiple levels of influence, ranging from individual characteristics to broader societal factors. The model is commonly visualized as intersecting circles, with individual factors in the innermost circle, surrounded by interpersonal factors (i.e., family, relationships); then organizational factors (i.e., workplace); then community and/or societal factors (i.e., neighborhood environments, policies). At its core, socioecological theory emphasizes the importance of considering both proximal and distal determinants of health (Wilderink et al., 2022), acknowledging that factors operating at different levels interact to shape health outcomes.

This approach may be one of the most common models of health used across multiple disciplines and is flexible for adaptation in both quantitative and qualitative methodological approaches. It's been used to support health promotion and understand the barriers and facilitators of health services within specific populations (Alwan et al., 2021; DiClemente et al., 2005; Mehtälä et al., 2014), the study of well-being (King et al., 2014), investigations of specific disease impacts across socioecological systems (Ohri-Vachaspati et al., 2015), empirical illustrations across subdomains of medical research (Kilanowski, 2017), resilience theory and systems research (Cretney, 2014; Masterson et al., 2017), and the study of SDoH itself (Kolak et al., 2020). Cretney (2014) explored socioecological resilience frameworks to highlight the interconnectedness of social and ecological systems, advocating for policies that consider these complex interactions. Wilderink et al. (2022) utilized the socioecological framework to investigate persistent socioeconomic health inequalities. This study emphasized the need for systemic interventions that address multiple levels of influence, such as individual behaviors, social norms, and broader economic and political factors. The socioecological perspective helps elucidate why certain health disparities persist despite various interventions, advocating for comprehensive strategies that consider the full complexity of these interactions.

When considering the SDoH, the socioecological framework has been extensively used to uncover and address inequalities, racial/ethnic health disparities, racism as a social determinant of health, and xenophobia and stigma. Racism, a public health crisis (Wamsley, 2021), persists across multiple scales, engaging differing mechanisms and health outcomes at individual, interpersonal, community, and structural levels. To address racism in pharmaceutical fields, for example, Nonyel et al. (2021) employed the socioecological model to suggest new interventions at curricular, interprofessional, institutional, community, and accreditation levels. In a different application, Alwan et al. (2021) engaged socioecological perspectives to understand challenges to healthcare access for undocumented persons in an "anti-immigrant" era, finding multiple factors at individual, organizational, and public policy levels. When considering stigma (e.g., stigma of opioid use disorder), research has found different associations across ecosocial scales of individual, inter-personal, and neighborhood levels (Lin et al., 2022a, b). In another study, multidi-mensional SDoH indices (generated by variables identified with a socioecological model) illuminated intersectional neighborhood experiences with differing dimen-sions of inequality (Kolak et al., 2020); areas with varying socioeconomic advan-tage, isolation, opportunities, and access to resources generated complex patterns across the United States that go beyond simplistic racial/ethnic or socioeconomic (SES) categories. The socioecological framework helps in understanding the inter-actions between human well-being and environmental factors, with some scholars advocating for participatory methods to assess these relationships. This approach acknowledges the complexity of human-environment interactions and the need for sustainable practices.

Fundamental Causes Theory: Engaging Context

Fundamental causes theory, proposed by Link and Phelan (1995), offers a macro-level perspective on health disparities by identifying the underlying social condi-tions that contribute to differential access to resources and opportunities. Born from the field of medical sociology, this theoretical approach focuses on the inherent association between SES and mortality. In its original formalization (Link & Phelan, 1995), a fundamental social cause of health inequality includes four components:

1. It influences multiple disease outcomes.
2. It affects disease outcomes through multiple risk factors.
3. Access to resources can be used to avoid risk or minimize consequences of a disease.
4. The association between a fundamental cause and health is reproduced over time via the replacement of intervening mechanisms.

Thus, the persistence of the effect on health leads the social factor to be called "fundamental," as summarized by Phelan and Link (2013). Research on fundamen-tal causes theory has expanded over the years to explore various dimensions of health inequalities, often integrating perspectives from epidemiology, sociology, and public health. This approach retains a strong focus on the mechanisms or "meta-mechanisms" through which SES impacts health, such as differential access to healthcare, exposure to stressors, and lifestyle factors (Lutfey & Freese, 2005), as well as "spillover," defined here as contextual resources that may occur through influence of an individual's social network (Phelan & Link, 2013). Studies have examined how SES influences a wide range of health outcomes, including chronic diseases, mental health, and mortality (Phelan et al., 2010). Fundamental causes theory has inspired interdisciplinary research, combining insights from social sciences, public health, and medicine. This includes looking at how social conditions interact with biological and environmental factors to affect health (Hatzenbuehler et al., 2013). This approach retains implications for public health policies aimed at reducing health inequalities, like interventions suggested to focus on reducing SES disparities and improving access to health-promoting resources (Phelan et al., 2010).

In public health pedagogy research, when discussing how to define the SDoH, Harvey (2020) advocates for a move beyond the inadequacies of behavioral theory to incorporate social theories of health inequality; among options, fundamental causes theory was heralded as an innovative and promising approach. Fundamental causes theory is seen as an early and influential framework to understand the mechanisms of SDoH with its focus on underlying factors impacting health inequities (Thimm-Kaiser et al., 2023). The term "fundamental causes" may be seen as synonymous as "structural drivers" of health equities in some applications. Other work linking SDoH and "fundamental causes" note that both access to resources (as traditionally argued by the theory) *and* psychosocial stress may drive poor health; either way, though, the drivers of health disparities relate to social position and social inequalities (Douglas, 2015). Empirical applications of fundamental causes theory demonstrate mixed findings, enabling new opportunities for the social theory to expand further. Mackenbach et al. (2017) tested four hypotheses of the fundamental causes theory with mortality data across twenty European populations; the first hypothesis was supported, lending support to the theory, with mixed results for the remaining ones. Contrary to expectations, differences in mortality between education groups were not larger when income inequality was greater, suggesting additional factors were contributing to changes in mortality decline. Additionally, more needs to be discerned in how "fundamental causes" are measured, as well as the spatial scale(s) in which the theory holds.

From a geographical standpoint, fundamental causes theory may highlight the spatial concentration of resources and opportunities which contribute to spatial disparities in health outcomes (Clouston & Link, 2021). While fundamental causes theory addresses how disease may cluster from social and physical exposures, it has been criticized for not addressing methodological issues (such as spatial clustering) or interpretations when brought across applications (Clougherty & Kubzansky, 2009). Additional approaches are needed to disentangle complex relationships between social and physical stressors, when they are both independent and interdependent. A spatially explicit approach may refine research engaging fundamental causes theory, especially in complex human-environmental interactions. Spatial structures in the data may result from (1) spatial heterogeneity, observed when underlying variation in outcomes is due to some external process (e.g., redlining, immigration patterns), and/or (2) spatial dependence, observed when phenomena influence their neighbors (e.g., polluting factory may impact nearby residents the most). Spatial spillover serves as a special case of spatial dependence, where persons or places may influence each other based on proximity. Connecting these concepts and methodological techniques with "spillover" topics in fundamental causality is promising. A geographic perspective can test and further extend theoretical frameworks, such as fundamental causes theory, by refining models of how changes in SES might influence health outcomes across different geographic areas, as well as how social inequalities are embedded and reproduced within place.

Need for Trauma-Informed Approaches

While social science theories and spatially explicit models may further understanding the mechanisms of the SDoH, acknowledging the deep structural factors and their impacts on health inequalities in a more profound manner remains essential. This impacts *how we do the work* of research itself, in addition to updating perspectives when developing measures, applying theoretical frameworks, or interpreting findings. Incorporating a trauma-informed approach into the study of health, place, and SDoH goes hand in hand with prioritizing equity in research processes, including the conceptual framework, research design, and data analysis. The term "trauma-informed" seeks to balance biological and contextual understanding of health disparities by actively healing trauma and minimizing further harm (Kennedy, 2024). Parker and Johnson-Lawrence (2022) argue that public health training should include trauma-informed approaches so that practitioners can recognize the pervasive impact of trauma on individuals and communities. Trauma-informed approaches prioritize safety, trustworthiness, choice, collaboration, and empowerment, understanding that trauma can stem from various experiences, including violence, abuse, neglect, and systemic oppression (Menschner & Maul, n.d.).

Building on research from historical trauma, collective trauma, and intergenerational trauma—traumatic events experienced by communities, particularly those related to displacement, segregation, and environmental injustices in specific locations—geographer Pain (2021) has emphasized the importance of place-based traumatic events by coining the term "geotrauma." By viewing space and place through a trauma-informed lens, researchers may better consider how minority identities are oppressed by socioecological systems over time and seek to pinpoint intervention opportunities for learning and healing. Trauma-informed approaches may also shed light on the spatial dimensions of trauma and its intersection with social and environmental factors (Schuurman et al., 2008). Geographic disparities in exposure to trauma-inducing events can worsen existing health inequities and perpetuate cycles of trauma within marginalized communities. By employing spatial analysis techniques, we can map patterns of trauma exposure, identify geographic hotspots of trauma prevalence, and develop targeted interventions to support affected individuals and communities. Thus to fully embrace a trauma-informed approach in public health, it is essential to integrate this perspective into the conceptual framework, research design, and data analysis phases of public health projects. Conceptually, a trauma-informed framework emphasizes understanding the widespread impact of trauma and recognizing signs and symptoms in individuals and communities. It involves integrating knowledge about trauma into policies, procedures, and practices to avoid re-traumatization.

In terms of research design, incorporating trauma-informed principles means ensuring that study protocols and methods are sensitive to participants' trauma histories and experiences. This includes designing studies that minimize potential triggers, providing informed consent processes that clearly explain the nature of the research, and offering support resources to participants. Data analysis within a trauma-informed framework involves interpreting findings with an awareness of how trauma might influence health outcomes and behaviors. Analyzing data through this lens can reveal patterns and correlations that might otherwise be overlooked, such as the relationship between trauma exposure and health disparities. Additionally, it involves using equitable data analysis techniques to present findings in ways that are understandable and useful to all stakeholders, including those from marginalized communities. This is particularly relevant in areas where historical and contemporary traumas, such as redlining and urban renewal, have left lasting impacts on the built environment and community well-being (Rothstein, 2018; Sugrue, 2014).

While trauma-informed principles offer a valuable framework for addressing the impacts of trauma on health, there are significant dangers in using it to generalize a population. Trauma experiences and responses are highly individual, influenced by a myriad of factors such as personal history, culture, socioeconomic status, and existing support systems. Generalizing trauma experiences can lead to several issues including the oversimplification of trauma, stigmatization, and misallocation of resources.

The implementation of trauma-informed practices and the use of SDoH in public health can significantly improve health outcomes for communities affected by trauma. However, there is a clear need for more empirical research to support these conceptual frameworks and validate their effectiveness in real-world settings. Collaboration across sectors, including public health, technology, and data science, is essential to develop more comprehensive and effective approaches to trauma-informed care (Champine et al., 2021; Chandra et al., 2022; Forkey et al., 2021; Phelos et al., 2022). By doing so, we contribute to building more equitable and resilient communities where everyone can thrive, regardless of their past experiences with trauma.

Conclusion

Integrating social theories with geographic perspectives can build rich and robust SDoH conceptual frameworks, with a focus on place, region, and scale providing a comprehensive lens for examining health disparities. By linking social theories to these geographic topics, we can enhance our understanding of the complexities between various factors that influence health outcomes. Encouraging exploration across disciplinary boundaries can move the science ahead, as well as roll out more useful interventions and policies to support community health. This approach re-roots SDoH into geographic thinking, diving back into a two hundred+ year discussion on the relationships between health and environments. Incorporating a trauma-informed view into our research practices further enriches this perspective, ensuring that we are sensitive to the lived experiences of individuals and communities affected by trauma.

Integrating social theories with geographic perspectives and incorporating trauma-informed principles can significantly enhance our understanding and address health disparities. A multidisciplinary approach fosters the development of more nuanced and effective public health interventions, contributing to the creation of equitable and resilient communities. By remaining open to new insights and promoting interdisciplinary collaboration, we can advance public health research and practice in ways that are truly beneficial to individuals and communities.

References

Abrams, E. M., & Szefler, S. J. (2020). COVID-19 and the impact of social determinants of health. *The Lancet Respiratory Medicine, 8*(7), 659–661. https://doi.org/10.1016/S2213-2600(20)30234-4

Alwan, R. M., Kaki, D. A., & Hsia, R. Y. (2021). Barriers and facilitators to accessing health services for people without documentation status in an anti-immigrant era: A socioecological model. *Health Equity, 5*(1), 448–456. https://doi.org/10.1089/heq.2020.0138

Bauer, G. F., Roy, M., Bakibinga, P., Contu, P., Downe, S., Eriksson, M., Espnes, G. A., Jensen, B. B., Juvinya Canal, D., Lindström, B., Mana, A., Mittelmark, M. B., Morgan, A. R., Pelikan, J. M., Saboga-Nunes, L., Sagy, S., Shorey, S., Vaandrager, L., & Vinje, H. F. (2020). Future directions for the concept of salutogenesis: A position article. *Health Promotion International, 35*(2), 187–195. https://doi.org/10.1093/heapro/daz057

Bowleg, L. (2012). The problem with the phrase women and minorities: Intersectionality—An important theoretical framework for public health. *American Journal of Public Health, 102*(7), 1267–1273. https://doi.org/10.2105/AJPH.2012.300750

Braveman, P., & Gottlieb, L. (2014). The social determinants of health: It's time to consider the causes of the causes. *Public Health Reports, 129*(1_Suppl. 2), 19–31. https://doi.org/10.1177/00333549141291S206

Bronfenbrenner, U. (1977). Toward an experimental ecology of human development. *American Psychologist, 32*(7), 513–531. https://doi.org/10.1037/0003-066X.32.7.513

Canadian Medical Association. (1965). Geography and health. *Canadian Medical Association Journal, 93*(25), 1322–1323.

Champine, R. B., Lang, J. M., & Mamidipaka, A. (2021). Equity-focused, trauma-informed policy can mitigate COVID-19's risks to children's behavioral health. *Policy Insights from the Behavioral and Brain Sciences, 8*(2), 103–110. https://doi.org/10.1177/23727322211031583

Chandra, A., Martin, L. T., Acosta, J. D., Nelson, C., Yeung, D., Qureshi, N., & Blagg, T. (2022). Equity as a guiding principle for the public health data system. *Big Data, 10*(Suppl. 1), S3–S8. https://doi.org/10.1089/big.2022.0204

Clougherty, J. E., & Kubzansky, L. D. (2009). A framework for examining social stress and susceptibility to air pollution in respiratory health. *Environmental Health Perspectives, 117*(9), 1351–1358. https://doi.org/10.1289/ehp.0900612

Clouston, S. A. P., & Link, B. G. (2021). A retrospective on fundamental cause theory: State of the literature, and goals for the future. *Annual Review of Sociology, 47*(1), 131–156. https://doi.org/10.1146/annurev-soc-090320-094912

Collins, A. B., Boyd, J., Cooper, H. L. F., & McNeil, R. (2019). The intersectional risk environment of people who use drugs. *Social Science & Medicine, 234*, 112384. https://doi.org/10.1016/j. socscimed.2019.112384

Cooper, H. L. F., Linton, S., Kelley, M. E., Ross, Z., Wolfe, M. E., Chen, Y.-T., Zlotorzynska, M., Hunter-Jones, J., Friedman, S. R., Des Jarlais, D., Semaan, S., Tempalski, B., DiNenno, E., Broz, D., Wejnert, C., & Paz-Bailey, G. (2016). Racialized risk environments in a large sample of people who inject drugs in the United States. *International Journal of Drug Policy, 27*, 43–55. https://doi.org/10.1016/j.drugpo.2015.07.015

Crenshaw, K. (1989). Demarginalizing the intersection of race and sex: A black feminist critique of antidiscrimination doctrine, feminist theory and antiracist politics. *The University of Chicago Legal Forum, 1989*, 139.

Cretney, R. (2014). Resilience for whom? Emerging critical geographies of socio-ecological resilience. *Geography Compass, 8*(9), 627–640. https://doi.org/10.1111/gec3.12154

D'Ignazio, C., & Klein, L. F. (2020). Seven intersectional feminist principles for equitable and actionable COVID-19 data. *Big Data & Society.* https://doi.org/10.1177/2053951720942544

Deguen, S., Amuzu, M., Simoncic, V., & Kihal-Talantikite, W. (2022). Exposome and social vulnerability: An overview of the literature review. *International Journal of Environmental Research and Public Health, 19*(6), 3534. https://doi.org/10.3390/ijerph19063534

DiClemente, R. J., Salazar, L. F., Crosby, R. A., & Rosenthal, S. L. (2005). Prevention and control of sexually transmitted infections among adolescents: The importance of a socio-ecological perspective—A commentary. *Public Health, 119*(9), 825–836. https://doi.org/10.1016/j. puhe.2004.10.015

Douglas, M. (2015). Beyond 'health': Why don't we tackle the cause of health inequalities? In K. E. Smith, C. Bambra, & S. E. Hill (Eds.), *Health inequalities* (pp. 109–123). Oxford University Press. https://doi.org/10.1093/acprof:oso/9780198703358.003.0008

Forkey, H., Szilagyi, M., Kelly, E. T., Duffee, J., The Council On Foster Care, Adoption, and Kinship Care, Council On Community Pediatrics, Council On Child Abuse and Neglect, Committee On Psychosocial Aspects Of Child and Family Health, Springer, S. H., Fortin, K., Jones, V. F., Vaden Greiner, M. B., Ochs, T. J., Partap, A. N., Davidson Sagor, L., Allen Staat, M., Thackeray, J. D., Waite, D., & Weber Zetley, L. (2021). Trauma-informed care. *Pediatrics, 148*(2), e2021052580. https://doi.org/10.1542/peds.2021-052580

Fuchs, A. (1959). Geography of ophthalmology. *Klinische Monatsblatter Fur Augenheilkunde Und Fur Augenarztliche Fortbildung, 135*, 579–583.

Fuchs, V. R. (2017). Social determinants of health: Caveats and nuances. *JAMA, 317*(1), 25–26. https://doi.org/10.1001/jama.2016.17335

Gallego-Osorio, C., Betancurth-Loaiza, D. P., Vélez-Álvarez, C., & Universidad de Caldas. (2021). Determinantes sociales de la salud y activos comunitarios: Importancia para el análisis de contexto. *Revista U.D.C.A Actualidad & Divulgación Científica, 24*(2). https://doi.org/10.31910/rudca.v24.n2.2021.1633

Goodchild, M. F. (1991). Geographic information systems. *Progress in Human Geography, 15*(2), 194–200. https://doi.org/10.1177/030913259101500205

Gudi-Mindermann, H., White, M., Roczen, J., Riedel, N., Dreger, S., & Bolte, G. (2023). Integrating the social environment with an equity perspective into the exposome paradigm: A new conceptual framework of the Social Exposome. *Environmental Research, 233*, 116485. https://doi.org/10.1016/j.envres.2023.116485

Hajat, A., Diez-Roux, A. V., Adar, S. D., Auchincloss, A. H., Lovasi, G. S., O'Neill, M. S., Sheppard, L., & Kaufman, J. D. (2013). Air pollution and individual and neighborhood socioeconomic status: Evidence from the Multi-Ethnic Study of Atherosclerosis (MESA). *Environmental Health Perspectives, 121*(11–12), 1325–1333. https://doi.org/10.1289/ehp.1206337

Hanchette, C. L. (2008). The political ecology of lead poisoning in eastern North Carolina. *Health & Place, 14*(2), 209–216. https://doi.org/10.1016/j.healthplace.2007.06.003

Harari, L., & Lee, C. (2021). Intersectionality in quantitative health disparities research: A systematic review of challenges and limitations in empirical studies. *Social Science & Medicine*, (1982), 277, 113876. https://doi.org/10.1016/j.socscimed.2021.113876

Harper, J. (2004). Breathless in Houston: A political ecology of health approach to understanding environmental health concerns. *Medical Anthropology, 23*(4), 295–326. https://doi.org/10.1080/01459740490513521

Harvey, M. (2020). How do we explain the social, political, and economic determinants of health? A call for the inclusion of social theories of health inequality within U.S.-Based Public Health Pedagogy. *Pedagogy in Health Promotion, 6*(4), 246–252. https://doi.org/10.1177/2373379920937719

Hatzenbuehler, M. L., Phelan, J. C., & Link, B. G. (2013). Stigma as a fundamental cause of population health inequalities. *American Journal of Public Health, 103*(5), 813–821. https://doi.org/10.2105/AJPH.2012.301069

Janiszewska-Zygier, A. (1966). Geography of glaucoma in Poland. *Klinika Oczna, 36*(1), 127–130.

Kearns, R. A. (1993). Place and health: Towards a reformed medical geography. *The Professional Geographer, 45*(2), 139–147. https://doi.org/10.1111/j.0033-0124.1993.00139.x

Kennedy, A. (2024). *What it means to be trauma-informed* (1st ed.). Routledge. https://doi.org/10.4324/9781003371533

Kilanowski, K. (2017) Breadth of the Socio-Ecological Model, Journal of Agromedicine, 22:4, 295–297, https://doi.org/10.1080/1059924X.2017.1358971

King, M. F., Renó, V. F., & Novo, E. M. (2014). The concept, dimensions and methods of assessment of human well-being within a socioecological context: a literature review. Social indicators research, 116, 681–698. https://doi.org/10.1007/s11205-013-0320-0

Koch, T. (2009). Social epidemiology as medical geography: Back to the future. *GeoJournal, 74*(2), 99–106. https://doi.org/10.1007/s10708-009-9266-9

Koch, T. (2022). Back to the future: Covid-19 and the recurring debate over social determinants of disease, and health. *Social Sciences & Humanities Open, 6*(1), 100298. https://doi.org/10.1016/j.ssaho.2022.100298

Kolak, M., Bhatt, J., Park, Y. H., Padrón, N. A., & Molefe, A. (2020). Quantification of neighborhood-level social determinants of health in the continental United States. *JAMA Network Open, 3*(1), e1919928. https://doi.org/10.1001/jamanetworkopen.2019.19928

Krieger, N. (2000). Epidemiology and social sciences: Towards a critical reengagement in the 21st century. *Epidemiologic Reviews, 22*(1), 155–163. https://doi.org/10.1093/oxfordjournals.epirev.a018014

Kruize, H., Droomers, M., Van Kamp, I., & Ruijsbroek, A. (2014). What causes environmental inequalities and related health effects? An analysis of evolving concepts. *International Journal of Environmental Research and Public Health, 11*(6), 5807–5827. https://doi.org/10.3390/ijerph110605807

Lin, Q., Kolak, M., Watts, B., Anselin, L., Pollack, H., Schneider, J., & Taylor, B. (2022a). Individual, interpersonal, and neighborhood measures associated with opioid use stigma: Evidence from a nationally representative survey. *Social Science & Medicine, 305*, 115034. https://doi.org/10.1016/j.socscimed.2022.115034

Lin, Q., Paykin, S., Halpern, D., Martinez-Cardoso, A., & Kolak, M. (2022b). Assessment of structural barriers and racial group disparities of COVID-19 mortality with spatial analysis. *JAMA Network Open, 5*(3), e220984. https://doi.org/10.1001/jamanetworkopen.2022.0984

Link, B. G., & Phelan, J. (1995). Social conditions as fundamental causes of disease. *Journal of Health and Social Behavior, 80–94.* https://doi.org/10.2307/2626958

Lutfey, K., & Freese, J. (2005). Toward some fundamentals of fundamental causality: Socioeconomic status and health in the routine clinic visit for diabetes. *American Journal of Sociology, 110*(5), 1326–1372. https://doi.org/10.1086/428914

Macgregor, G. (1961). Social determinants of health practices. *American Journal of Public Health and the Nations Health, 51*(11), 1709–1714.

Mackenbach, J. P., Looman, C. W. N., Artnik, B., Bopp, M., Deboosere, P., Dibben, C., Kalediene, R., Kovács, K., Leinsalu, M., Martikainen, P., Regidor, E., Rychtaříková, J., & de Gelder,

R. (2017). 'Fundamental causes' of inequalities in mortality: An empirical test of the theory in 20 European populations. *Sociology of Health & Illness, 39*(7), 1117–1133. https://doi.org/10.1111/1467-9566.12562

Masterson, V. A., Stedman, R. C., Enqvist, J., Tengö, M., Giusti, M., Wahl, D., & Svedin, U. (2017). The contribution of sense of place to social-ecological systems research: A review and research agenda. *Ecology and Society, 22*(1) https://www.jstor.org/stable/26270120

Mehtälä, M. A. K., Sääkslahti, A. K., Inkinen, M. E., & Poskiparta, M. E. H. (2014). A socio-ecological approach to physical activity interventions in childcare: A systematic review. *The International Journal of Behavioral Nutrition and Physical Activity, 11*, 22. https://doi.org/1 0.1186/1479-5868-11-22

Menschner, C., & Maul, A. (n.d.). Key ingredients for trauma-informed care implementation. In *Center for health care strategies*. Retrieved July 29, 2024, from https://www.chcs.org/resource/key-ingredients-for-successful-trauma-informed-care-implementation/

Mittelmark, M. B., & Bull, T. (2013). The salutogenic model of health in health promotion research. *Global Health Promotion, 20*(2), 30–38. https://doi.org/10.1177/1757975913486684

Moise, I. K. (2020). Variation in risk of COVID-19 infection and predictors of social determinants of health in Miami–Dade County, Florida. *Preventing Chronic Disease, 17*, 200358. https://doi.org/10.5888/pcd17.200358

Moise, I. K., Ortiz-Whittingham, L. R., Owolabi, K., Halwindi, H., & Miti, B. A. (2024). Examining the role of social determinants of health and COVID-19 risk in 28 African countries. *COVID, 4*(1), 87–101. https://doi.org/10.3390/covid4010009

Morrill, R. L., & Earickson, R. (1968). Variation in the character and use of chicago area hospitals. *Health Services Research, 3*(3), 224–238.

Mullins, L. S. (1966). Sources of information on medical geography. *Bulletin of the Medical Library Association, 54*(3), 230–242.

Neumann, R. P. (2009). Political ecology: Theorizing scale. *Progress in Human Geography, 33*(3), 398–406. https://doi.org/10.1177/0309132508096353

Nonyel, N. P., Wisseh, C., Riley, A. C., Campbell, H. E., Butler, L. M., & Shaw, T. (2021). moving from injustice to equity: A time for the pharmacy profession to take action: Conceptualizing social ecological model in pharmacy to address racism as a social determinant of health. *American Journal of Pharmaceutical Education, 85*(9), 959–965.

Obinna, D. N. (2021). Confronting disparities: Race, ethnicity, and immigrant status as intersectional determinants in the COVID-19 era. *Health Education & Behavior, 48*(4), 397–403. https://doi.org/10.1177/10901981211011581

Ohri-Vachaspati P, DeLia D, DeWeese RS, Crespo NC, Todd M, Yedidia MJ (2015). The relative contribution of layers of the Social Ecological Model to childhood obesity. Public Health Nutrition. 18:11, 2055–2066. https://doi.org/10.1017/S1368980014002365

Openshaw, S., Charlton, M., Craft, A. W., & Birch, J. M. (1988). Investigation of leukemia clusters by use of a geographical analysis machine. *The Lancet, 331*(8580), 272–273. https://doi.org/10.1016/S0140-6736(88)90352-2

Pain, R. (2021). Geotrauma: Violence, place and repossession. *Progress in Human Geography, 45*(5), 972–989. https://doi.org/10.1177/0309132520943676

Parker, S., & Johnson-Lawrence, V. (2022). Addressing trauma-informed principles in public health through training and practice. *International Journal of Environmental Research and Public Health, 19*(14), 8437. https://doi.org/10.3390/ijerph19148437

Paul, R., Arif, A. A., Adeyemi, O., Ghosh, S., & Han, D. (2020). Progression of COVID-19 from urban to rural areas in the United States: a spatiotemporal analysis of prevalence rates. *The Journal of Rural Health*. https://doi.org/10.1111/jrh.12486

Pérez-Wilson, P., Marcos-Marcos, J., Morgan, A., Eriksson, M., Lindström, B., & Álvarez-Dardet, C. (2021). 'A synergy model of health': An integration of salutogenesis and the health assets model. *Health Promotion International, 36*(3), 884–894. https://doi.org/10.1093/heapro/daaa084

Phelan, J. C., & Link, B. G. (2013). Fundamental cause theory. In W. C. Cockerham (Ed.), *Medical sociology on the move: New directions in theory* (pp. 105–125). Springer. https://doi. org/10.1007/978-94-007-6193-3_6

Phelan, J. C., Link, B. G., & Tehranifar, P. (2010). Social conditions as fundamental causes of health inequalities: Theory, evidence, and policy implications. *Journal of Health and Social Behavior, 51*(Suppl. 1), S28–S40. https://doi.org/10.1177/0022146510383498

Phelos, H. M., Kass, N. M., Deeb, A.-P., & Brown, J. B. (2022). Social determinants of health and patient-level mortality prediction after trauma. *Journal of Trauma and Acute Care Surgery, 92*(2), 287–295. https://doi.org/10.1097/TA.0000000000003454

Phillips, T. A. (1959). Leukæmia and geography. *The Lancet, 274*(7104), 659–661. https://doi. org/10.1016/S0140-6736(59)91427-8

Reading, C. L., & Wien, F. (2009). *Health inequalities and social determinants of aboriginal peoples' health*. National Collaborating Centre for Aboriginal Health.

Rhodes, T. (2002). The 'risk environment': A framework for understanding and reducing drug-related harm. *International Journal of Drug Policy, 13*(2), 85–94. https://doi.org/10.1016/S0955-3959(02)00007-5

Richardson, D. S., Richardson, D., Volkow, N. D., Kwan, M.-P., Kaplan, R. M., Goodchild, M. F., & Croyle, R. T. (2013). Spatial turn in health research. *Science*. https://doi.org/10.1126/science.1232257

Richmond, C., Elliott, S. J., Matthews, R., & Elliott, B. (2005). The political ecology of health: Perceptions of environment, economy, health and well-being among 'Namgis First Nation. *Health & Place, 11*(4), 349–365. https://doi.org/10.1016/j.healthplace.2004.04.003

Rollston, R., & Galea, S. (2020). COVID-19 and the social determinants of health. *American Journal of Health Promotion: AJHP, 34*(6), 687–689. https://doi.org/10.1177/0890117120930536b

Rothstein, R. (2018). *The color of law: A forgotten history of how our government segregated America (First published as a Liveright paperback 2018)*. Liveright Publishing Corporation, A Division of W.W. Norton & Company.

Schuurman, N., Hameed, S. M., Fiedler, R. S., Bell, N., & Simons, R. K. (2008). The spatial epidemiology of trauma: The potential of geographic information science to organize data and reveal patterns of injury and services. *Canadian Journal of Surgery*.

Smithsonian Institution. (2024). Guide to the Gordon Macgregor papers, 1951–1966. *National Anthropological Archives*. Retrieved from https://sova.si.edu/record/naa. xxxx.0302?q=Poverty&t=C

Sugrue, T. J. (2014). *The origins of the urban crisis: Race and inequality in postwar detroit (First Princeton classics edition)*. Princeton University Press.

Thimm-Kaiser, M., Benzekri, A., & Guilamo-Ramos, V. (2023). Conceptualizing the mechanisms of social determinants of health: A heuristic framework to inform future directions for mitigation. *The Milbank Quarterly, 101*(2), 486–526. https://doi.org/10.1111/1468-0009.12642

Tobler, W. R. (1967). Computer use in geography. *Behavioral Science, 12*(1), 57–58. https://doi. org/10.1002/bs.3830120108

U.S. Dept of Health and Human Services. (n.d.). *Social determinants of health—Healthy people 2030*. Retrieved July 30, 2024, from https://health.gov/healthypeople/priority-areas/social-determinants-health

Vinje, H. F., Langeland, E., & Bull, T. (2017). Aaron Antonovsky's development of salutogenesis, 1979–1994. In M. B. Mittelmark, S. Sagy, M. Eriksson, G. F. Bauer, J. M. Pelikan, B. Lindström, & G. A. Espnes (Eds.), *The handbook of salutogenesis* (pp. 25–40). Springer International Publishing. https://doi.org/10.1007/978-3-319-04600-6_4

Wamsley, L. (2021, April 8). *CDC director declares racism a "Serious Public Health Threat"*. NPR. Retrieved from https://www.npr.org/2021/04/08/985524494/cdc-director-declares-racism-a-serious-public-health-threat

Webb Hooper, M., Marshall, V., & Pérez-Stable, E. J. (2022). COVID-19 health disparities and adverse social determinants of health. *Behavioral Medicine (Washington, D.C.), 48*(2), 133–140. https://doi.org/10.1080/08964289.2021.1990007

Wild, C. P. (2005). Complementing the genome with an "Exposome": The outstanding challenge of environmental exposure measurement in molecular epidemiology. *Cancer Epidemiology, Biomarkers & Prevention, 14*(8), 1847–1850. https://doi.org/10.1158/1055-9965.EPI-05-0456

Wilderink, L., Bakker, I., Schuit, A. J., Seidell, J. C., Pop, I. A., & Renders, C. M. (2022). A theoretical perspective on why socioeconomic health inequalities are persistent: Building the case for an effective approach. *International Journal of Environmental Research and Public Health, 19*(14), 8384. https://doi.org/10.3390/ijerph19148384

Wind, K. S. (2021). *What causes health? Revisiting the Social Determinants of Health (SDH) through a Salutogenic Lens and Self-Reported Health (SRH) as the main outcome: A realist evaluation [Ph.D., University of Toronto (Canada)].* Retrieved from https://www.proquest.com/docview/2608509766/abstract/584A42E1307A49B4PQ/1

World Health Organization. (2010). *A conceptual framework for action on the social determinants of health* (p. 76).

Chapter 3
Defining the Social-Spatial Determinants of Health Through Transformative Critical Praxis

Bryce Puesta Takenaka, Ijeoma Opara, and Trace Kershaw

Introduction

Conversations on health and where health inequities are occurring (or absent) remain a common denominator for many of us in the field of public health. Often, the question that is asked among health populations: "Why do certain populations who reside in certain neighborhoods have worse health outcomes than others? Which populations make up socially vulnerable neighborhoods compared to other groups?" The abundance of evidence that attempts to answer such questions suggests that our environment plays a role in creating, maintaining, and reinforcing adverse health outcomes and health inequities. In this chapter, we will first describe the existing conceptualizations of the social and spatial determinants of health. Second, we will explore the current applications of the critical praxis of the ecosocial theory of disease distribution and intersectionality. Lastly, we will propose an integrated analytical framework to advance our definition of social-spatial determinants of health (SSDoH). We question how SSDoH may challenge or even improve current deterministic assumptions of "causes of health" that are perched downstream. In doing so, we demonstrate how both ecosocial theory and intersectionality may together more accurately interrogate the structural underpinnings of how we understand health and health inequities.

B. P. Takenaka (✉) · I. Opara · T. Kershaw
Department of Social & Behavioral Sciences, Yale School of Public Health,
New Haven, CT, USA
e-mail: bryce.takenaka@yale.edu

© The Author(s) 2026
M. A. Kolak, I. K. Moise (eds.), *Place and the Social-Spatial Determinants of Health*, Global Perspectives on Health Geography,
https://doi.org/10.1007/978-3-031-88463-4_3

The Social Determinants of Health

The purpose of the inaugural Commission on the Social Determinants of Health (CSDH) was to upheave the global movement to address the social drivers, apart from medical care factors, that shape health inequities across a range of settings and populations (World Health Organization [WHO], 2008; Marmot, 2000). The social determinants of health (SDoH) are defined as the "conditions in which people are born, grow, live, work, and age" that are "shaped by the distribution of money, power, and resources globally, nationally, and locally" (Marmot et al., 2008). Common examples of the SDoH may include, but are not limited to: education, employment, income and wealth, the physical environment, public safety, transportation, health systems and services, and social policies. Later, the WHO expanded this definition, adding "… and the wider set of forces and systems shaping the conditions of daily life," including the "economic policies and systems, development agendas, social norms, social policies, and political systems" (WHO, 2020). There is no doubt that the SDoH framework has added value to understanding and addressing multiple forms of health inequality. However, we have yet to see exponential results in closing long-standing health gaps, especially among communities that continue to be neglected and oppressed.

This is not to discredit the SDoH nor existing spatial perspectives of public health, which have an extensive history of promoting public health but are yet to be clearly defined. Here, we start with spatial epidemiology as a well-suited proxy for spatial approaches of public health, which is "the description and analysis of geographic variations in disease with respect to demographic, environmental, behavioral, socioeconomic, genetic, and infectious risk factors" (Kirby et al., 2017). Spatial epidemiological studies and technology such as geographic information systems (GIS) have risen in utility over the last two decades. Such approaches to public health have provided value through the creative and effective deployment of spatial health data and advancing our understanding of clustered health phenomena through spatial analysis—from measuring the distribution of healthy food options and food deserts, to the examination of associations between racialized housing policies and chronic asthma–related outcomes, as well as exploring rates of SARS-CoV-2 vaccination coverage between socially vulnerable neighborhoods. The spatialities of health have guided scholars to understand the environmental determinants of health.

The SDoH has been referred to as the "causes of the causes" (Braveman & Gottlieb, 2014), yet the framing of what SDoH means has been misused and been taken out of context due to its ambiguity in deterministic language. Common examples, such as socioeconomic status and education, should not be interpreted as "root causes" of health but rather be interpreted as social positions. Taking social positions like "socioeconomic status" and "education" into account is necessary for improving health. However, these arrangements are not created by coincidence, but rather driven by forces that operate further upstream that may not be captured through SDoH framing (Lundberg, 2020). Thus, there is a need to shift more attention to why and how opportunities, advantages, and disadvantages are created and maintained in the first place.

The Built Environment

Individual and population health are inextricably linked to our environment, particularly the physical (built) environment. The interpretation of the built environment varies across the literature. For this chapter, we will refer to the built environment as a collection of human-designed objects that include buildings and cities, transportation systems, urban design localities, and landscapes that shape the opportunities that promote or hinder health equity. Among other physical features, an additional framing of the built environmental processes on health can consider walkable and attractive routes (e.g., sidewalks, paths) to and from certain localities that shape health (e.g., hospitals, retail stores, schools, parks). These features can also be reduced to the characteristics of one's house, such as the type of unit (e.g., high-rise, multiple dwelling units, congregate setting) and the quality (e.g., quality of maintenance and upkeep, amenities). The built environment and other SDoH have been widely approached through a socioecological lens, from micro (individual)-, meso (community)-, and macro-level (policy) stages (McLeroy et al., 1988). However, important structural-level forces (e.g., white supremacy, racism, heterosexism) and their role within our environment are often omitted (Czyzewski, 2011).

External forces often limit resources and create unsupportive and unhealthy environments that influence neighborhood-level built environments. One prominent example of this structural discourse is the historical use of housing policies rooted in racism and racial capitalism. The US federal government established the Home Owners' Loan Corporation (HOLC), created in 1933 to improve the housing market by providing refinancing aid, low-interest loans to recover foreclosed homes, and purchasing new mortgages with extended repayment timelines to qualified homeowners (Mitchell & Franco, 2018). Shortly after in 1934, the Federal Housing Administration (FHA) produced an Underwriting Manual outlining the guidelines to "preserve" property values to ensure secure bank comfort in distributing loans, stating "[i]f a neighborhood is to retain stability, it is necessary that properties shall continue to be occupied by the same social and racial classes" (Federal Housing Administration Underwriting Manual, 1938). Qualities of neighborhoods were also "investigated to determine whether incompatible racial and social groups are present." This intentional use of color-coded maps to determine whether residents are unfit or "incompatible" is referred to as redlining (Rothstein, 2017). While HOLC ceased its operations in 1951, institutional racism and racialized practices continued, such as predatory lending, zoning, restrictive covenants, and blockbusting, to name a few (Taylor, 2018). According to recent studies, historical redlining practices where neighborhoods received lower graded HOLC scores have manifested into present-day neighborhood disadvantages, such as lower home values, more vacancies, increased poverty, and socioeconomic immobility (Aaronson et al., 2021; Lynch et al., 2021).

Similar colonial parallels of displacement and segregation are apparent with the August 2023 wildfires that ravaged Lāhainā, Maui Komohana (west), killing an estimated 100 people (Bonilla, 2023). This disaster led to a compounding effect of numerous health implications, such as the effects of smoke inhalation and toxic debris, trauma-related tolls, and reckoning on marine ecosystems (Taparra et al., 2023). Scientists suggested climate change as a major reinforcer of the fires. However, interrogating this disaster with a simplistic SDoH framing does not tell the complete story. Instead, this event should be contextualized into its settler colonial roots, involving advantageous Western capitalism that enabled the inequitable diversion of water from Lāhainā and homesteads of Native Hawaiians to maintain plantations and luxury estates (Klein & Sproat, 2023). The legacy of water mismanagement has jeopardized the ecosystems of Hawai'i, which has made Lāhainā vulnerable to climate disasters and enabling continued harm to Native Hawaiians. The afterlives of plantations to tourism in Hawai'i has led to exorbitant invasive plants, like guinea grass and buffelgrass that made Lāhainā a flammable area, coupled with the continued displacement of Native Hawaiians due to unaffordable costs of living (Magbual, n.d.; Taparra et al., 2023). According to the 2020 US Census Bureau, only 47% of Native Hawaiians live in Hawai'i, with a 29% increase in migration to the continent since 2010 (US Census Bureau, 2023). This forced displacement and disposession may be attributable to unequal housing opportunities, as suggested by the Hawai'i Appleseed Center for Law & Economic Justice: data from 2018 showed that in Maui alone, about 52% of homes were sold to non-residents, with 60% of apartments and condos to out-of-state investors or second homeowners (Geminiani & DeLuca, 2018).

Similar to tourist-heavy cities like San Francisco with similar housing trajectories, the uptake of popular vacation rental units (VRU) has led to higher costs of housing–149% over the national average (Dubetz et al., 2022). Egregious cost of housing and living prices are partial products of larger neocolonial-driven pillars in Hawai'i that curate forced displacement of Indigenous Pacific Islanders. These examples illustrate the ways in which business and political elites organize neighborhoods to exclude certain groups to determine how the residential and industrial planes would desirably evolve, hence manufacturing the built environment. While much of built environment research has centered on defining population distributions and uncovering health inequalities, this chapter highlights the need to broaden such work with critical analytic frameworks to understand the "causes of the causes."

Current Perspectives on Neighborhoods and Health

Spatial epidemiological research on the built environment has some conceptual and methodological overlaps with neighborhood effects research. The difference here is that rather than a lens primarily focusing on disease distribution and its social factors, neighborhood effects research often tells a story of structures that dictate where people live and how neighborhoods become "resource-deprived" through an environmental justice lens. It is well understood from the SDoH framework that where you live impacts your health (Diez Roux, 2001). Similarly, neighborhood effects research assumes where you live affects your life chances (Slater, 2013). However, neighborhood effects welcome an explanatory lens to examine the structural implications of these assumptions, centering the question: *why do people live where they are? how are places created, arranged, and organized in the first place?*

Investigating the role of capitalist urbanization (i.e., arrangement of private property rights) on neighborhoods, Tom Slater illustrates how the forced concentration of poor urban dwellers in specific neighborhoods in Manchester shaped their life chances but also how these residents mutually shaped their neighborhood under the institutional apparatus of capitalist urbanization (Slater, 2013). Here, neighborhoods can be seen as merely political instruments of accusation that imply that residents alone who occupy these impoverished places are "ultimately responsible for their own social and economic situation," failing to capture external forces that operate outside of these neighborhoods. However, efforts to understand and address the *why* and *how* are limited if we continue to attempt to solve structural issues through an SDoH lens (Rasanathan, 2018). Neighborhoods are not inherently "good" or "bad" but operate as intended based on sociopolitical systems of injustice. Public health scholars need to not only reject the problematic causal assumptions of perpetuated stigmatization that create fables of poor neighborhoods, but also invert such statements and questions that seek to disentangle why and how life chances affect where you live (Slater, 2013). Studying neighborhood effects may help challenge us to ask deeper, structurally rooted questions.

Alongside the physical and built attributes of the environment, such as schools, parks, and workplaces, neighborhoods are hubs that also comprise and enact complex social and spatial processes and meaning. However, what is considered a neighborhood? And how do you define something so abstract? Over the years, health researchers have cycled through multiple definitions of neighborhoods. Within the domains of public health, neighborhoods are considered "geographical places that can have social and cultural meaning to residents and non-residents alike and are subdivisions of large places" (Duncan & Kawachi, 2018). Other disciplines, such as social geography, urban planning, and sociology, have referred to neighborhoods as "subsections of a larger community–a collection of both people and institutions occupying a spatially defined area influenced by ecological, cultural, and sometimes political forces" (Weiss et al., 2007, p. 154–159). Despite geographic advances in recognizing the social constructions and spatial complexity of neighborhoods, many interpretations in public health continue to refer to neighborhoods as exclusively static administrative geographic units of residential and population (Arcaya et al., 2016) and very rarely about the ways social interaction and symbolic meaning and significance factor in Erfani (2022), Gould and White (2012). Neighborhoods, like places, may actually transcend through multiple accumulated past experiences and continuous realities of re-shaping and sharing, which may not obey the constraints of physical and economic materialities, thus allowing individuals the freedom to establish their own world, full of value and meaning (Cresswell, 2014; Tuan, 1971). The ubiquity of interpretations reflects how places shape people and people shape places, a re-conceptualized lens to help us see neighborhoods beyond their arbitrary and singular confines.

The impact of neighborhoods on health is nothing new. Among some of the most prominent thought leaders of place and health was William Edward Burghardt (W. E. B.) Du Bois. W. E. B. Du Bois was a sociologist, historian, and abolitionist, who led some of the earliest investigations on racialized neighborhoods. In *The Philadelphia Negro: A Social Study,* he investigated associations of concentrated poverty and spatial segregation as the result of institutional racism that excluded Black residents living in the seventh ward of Philadelphia in the late nineteenth century (Du Bois & Eaton, 1996). Joined with other conceptualizations of neighborhoods, there has been a growing recognition of neighborhoods being more than just their visible features but also possessing symbolic meaning and significance that are created and reproduced within and throughout neighborhoods (Keene & Padilla, 2014). This concept of spatial stigma was developed from the scholarship around structural stigma, which refers to the processes of how different localities and their residents are stereotyped (Hatzenbuehler & Link, 2014).

Rooted in white supremacy and heteropatriarchal ideologies (Mitchell et al., 2021), spatial stigma overlaps with our previous examples of racialized policies of redlining to maintain desirable white neighborhoods and the neocolonial procedures of disinvestment in native homesteads of Hawai'i. Indeed, spatial stigma operates under similar structural paradigms that create and maintain inequality. Comparably, "territorial stigmatization" as defined by Loïc Wacquant considers a type of violence that proliferates racialized urban poverty into spatial processes generating excluded "ghettos" and degraded urban spaces (Wacquant, 2008). These so-called "blemish of places" would then become racialized neighborhoods that moderate the perceptions of disorder, being more apparent in predominantly neighborhoods of color than in white affluent neighborhoods. As a result, these neighborhoods become subjected to disinvestment and painted with false narratives that isolate and exclude those residing in these neighborhoods (Aaronson et al., 2021; Lynch et al., 2021). We can also critique the ways spatial stigma inversely occurs in unique cultural localities, such as Hawai'i. The advanced marginality of Indigenous Pacific Islanders in Hawai'i may be seen as a result of positive or desirable reputation (Tran et al., 2020) that reinforced the development of plantations and luxury resorts, and disinvestment in important ecosystems, such as kalo pondfields and abundant wetlands.

Spatial stigma demonstrates how a place becomes stigmatized by those who occupy it, leading to a cycle of disinvestment, exclusion, and abandonment from symbolic marginalization (Keene & Padilla, 2014; Wacquant, 2008). With the growing literature around theoretical constructs such as spatial stigma, researchers may move on from summations that "where you live affects your life chances" toward more complex, structural views that inquire why people live in these places in the first place (Slater, 2013). With that, understanding and considering neighborhoods as a social determinant of health integrate social-spatial theories and praxis to address health inequalities beyond the individual level (Diez Roux, 2001).

Gaps in Current Social-Spatial Health

The maturation of SDoH framing and understanding to address health inequities has been greatly attributable to the integration of both spatial health and neighborhood effects scholarship. Overall, we have been able to discern the specificity of how dynamic and unbound neighborhood context and compositions are in relation to the health of residents and non-residents alike. Despite the nuances in the current theoretical understanding of neighborhoods and their social and spatialities on health, major conceptual and methodological gaps in place-health research are apparent. For this chapter, we focus on the conceptual definition of the SSDoH, as proposed methodological extensions and empirical illustrations are discussed elsewhere.

One core agent that reinforces the social-spatial attributes of neighborhoods and health is the examination of "power" (Foucault, 1978). The utility of power, from execution to subjugation, was conceptualized as the "techniques for achieving the subjugation of bodies and the control of populations" (Foucault, 1978). Power may help us recognize the ways institutions legitimize and produce social arrangements that manage the norms that conform to secure a "vital population" (Roach, 2009). Again, we refer to the ways power has been exercised through HOLC redlining practices that excluded communities of color from achieving housing opportunities, as well as the colonial violence that created a vulnerable landscape and unsustainable economy in Hawai'i. Power is also not a traditional apparatus of domination from top to bottom, but should be seen on a multiplicative scale as interlocking systems of oppression (Collins, 2022). Power becomes embodied through the assimilation of societal arrangements (what we know as the SDoH) and how they are distributed, produced, and reproduced through place. This embodiment of place and power is important since place and identity are attached through symbolic meanings, which mutually shape one's environment (Petteway et al., 2019b).

There has also been a long disconnect in the production of spatial knowledge of *who* gets to tell these stories and define "place" (Petteway et al., 2019a). More specifically, this reflects how dominant modes of knowledge production, particularly spatial knowledge, render individuals with less agency in their own place-making, and obscure the actual systems that are responsible for embodied health inequities. This concept of embodied inequalities is familiar to public health, as embodiment has been described by Krieger (1994) through the ecosocial theory of disease distribution. Ecosocial theory grounds the SDoH into rich spatiotemporal scales that transcend historical, generational, and life course perspectives—challenging the contemporary definition of SDoH. As power operates and flows through multiple axes, geographic health inequities are produced, reproduced, and become complex hubs. Developed through intersectional geographic perspectives and now growing in contemporary SDoH, spatial public health has looked at place alongside identities of race and ethnicity, gender, sexuality, age, and class.

The existing scholarship in neighborhoods and health research has paved the way to make place-based health inequities visible. However, the current social-spatial health paradigms have yet to be standardized to a common interpretation. A great challenge with the current SDoH framework is its lack of precision and its omission of complex pathways that create, manipulate, and uphold patterns of geographic health inequities. Moreover, we turn to literature, within and beyond spatial public health, the built environment, and neighborhood effects research, to urge a shift in conventional SDoH thinking and defining of health through a wider spatial lens. We also challenge the hesitancy and silence that becomes enacted from within an oversimplified SDoH framework, impeding the efforts for radical change in redistributing power.

A Place for Interdisciplinary Critical Praxis

Ecosocial Theory of Disease Distribution

Various attempts to explain and intervene in the societal inequities of health have been made through the many couplings of public health theoretical frameworks (Marmot et al., 2008; McLeroy et al., 1988). However, most theoretical frameworks in public health scholarship only do so much as report on the descriptive societal processes of health outcomes that often overcast opportunities for explanatory questioning and examination. Here, we consider Nancy Krieger's ecosocial theory of disease distribution, which encompasses the engagement of "the societal, biological, and ecological processes at the heart of embodying (in)justice" (Krieger, 2021)—that is, how interactions of the external social and physical world disrupt our literal bodies' physical, psychological, and emotional functions.

Unlike other public health theoretical frameworks, ecosocial theory highlights how the conditions in which peoples' embodied realities occur are controlled and maintained beyond standard classifications of social ecology, but are fundamentally interdependent on intergenerational material and symbolic arrangements of power and privilege (Krieger, 2001). The concepts of embodiment guide us to understand these realities through various spatiotemporal scales over the life course, which has enabled scholars to explore the ways embodiment occurs within and between places (neighborhoods) and how these neighborhoods of existing, living, playing, working, and aging are concretized in the body (McKittrick, 2011; Petteway et al., 2019b). Additionally, embodiment describes the realities crafted through our everyday lives that occur in certain localities at different points in time, suggesting how embodied truths are "a story about a place—what it is, where it is, how it is, and for whom" (Petteway et al., 2019a). These embodied stories in relation to place have informed the last few decades of place-embodied scholarship.

For instance, Petteway et al. (2019a, b) explore the geographies of embodiment as part of the People's Social Epi Project (PSEP), which is a multimethod intergenerational community-based participatory research (CBPR) initiative looking at the intersection of place, embodiment, and health. A combination of mapping methodologies, such as photovoice, activity space mapping, and participatory GIS, pinpoints person-centered place-health experiences. First, photovoice was used to document participants' daily places and exposures within certain places they self-perceive to impact their health. These images were narrated, giving each image its own story. The second method, activity space mapping, involved participants geolocating and rating their photovoice images, as well as other non-photographed points on a map. Then, X-ray mapping, an exercise of "body mapping" and "body storytelling," was used to qualitatively identify precise pathways the environment has on the body. Finally, participatory GIS allowed participants to digitally map their own data via a web-based platform to digitally share with others, as well as the broader community and city leaders. Extending outside participants' administrative boundaries and within overlapping embodied places—*where* and *how* place attributes were affecting their bodies—was captured using multi-intervention approaches to capture the subjective meaning of place and its role in embodied health inequalities.

This study is an example of the increased need to divorce from administrative boundaries of place toward more fluid and relational approaches that give communities back the pen to tell their stories of embodied places. Highlighting the importance of people's mobility and unique activity spaces reflects the call for an urgent shift in person-centered spatial reasoning of current and historical configurations that give place meaning. In place-embodied health research, these examples crystallize the link between place and physiological being. Place embodiments are shaped through continual societal arrangements of power and opportunity, drawing on the implications of how local sociopolitical processes and practices create these daily conditions and environmental exposures that become a part of a person. Related to "place identity," which has evolved into the understanding that social sense-making that varies by place or neighborhood becomes a hub of being and belonging (Wacquant, 2008), has been reflective of how subjective and experiential facets of geographical health inequities take place. As with the existing place-based health literature, there is a possible apparatus in which the allocation of place-based experiences becomes biologically embodied.

Intersectionality

Health inequity and its midstream SSDoH in the United States have been widely influenced by entrenched sociopolitical systems (e.g., white supremacy, heteropatriarchy) (hooks, 1982). With the growth in identifying and attempting to eliminate these health injustices, more public health scholars are considering the ways in which people's lived experiences are shaped through multiple forms of oppression and maintained by asymmetries in power. Intersectionality, coined by Black feminist legal scholar Kimberlé Crenshaw (1991), is a critical theoretical framework intended to describe how multiple and interlocking systems of power, privilege, and oppression shape realities. However, the roots of intersectional thinking may date back to abolitionist Sojourner Truth's (1851) examination of the intersection of race and gender in her "Ain't I a Woman?" speech at the 1851 Women's Rights Convention in Akron, Ohio. The scholarship of intersectionality has evolved through the documentation of similar experiences, including the stereotypes of Black women upon various oppressive systems, tokenism, subordination, and erasure and silence within a white world (Collins, 1989; Combahee River Collective, 1977). Although developed within Black feminist theory to originally understand the intersection of Black women's race, gender, and class positions (Collins, 1989), intersectionality has been extendable to understanding a broad landscape of oppression and how it marginalizes populations based on racial/ethnic, sex, gender, sexual orientation, socioeconomic position, location, and other social identities (Bowleg, 2012). Such social positions at the individual level are suggested to reflect much larger and deeper institutional and structural empires of oppression (e.g., racism, colonialism, white supremacy) (Combahee River Collective, 1977; Czyzewski, 2011).

Similar to the ecosocial theory of disease distribution, intersectionality also emphasizes the importance of interacting structural power relations (Krieger, 2021). However, unlike ecosocial theory, intersectionality praxis does not concern itself with theorizing an individual's biological embodiment pathways with reflection on population health trends. Instead, intersectionality was originally born outside of the academy to advance advocacy and justice-oriented action (Crenshaw, 1991). Another special element of intersectional analysis in public health research is that it rejects the atomization of health inequities from a single-axis pathway of a person's identity but rather a "matrix"—where multiple dimensions of power are related and mutually situated to marginalize folks differently based on those identities, which shape health inequity outcomes (Bowleg, 2008; Collins, 2022). Along these lines, Patricia Hill Collins argues how social contexts (e.g., race, racism, and antiracism) hold different interpretations depending on an individual's or group's position (Collins, 2022) and suggests that "systems of power may be theoretically intersectional, yet in practice, some forms of oppression will be especially salient during particular periods of time and in particular social contexts" (p., 203).

Intersectionality extends the ability to illuminate how certain experiences of communities come about in certain places because groups are differently situated in a given locality within the nexus of power (Collins, 2022). Health research terminologies and messages need to consider the role of "place as an aspect of intersectionality" (Bambra, 2022). Moreover, an intersectional perspective in relation to place provides an opportunity to understand and more clearly define SSDoH through a compositional-contextual lens outside of their siloed customary social factors (e.g., social economic status, race/ethnicity, gender, age) and the physical make-up of neighborhoods (e.g., schools, parks, greenspace, healthcare facilities). We can better recognize the position of these conditions and factors beyond standard categories, and consider how they reinforce each other. An intersectional lens may aid further considerations of how places produce unique intersecting place-embodied experiences. Although intersectionality praxis is still in its infancy in public health research, it underscores the promise to converge the interlocking structures of SSDoH that shape health inequities (Bowleg, 2021; May, 2015).

Ecosocial and Intersectionality for Social-Spatial Determinants of Health

The social-spatial advancements of SDoH research to understand and address health inequity have developed over the years. More scholars are now pivoting their vantage point toward structural-level facets within an SDoH framework. However, as we have ascertained, upstream structural factors are frequently mistaken with socioeconomic positions, and often omit what puts people in these positions and living conditions in the first place. This framing is problematic, implying that the social risk factors for health are perched at the social level and that risk efforts for intervention and action remain at the individual level. Supported by a spatial view of public health, the built environment, and neighborhood effects research, an opportunity presents itself to understand how health inequities occur in the first place, and even why people who embody health inequities reside and remain trapped in disadvantaged neighborhoods. The deterministic language around current SDoH and spatial framings has helped identify health inequity and pinpoint its descriptive attributes, but continues to fall short in theorizing and critically analyzing deeper systems of power and domination, as well as the practice of addressing those injustices.

Adding to the many theoretical approaches engaged with the SDoH and spatial perspectives, we describe the independent and similar attributes of how the ecosocial theory of disease distribution and intersectionality has been used to address the macrosocial-spatial causes and power asymmetries that shape health inequities. Ecosocial theory and intersectionality may offer a valuable extension to how both are used independently. In fact, we are not the first to recognize the possibilities of using both ecosocial theory and intersectionality together (Merz et al., 2023). However, extending similar calls to action, we add to the ways both analytical frameworks can help with making sense of the role of place as both an embodied reality and an identity linked to power, which in turn may illuminate the dynamic structures of the sociopolitical economy driven through white supremacy, institutional discrimination, structural stigma, and transgenerational trauma (Stonington et al., 2018).

We propose a framework in which to integrate ecosocial and intersectionality perspectives of the SSDOH (see Fig. 3.1). The overarching intersecting structures of oppression (also coined as the "matrix of domination") (Collins, 2022) include but are not limited to settler colonialism, racism, patriarchy, capitalism, heterosexism, ableism, nationalism, and other forms of oppression (López et al., 2017). Within the historical context and intersecting structures is the matrix of arranged hierarchical

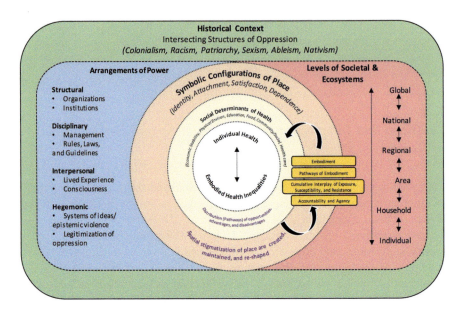

Fig. 3.1 Ecosocial and intersectionality perspective of the SSDoH, with adaptations from Collins (2022), Krieger (2021) and López et al. (2017)

power at multiple levels—macro structures, meso practices or disciplinary powers, and micro individual power. Hegemonic (or cultural) domains of power are considered the "glue" that binds all levels of power through principles and practices of ideas (Collins & Bilge, 2020). Coinciding the matrix of domination are the levels, pathways, and power drawn from Krieger's (2021) ecosocial theory of disease distribution. Each construct operates at different levels of society and the corresponding processes connected between and within them. This connection discerns how different forms of oppression, such as racism and classism, shape the way in which place is configured, and its position within the pathway of place embodiment and embodied health inequalities. Lastly, the spatial configuration of place and the arrangement of power influence the distribution of the social determinants, where differences in opportunities, advantages, and disadvantages arise.

Recognizing that the determinants of health are more than static and siloed social, spatial, and structural pathways to health inequity, we expand the current definition of SDoH using ecosocial theory and intersectionality. This critical praxis provides public health scholars with the opportunity to diverge from authoritative epistemologies that use deterministic social-spatial language toward expanded thinking and understanding of health inequities. Therefore, we suggest that the definition of SSDoH can be pursued and evolved as *interlocking and mutually reinforcing arrangements of social, economic, political, intergenerational, and ecological advantages and disadvantages where people are born, live, work, play, and age that enacts as a process in shaping embodied health realities across objective and subjective place over time*, while we acknowledge that the ways we present the analytic praxis to delineate structural determinants will vary across scholars and remain open to interpretation. However, we hope this reframing of the present SDoH framework and spatial paradigms inspires alternative conceptualizations and praxis for place-based health.

Conclusion

In this chapter, we demonstrated how place is inextricably linked to health. Guided by an ecosocial theory and intersectionality praxis, we welcome alternative ways to reconsider how place (physically and symbolically) operates as a determinant of health. Challenging current definitions of the social and spatial aspects that influence health and health inequity, we examined how these analytical frameworks may illuminate the systemic forms of power that often go unnoticed in SDoH and spatial health paradigms. The SDoH frameworks currently used are intended to provide the principles that show us what public health is and operate at multiple social levels that affect health. While these conceptualizations have been well intended to attain health equity and justice, they have often presented limited and amorphous views of what influences health by focusing on traditional social factors, which certainly contribute to health, but are not the root causes of health inequity.

Inconsistencies and misguided deterministic framing of SDoH factors such as socioeconomic status (e.g., income, wealth, and education) may omit the deeper systems in which power operates (e.g., white supremacy, colonialism, racism). Similarly, SDoH restricts concepts of place (neighborhoods) to its quantified administrative units and the physical institutions and localities, rather than complex terrains that take up symbolic meaning. Through an ecosocial lens, we understand that what leads to health is more dynamic and made up of integrated experiences that become embodiments (biologically embedded) from place (physical and social world) (Krieger, 2021). Embodied inequalities are socially arranged by power and privilege, not by inherent social, economic, and political conditions. This may help scholars think beyond an individual-to-policy view of SDoH, as well as understand *place* outside of arbitrary administrative boundaries. Place and how people embody health inequities are not static to their residence but are rather comprised of peoples' daily activities, which are informed by structural distributions of power and opportunity. As we think more relationally about the spatiality of embodiment, researchers can begin to focus on where these embodied realities take place and how they are experienced. In contrast to ecosocial theory, intersectionality compliments the way we think about how identities (in place and time) are reflections of larger interlocking roots of power and oppression.

As an extension of ecosocial theory, scholars can move beyond the static notions of place to also consider other social positions and relations to power (Collins, 2022; Combahee River Collective, 1977). With more focus on the embodied intersectional positions within different locations linked to oppression, scholars are better situated to shift attention to the structural determinants that inform the

SSDoH. Intersectionality grounds ecosocial thinking in systems of power that are interconnected and mutually reinforcing. For this analytical praxis to maintain its radical edge, researchers should continuously ground their questioning and reflexivity in its Black feminist roots—centering on power and justice (Bowleg, 2021). This dual lens may allow for multiple-axis perspectives in complement to the social-spatial factors created and maintained through historical and evolved pathways on health.

References

Aaronson, D., Hartley, D., & Mazumder, B. (2021). The effects of the 1930s HOLC "Redlining" Maps. *American Economic Journal: Economic Policy, 13*(4), 355–392. https://doi.org/10.1257/pol.20190414

Arcaya, M., Tucker-Seeley, R., Kim, R., Schnake-Mahl, A., So, M., & Subramanian, S. (2016). Research on neighborhood effects on health in the United States: A systematic review of study characteristics. *Social Science & Medicine, 1982*(168), 16–29. https://doi.org/10.1016/j.socscimed.2016.08.047

Bambra, C. (2022). Placing intersectional inequalities in health. *Health & Place, 75*, 102761. https://doi.org/10.1016/j.healthplace.2022.102761

Bonilla, Y. (2023). Opinion | a legacy of colonialism set the stage for the maui wildfires. *The New York Times*. Retrieved from https://www.nytimes.com/2023/08/27/opinion/maui-wildfire-colonialism.html

Bowleg, L. (2008). When Black+ lesbian+ woman≠ Black lesbian woman: The methodological challenges of qualitative and quantitative intersectionality research. *Sex Roles, 59*, 312–325.

Bowleg, L. (2012). The problem with the phrase women and minorities: Intersectionality—An important theoretical framework for public health. *American Journal of Public Health, 102*(7), 1267–1273. https://doi.org/10.2105/AJPH.2012.300750

Bowleg, L. (2021). Evolving intersectionality within public health: From analysis to action. *American Journal of Public Health, 111*(1), 88–90. https://doi.org/10.2105/AJPH.2020.306031

Braveman, P., & Gottlieb, L. (2014). The social determinants of health: It's time to consider the causes of the causes. *Public Health Reports, 129*(Suppl. 2), 19–31.

Combahee River Collective. (1977). *A black feminist statement*.

Collins, P. H. (1989). The social construction of black feminist thought. *Signs: Journal of Women in Culture and Society, 14*(4), 745–773.

Collins, P. H. (2022). *Black feminist thought: Knowledge, consciousness, and the politics of empowerment*. Routledge.

Collins, P. H., & Bilge, S. (2020). *Intersectionality*. John Wiley & Sons.

Crenshaw, K. (1991). Mapping the margins: Intersectionality, identity politics, and violence against women of color. *Stanford Law Review, 43*(6), 1241–1299. https://doi.org/10.2307/1229039

Cresswell, T. (2014). *Place: An introduction*. John Wiley & Sons.

Czyzewski, K. (2011). Colonialism as a broader social determinant of health. *The International Indigenous Policy Journal, 2*(1).

Diez Roux, A. V. (2001). Investigating neighborhood and area effects on health. *American Journal of Public Health, 91*(11), 1783–1789.

Du Bois, W. E. B., & Eaton, I. (1996). *The Philadelphia Negro: A social study*. University of Pennsylvania Press. Retrieved from https://www.jstor.org/stable/j.ctt3fhpfb

Dubetz, A., Horton, M., & Kesteven, C. (2022). Staying power: The effects of short-term rentals on California's tourism economy and housing affordability.

Duncan, D. T., & Kawachi, I. (2018). Neighborhoods and health: A progress report. In *Neighborhoods and health*. Oxford University Press. https://doi.org/10.1093/oso/9780190843496.003.0001

Erfani, G. (2022). Reconceptualising sense of place: Towards a conceptual framework for investigating individual-community-place interrelationships. *Journal of Planning Literature, 37*(3), 452–466. https://doi.org/10.1177/08854122221081102

Federal Housing Administration Underwriting Manual. (1938). Retrieved October 4, 2023, from https://www.huduser.gov/portal/sites/default/files/pdf/Federal-Housing-Administration-Underwriting-Manual.pdf

Foucault, M. (1978). Right of death and power over life. In *Caring labor: An archive*.

Geminiani, V., & DeLuca, M. (2018). *Hawai'i appleseed center for law and economic justice*.

Gould, P., & White, R. (2012). *Mental maps*. Routledge.

Hatzenbuehler, M. L., & Link, B. G. (2014). Introduction to the special issue on structural stigma and health. *Social Science & Medicine, 103*, 1–6. https://doi.org/10.1016/j.socscimed.2013.12.017

hooks, B. (1982). *Ain't I a woman: Black women and feminism*.

Keene, D. E., & Padilla, M. B. (2014). Spatial stigma and health inequality. *Critical Public Health, 24*(4), 392–404. https://doi.org/10.1080/09581596.2013.873532

Kirby, R. S., Delmelle, E., & Eberth, J. M. (2017). Advances in spatial epidemiology and geographic information systems. *Annals of Epidemiology, 27*(1), 1–9. https://doi.org/10.1016/j.annepidem.2016.12.001

Klein, N., & Sproat, K. (2023). *Why was there no water to fight the fire in Maui?* The Guardian.

Krieger, N. (1994). Epidemiology and the web of causation: Has anyone seen the spider? *Social Science & Medicine, 39*(7), 887–903. https://doi.org/10.1016/0277-9536(94)90202-X

Krieger, N. (2001). Theories for social epidemiology in the 21st century: An ecosocial perspective. *International Journal of Epidemiology, 30*(4), 668–677. https://doi.org/10.1093/ije/30.4.668

Krieger, N. (2021). 1C1 from embodying injustice to embodying equity: Embodied truths and the ecosocial theory of disease distribution. In N. Krieger (Ed.), *Ecosocial theory, embodied truths, and the people's health*. Oxford University Press. https://doi.org/10.1093/oso/9780197510728.003.0001

López, N., Erwin, C., Binder, M., & Chavez, M. (2017). Making the invisible visible: Advancing quantitative methods in higher education using critical race theory and intersectionality. *Race Ethnicity and Education, 21*, 1–28. https://doi.org/10.1080/13613324.2017.1375185

Lundberg, O. (2020). Next steps in the development of the social determinants of health approach: The need for a new narrative. *Scandinavian Journal of Public Health, 48*(5), 473–479. https://doi.org/10.1177/1403494819894789

Lynch, E. E., Malcoe, L. H., Laurent, S. E., Richardson, J., Mitchell, B. C., & Meier, H. C. S. (2021). The legacy of structural racism: Associations between historic redlining, current mortgage lending, and health. *SSM—Population Health, 14*, 100793. https://doi.org/10.1016/j.ssmph.2021.100793

Magbual, N. J. (n.d.). Paradise for tourists, a struggle for natives: Native Hawaiian homelessness in the Hawaiian Islands.

Marmot, M. (2000). Social determinants of health: From observation to policy. *The Medical Journal of Australia, 172*(8), 379–382. https://doi.org/10.5694/j.1326-5377.2000.tb124011.x

Marmot, M., Friel, S., Bell, R., Houweling, T. A. J., & Taylor, S. (2008). *Closing the gap in a generation: Health equity through action on the social determinants of health: Final report of the commission on social determinants of health*. World Health Organization.

May, V. M. (2015). *Pursuing intersectionality, unsettling dominant imaginaries*. Routledge.

McKittrick, K. (2011). On plantations, prisons, and a black sense of place. *Social & Cultural Geography, 12*(8), 947–963.

McLeroy, K. R., Bibeau, D., Steckler, A., & Glanz, K. (1988). An ecological perspective on health promotion programs. *Health Education Quarterly, 15*(4), 351–377. https://doi.org/10.1177/109019818801500401

Merz, S., Jaehn, P., Mena, E., Pöge, K., Strasser, S., Saß, A.-C., Rommel, A., Bolte, G., & Holmberg, C. (2023). Intersectionality and eco-social theory: A review of potentials for public health knowledge and social justice. *Critical Public Health, 33*(2), 125–134. https://doi.org/1 0.1080/09581596.2021.1951668

Mitchell, B., & Franco, J. (2018, March 20). *HOLC "redlining" maps: The persistent structure of segregation and economic inequality.* NCRC. Retrieved from https://ncrc.org/holc/

Mitchell, U. A., Nishida, A., Fletcher, F. E., & Molina, Y. (2021). The long arm of oppression: How structural stigma against marginalized communities perpetuates within-group health disparities. *Health Education & Behavior: The Official Publication of the Society for Public Health Education, 48*(3), 342–351. https://doi.org/10.1177/10901981211011927

Petteway, R. J., Mujahid, M., & Allen, A. (2019a). Understanding embodiment in place-health research: Approaches, limitations, and opportunities. *Journal of Urban Health: Bulletin of the New York Academy of Medicine, 96*(2), 289–299. https://doi.org/10.1007/s11524-018-00336-y

Petteway, R. J., Mujahid, M., Allen, A., & Morello-Frosch, R. (2019b). The body language of place: A new method for mapping intergenerational "geographies of embodiment" in place-health research. *Social Science & Medicine, 223*, 51–63. https://doi.org/10.1016/j.socscimed.2019.01.027

Rasanathan, K. (2018). 10 years after the commission on social determinants of health: Social injustice is still killing on a grand scale. *The Lancet, 392*(10154), 1176–1177. https://doi.org/10.1016/S0140-6736(18)32069-5

Roach, T. (2009). *Sense and sexuality: Foucault, Wojnarowicz, and biopower.* Nebula.

Rothstein, R. (2017). The color of law: A forgotten history of how our government segregated America. *Liveright Publishing*.

Slater, T. (2013). Your life chances affect where you live: A critique of the cottage industry of neighbourhood effects research. *International Journal of Urban and Regional Research*.

Stonington, S. D., Holmes, S. M., Hansen, H., Greene, J. A., Wailoo, K. A., Malina, D., Morrissey, S., Farmer, P. E., & Marmot, M. G. (2018). Case studies in social medicine—Attending to structural forces in clinical practice. *The New England Journal of Medicine, 379*(20), 1958–1961. https://doi.org/10.1056/NEJMms1814262

Taparra, K., Purdy, M., & Raphael, K. L. (2023). From ashes to action—Indigenous health perspectives on the Lāhainā fires. *New England Journal of Medicine.* https://doi.org/10.1056/NEJMp2309966

Taylor, K.-Y. (2018). How real estate segregated America. *Dissent, 65*(4), 23–32.

Tran, E., Blankenship, K., Whittaker, S., Rosenberg, A., Schlesinger, P., Kershaw, T., & Keene, D. (2020). My neighborhood has a good reputation: Associations between spatial stigma and health. *Health & Place, 64*, 102392. https://doi.org/10.1016/j.healthplace.2020.102392

Truth, S. (1851). *Ain't I a woman?*

Tuan, Y. F. (1971). Geography, phenomenology, and the study of human nature. *Canadian Geographies/Géographies Canadiennes, 15*(3), 181–192. https://doi.org/10.1111/j.1541-0064.1971.tb00156.x

US Census Bureau. (2023). *Chuukese and Papua New Guinean populations fastest growing Pacific Islander Groups in 2020.* Census.Gov. Retrieved from https://www.census.gov/library/stories/2023/09/2020-census-dhc-a-nhpi-population.html

Wacquant, L. (2008). *Territorial stigmatization in the age of advanced marginality* (pp. 43–52). Brill. https://doi.org/10.1163/9789087902667_006

Weiss, L., Ompad, D., Galea, S., & Vlahov, D. (2007). Defining neighborhood boundaries for urban health research. *American Journal of Preventive Medicine, 32*(Suppl. 6), S154–S159. https://doi.org/10.1016/j.amepre.2007.02.034

WHO. (2020). *Social determinants of health.* Retrieved from https://www.who.int/health-topics/social-determinants-of-health#tab=tab_1

World Health Organization (WHO). (2008). *Social determinants of health.* Retrieved from https://www.who.int/westernpacific/activities/taking-action-on-the-social-determinants-of-health

Chapter 4
Residual Effects of Historical Place-Based Discrimination on Health

Ayodeji Iyanda

Introduction

In recent years, there has been a growing focus on the impacts of place-based discrimination on the health of minority populations. In this chapter, I use the term "historical place-based discrimination" (H-PBD) to describe the enduring spatial segregation and exclusion practices that have deprived minority populations of social, economic, and political opportunities. Place-based and racial-based discrimination are intertwined and perpetuated through institutionalized policies and social practices, ultimately leading to disparities in health outcomes. To bring to the fore, I present the concept from the spatial-social health determinants of health perspective (Fig. 4.1). I aim to emphasize that the concept of the H-PBD results from power dynamics and social structures that have shaped the distribution of resources, opportunities, and hazards in specific areas, consequently contributing to health disparities among different populations. Addressing the underlying structural and institutional factors perpetuating spatial health inequalities requires a comprehensive approach due to the complex nature of the H-PBD. Numerous studies have explored this issue in developed and developing countries, underscoring the detrimental implications of spatial segregation and exclusion on health and overall well-being (Do et al., 2019; Huang et al., 2023). In this chapter, I not only illustrate the impact of the H-PBD from a critical geography perspective but also provide case studies to bolster my argument on how this issue correlates with health inequity (a term I used interchangeably for discrimination), employing life expectancy (LE) as a health outcome.

In the remainder of the chapter, I borrow from the critical geography framework in explaining H-PBD. I also show global examples of historical and current

A. Iyanda (✉)
Division of Social Sciences, Prairie View A&M University, Prairie View, TX, USA
e-mail: aeiyanda@pvamu.edu

M. A. Kolak, I. K. Moise (eds.), *Place and the Social-Spatial Determinants of Health*, Global Perspectives on Health Geography,
https://doi.org/10.1007/978-3-031-88463-4_4

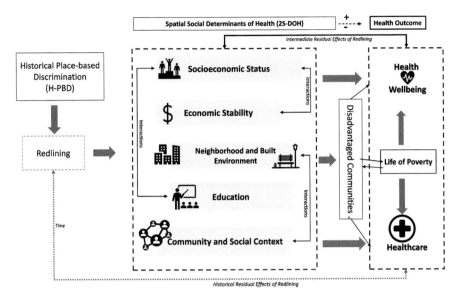

Fig. 4.1 A Conceptual Framework of Historical Place-based Discrimination

discrimination practices in minority health from the literature. I touch on various measures and techniques in identifying historical features of the H-PBD in social research. In what follows, I present case studies of the H-PBD in the United States (US). In the concluding section, I bring my argument to a close by submitting to the fact that we need policies and interventions designed encompassingly in understanding how discrimination based on race, ethnicity, gender, sexuality, and other forms of identity interact to create unique experiences of health inequity for different populations.

Theoretical Perspective Based on Critical Geography

The concept of H-PBD is deeply rooted in critical geography, which emphasizes the significance of power relations and social structures in shaping spatial inequalities. Critical geography acknowledges the influence of historical and contemporary policies and practices in creating an uneven distribution of resources, opportunities, and hazards among diverse populations based on their geographic location and social identity. This unequal distribution has profound consequences for the health outcomes of marginalized groups, including heightened rates of chronic illnesses, mental health issues, and LE.

The emergence of critical geography in the 1970s and 1980s challenged the traditional focus of geography, primarily centered on the physical and natural environment. Critical geography called for a more comprehensive approach that considered social, economic, and political factors that influence the distribution of resources and opportunities in different spatial contexts (see Soja, 2013). The framework is concerned with understanding the role of power and inequality in shaping the built environment, as well as social and spatial disparities in health outcomes. This perspective provides a convincing framework for examining the impact of H-PBD on health outcomes. As Soja (2013) argued, spatial justice is central to this approach by emphasizing the fair and equitable distribution of resources and opportunities across geographic space. Niagra and colleagues have highlighted how spatial disparities in access to resources such as healthcare, healthy food, and safe housing contribute to health disparities (Nigra et al., 2020). These studies underscore the significance of examining the unequal distribution of resources within the context of health outcomes.

Power relations, a critical concept in critical geography, encompasses the exercise and contestation of power in social and spatial contexts (Sandoval et al., 2017). Power relations are indispensable in shaping the spatial distribution of resources, opportunities, and hazards in the historical context of power, boundaries, and differences. Policies and practices that historically excluded minority populations from accessing quality education and employment opportunities also perpetuated social and spatial inequalities, which subsequently impacted health outcomes (Seitles, 1998).

Social movement is another fundamental concept within critical geography, representing collective efforts to challenge dominant power relations and promote social change. Social movements have played a crucial role in advancing spatial justice and contesting institutionalized policies and practices perpetuating health disparities among minority populations. Grassroots initiatives focusing on promoting healthy living in low-income neighborhoods, advocating for affordable housing, ensuring food access, and creating green spaces have effectively contributed to spatial justice and improved health outcomes (Alkon et al., 2020; Jennings et al., 2019; Parr, 2004), and have also been applied to the medical geography of discrimination of various health issues.

Furthermore, several studies in urban health have demonstrated the importance of the built environment in shaping health outcomes and the need for policies and interventions that promote equitable access to safe and healthy environments (Northridge & Freeman, 2011; Suglia et al., 2016). Admittedly, the concentration of hazardous waste facilities and other environmental hazards in low-income and minority neighborhoods is not coincidental but results from systemic discrimination and racism (Bowen et al., 1995; Bullard, 1993). These hazards have been linked to various adverse health outcomes including cancer, asthma, and developmental disorders (Sly et al., 2016; Suk et al., 2016).

Critical geography emphasizes the importance of community engagement in addressing the impact of the H-PBD on health outcomes. Community-led interventions that promote healthy food options and safe recreational spaces can help mitigate the impacts of the H-PBD on health outcomes (Egan et al., 2021). Furthermore, community engagement can help identify the underlying structural factors contributing to health disparities and inform policy changes in order to promote health equity. Therefore, community-based research is critical to understanding the link between H-PBD and health disparities, as demonstrated later in the article. Undoubtedly, the transitory impacts of the H-PBD on health disparities are significant for a better understanding of community well-being. As it is, the residual impacts of the H-PBD (Fig. 4.1) require effective policy interventions and community-led solutions that recognize the contrast between past and present realities.

Global Evidence of Historical Place-Based Discrimination in Health

Despite variations in the specific health outcomes demonstrated in health research in different locations, a consistent underlying factor always emerges: institutionalized policies and social practices play a significant role in perpetuating spatial disparities in health, as observed from both regional and international perspectives. For example, in many developed countries, government policies such as redlining and discriminatory lending practices have led to disinvestment in minority neighborhoods, limiting access to healthy food, quality health care, and safe housing (Braveman et al., 2011; Huang & Sehgal, 2022). Similarly, in developing countries, historical patterns of exclusion from political and economic power have led to the concentration of poverty and lack of access to essential services in minority neighborhoods (Cobbinah et al., 2013; Mosse, 2010).

Evidence from the Developed World

Research conducted in developed countries such as the US and Australia has revealed a range of health impacts linked to the H-PBD. Residential segregation and disinvestment in minority neighborhoods, for instance, have been associated with higher rates of chronic diseases, including diabetes, hypertension, and cardiovascular disease, and increased exposure to environmental hazards such as air pollution and toxic waste (Hicken et al., 2023). A recent study in Australia discovered that Indigenous Australians encounter significant spatial disparities in health outcomes caused by historical and contemporary discriminatory practices, such as limited access to healthcare services, inadequate housing conditions, and lower levels of education and employment opportunities (Paradies, 2016). The research emphasized the necessity of addressing the H-PBD to improve health outcomes for Indigenous Australians.

One of the several ways to measure the H-PBD is through redlining. Several studies have explored the impact of historical redlining on asthma prevalence and found that neighborhoods redlined in the 1930s had higher rates of asthma prevalence and asthma-induced hospitalization in the twenty-first century (Nardone et al., 2020; Schuyler & Wenzel, 2022). Earlier studies conducted by Lillie-Blanton and Laveist (1996) found that neighborhoods segregated by race and income in the twentieth century continue to experience higher rates of premature mortality in the twenty-first century.

Experiences of discrimination at the individual level, including racial and ethnic discrimination, have contributed to adverse health outcomes, such as increased risks of chronic diseases, mental health disorders, and mortality (Paradies, 2016; Vargas et al., 2021). Vargas et al. (2021) found that experiences of discrimination were linked to higher risks of hypertension among African Americans, while Paradies et al. (2016) discovered that racism incidents increased the risks of mental health disorders, such as depression and anxiety, among Indigenous Australians. English et al. (2020) demonstrated that neighborhoods with a history of racial discrimination had higher rates of depressive symptoms among residents. Similarly, research in the US has found that Black Americans who live in counties with a history of lynching have higher rates of cardiovascular disease and mortality than those who live in counties without a history of lynching (Kramer et al., 2017; Probst et al., 2019).

Addressing individual-level experiences of discrimination and structural-level factors is vital to promoting health equity. For example, Tung et al. (2021) found that implementing policies to reduce residential segregation and promote economic and social integration can improve health outcomes, particularly among low-income and racial/ethnic minority groups. Similarly, Thornton et al. (2016) observed that policies that increase access to affordable housing and improve neighborhood conditions could help reduce health disparities.

Evidence from the Developing World

Studies have documented overwhelming evidence of the H-PBD concerning health in developing countries. I draw examples from three continents: Africa, Asia, and Latin America, specifically focusing on South Africa, India, and Brazil. These trios have gone through notable historical discrimination.

The legacy of apartheid has resulted in significant health disparities between different racial groups in South Africa (Bell et al., 2022; Das-Munshi et al., 2016). The spatial segregation of different racial groups during apartheid has resulted in limited access to healthcare services, environmental hazards, and social and economic disadvantage, including green infrastructure for black and colored populations (Stull et al., 2016; Venter et al., 2020). Researchers have linked H-PBD (e.g., apartheid) to high HIV/AIDS, tuberculosis, and other communicable diseases in these populations (Ndinda et al., 2018; Pillay-van Wyk et al., 2016).

Evidence of the H-PBD also exists in Asia. The caste system in India has resulted in significant health disparities between different caste groups (Maity, 2017; Thapa et al., 2021). For example, Dalit populations, who are at the bottom of the caste hierarchy, experience higher rates of poverty, malnutrition, and poor health outcomes compared to higher-caste folks. The spatial segregation of Dalit populations has also resulted in limited access to healthcare services and environmental hazards (Ambade et al., 2022; Bharathi et al., 2018).

Supporting the findings from Africa and Asia, substantial evidence points to historical discrimination in Latin America. A prime illustration can be observed in Brazil, which harbors the largest population of individuals with African heritage on the continent. Extensive research has revealed that the socioeconomic opportunities (SEPs) and health conditions in the country substantiate the existence of deeply ingrained historical slavery within societal structures, which manifests as racism across various domains such as employment, healthcare, education, and housing (Bastos et al., 2014; Chor & Lima, 2005; Ikawa & Mattar, 2008; Mitchell-Walthour, 2017; Pavao et al., 2012).

Measures or Indicators of Historical Place-Based Discrimination

In the US, H-PBD has been measured using various indicators, including redlining, residential segregation, environmental injustice, and disparities in access to health-care, education, employment, and income (Krieger, 2012). Redlining refers to deny-ing mortgage loans or insurance based on race or neighborhood racial composition, contributing to the segregation of urban areas. In contrast, residential segregation physically separates racial and ethnic groups into disparate neighborhoods, influ-enced by historical discrimination and institutionalized policies such as redlining (Massey, 2020; Massey & Denton, 1993).

Bullard (1993) and Cutter (1995) define environmental injustice as the dispro-portionate burden of environmental hazards on minority communities. Discrimination has also impacted healthcare access for minorities, including disparities in health insurance coverage, a lack of healthcare facilities in minority neighborhoods, and discrimination by healthcare providers. Disparities in SEPs including education, employment, and income, are also used as indicators of historical discrimination. Discrimination in employment and income opportunities include inequities in hir-ing and promotion, wage gaps, and discrimination in access to credit and financing (Huffman & Cohen, 2004). In most developing regions, residential segregation based on race, ethnicity, or SEPs is a crucial indicator of place-based discrimination (Burgos et al., 2017; van Ham et al., 2021). Due to discriminatory practices, access to basic services such as healthcare, education, water, and sanitation can also be limited for certain groups. Furthermore, limited access to land and property owner-ship, a crucial indicator of wealth, is an important factor in economic and social mobility, with marginalized communities facing historically discriminatory prac-tices (van Ham et al., 2021).

Approaches for Revealing Place-Based Discrimination

Addressing the disparities from place-based discrimination requires a comprehen-sive and multifaceted approach that promotes health equity and addresses the root causes of discrimination (Krieger, 2012). Community-based participatory research (CBPR), machine learning (ML), and spatial analysis are promising approaches for understanding the spatial and social determinants of health and identifying interventions tailored to the community's specific needs and priorities.

Machine learning and spatial analysis may facilitate understanding the spatial and social determinants of health and identifying interventions tailored to the community's specific needs and priorities. Studies have used geospatial techniques to determine the distributional characteristics of syringe exchange programs in New York City (Cooper et al., 2012). They found that areas with high rates of HIV infection had lower access to syringe exchange programs. This information was used to inform the development of new syringe exchange programs in areas with increased rates of HIV infection, which helped to reduce the spread of HIV. Lotfata et al. (2023) similarly used machine learning algorithms to identify the SDoH most predictive of asthma prevalence in the US. The study concluded that the spatial information generated using ML could be used to develop targeted interventions that address the specific SDoH most strongly associated with asthma prevalence. More recent research used machine learning algorithms to assess disparities in accessing healthcare, finding that healthcare utilization was primarily driven by upstream SDoH (Koski et al., 2022; Tan et al., 2020).

Another promising approach to addressing health disparities is CBPR. CBPR involves a collaborative partnership between researchers and community members to identify research questions, design studies, and implement interventions tailored to the community's needs and priorities (Ziegler et al., 2019). This approach can help build trust and establish shared ownership of the research process, improving the relevance and effectiveness of intervention programs. For example, the Community Engagement Core of the National Institute of Environmental Health Sciences has implemented a CBPR approach to address environmental health disparities in minority communities in the US (Israel et al., 2005). This program involves partnering with community organizations to identify environmental hazards and develop interventions to reduce exposure and improve health outcomes. The Healthy Environments Partnership is another approach that collaborates with community organizations and academic researchers to address environmental health disparities in low-income neighborhoods in Detroit, Michigan. This program involves community members in all aspects of the research process, from study design to data analysis and interpretation.

By engaging with the community in a collaborative and participatory manner, researchers and community stakeholders identified environmental hazards and developed interventions to reduce exposure and improve health outcomes (Israel et al., 2005; Ziegler et al., 2019). Including a health equity lens in zoning and land-use policies is another approach that has been used to address the impacts of H-PBD. This approach has been successfully implemented in several cities, including Denver, Los Angeles, Chicago, Minneapolis, and Seattle, where planners designed new zoning policies to promote equitable access to amenities such as urban green infrastructure (e.g., parks and greenspace) and grocery stores (e.g., supermarkets and Fruit stores; see Dubowitz et al., 2015). These policies address the historical legacy of redlining and other discriminatory practices that have resulted in unequal access to resources and life-changing opportunities in marginalized communities.

Case Studies of Historical Place-Based Discrimination

Between 1940 and 1960, the African-American population of the US underwent a massive relocation from the Southern states to urban areas of the Northeast. During this period, 1.24 million African Americans left the south, and by 1960, most African Americans lived in urban areas (Sugrue, 1995). Southern states weaponized specialized laws against racial integration. On the other hand, the Northern states institutionalized laws and restrictive covenants to achieve segregation and ghettoization (Massey & Denton, 1993). Hence, Black American communities faced discrimination in real estate and banking using devices such as deed restrictions, restrictive covenants, racial redlining, and other discriminatory practices that led to neighborhood decline (Massey, 2020). The practice of redlining contributed to the segregation of African Americans in urban areas, as they were forced to live in neighborhoods with limited economic opportunities, poor housing conditions, and limited access to quality education and healthcare (Massey & Denton, 1993).

Methodology

This study employed a powerful machine learning tool called forest-based classification and regression, which operates within the ArcGIS framework. Using this tool, I trained and analyzed county-level variables from the County Health Rankings database (2022), enabling me to draw meaningful insights. Prior to data training, exploratory regression was conducted to assess all potential combinations of identified 32 candidate explanatory variables to predict LE.[1] This evaluation aimed to identify Ordinary Least Squares (OLS) models that most effectively elucidate LE, specifically within the social-spatial disparity framework.

The exploratory regression tests all variable combinations for redundancy, completeness, significance, bias, and model performance using different parameters such as maximum coefficient p-values of 0.05, maximum variance inflation factor (VIF) value of 7.5, minimum acceptable Jarque Bera p-value of 0.1, and minimum acceptable spatial autocorrelation p-value of 0.1. The process led to the six best

---

[1] Predictor Variables used from *County Health Rankings* (2022): % adult population to access to locations for physical activity; % adult reporting no leisure time PA; % Adult obesity; % Adult smoking; % Rural area; Residential segregation-Black/White; Residential segregation-nonwhite/White; % Drug poisoning; Median household income; Median household income-Black; Median household income-Hispanic; Household income-White; School segregation index; Ratio of population and other physicians; %No insurance (Children < 19); % Uninsured people; % Under 65 uninsured; %Sleep < 7 h; Motor vehicle crash per 100,000; % Low income residents living far from grocery store; % Pop. with low access to food; % Driving > 30 mins to work; % Households with severe housing problem; Average daily density of PM2.5; Violent Crimes*100k; Income inequality 80th percentile/20 percentile; Ratio pop/Mental health providers; MH providers*100k; PC physicians*100k; % Excessive drinking; % Access to physical activity; % Adult >18 yr with no insurance.

combinations based on the specified parameters. Subsequently, I applied the trained data to create predictive models for LE in the US, specifically focusing on three historical cities: Philadelphia, Baltimore, and Trenton. These models were developed based on the same associated explanatory variables, allowing me to make informed projections. The essential variables used in this analysis are highlighted in Table 4.1, which displays the six most influential factors for modeling LE in the study areas.

Within this framework, I defined adult smoking as a lifestyle variable. This variable helped me evaluate the impact of smoking habits on LE. Additionally, I identified rurality as a spatial determinant of health, acknowledging its significance in understanding the disparities in health outcomes between urban and rural areas at the county level. I classified and interpreted the remaining four variables as residual variables of H-PBD. Note that *residual* in this context refers to the long impact of historical discrimination and not the mathematical interpretation of it. These variables encapsulate the enduring impacts of past discriminatory practices that shaped these cities' socioeconomic and healthcare landscapes.

According to the National Community Reinvestment Coalition (NCRC), most of the neighborhoods that the Home Owners' Loan Corporation (HOLC) graded as high-risk or "hazardous" eight decades ago (74%) are low-to-moderate income (LMI) today. Additionally, most of the HOLC-graded "hazardous" areas (nearly 64%) are minority neighborhoods in the present day. By including these residual variables, I aimed to account for the long-standing systemic inequalities that continue to impact the health and well-being of three specific study regions in the Northeastern part of the US: Philadelphia, Pennsylvania; Baltimore, Maryland; and Trenton, New Jersey.

Results and Discussion

The combination of the six chosen variables in the spatial model collectively predicted LE across 7147 counties within the US, revealing notable spatial disparities. The average predicted LE value was 77.45 years, with a standard deviation of ±2.47. The range of LE values spanned from 70.7 to 84.7 years.

Case Study 1: Philadelphia, Pennsylvania

The spatial regression indicates compelling evidence of the historical impact of discrimination, as the model y predicted LE in Philadelphia's poorest areas, such as Stenton and Kensington (Fig. 4.2a, b). This is supported by various socioeconomic indicators highlighting these communities' disparities. Areas designated as "Best" and "Desirable" coincide with higher life expectancy, indicating a more favorable socioeconomic and health environment in those areas.

Table 4.1 Predictors of LE and their importance

Variables	Variable definition	Source and year	Importance percentage (%)	Variable statistics Mean (Min–Max)
Outcome variable				
Life expectancy	The average number of years a person can expect to live. The County Health Rankings used data from 2018 to 2020 for this measure	National Center for Health Statistics - Mortality Files, 2018–2020	–	
Predictors				
Adult smoking	Percentage of adults who are current smokers (age-adjusted)	Behavioral Risk Factor Surveillance System (BRSF), 2020	61.4	16.9 (8.7–29.3)
Rurality	Percentage of rural areas	Census Population Estimates, 2010	12.52	5.7 (0–54.6)
School segregation index	Measures how evenly representation of racial and ethnic groups in the student population is spread across schools using Theil's Index, a segregation index ranging between 0 and 1. Higher values represent high segregation.	National Center for Education Statistics, 2021–2022	8.99	0.24 (0.03–0.48)
Air Pollutant (PM2.5)	Average daily density of fine particulate matter in micrograms per cubic meter (PM2.5)	Environmental Public Health Tracking Network, 2019	8.37	9.7 (3.4–17.8)
Residential segregation-Black/White	The index of dissimilarity where higher values indicate greater residential segregation between Black and White county residents, ranges between 0 (complete integration) and 100 (complete segregation)	American Community Survey, 5-year estimates, 2017–2021	7.11	61.5 (27.5–84.8)
Redlining	The practice of denying borrowers access to credit based on the location of properties in the minority or economically disadvantaged	National Community Reinvestment Coalition, n.d.	1.60	0.34 (0–0.69)

60 A. Iyanda

The history of African Americans in Philadelphia dates back to the seventeenth century when they arrived as enslaved Africans. Over time, the population grew to include free Black residents actively participating in the abolitionist movement and the underground railroad. Despite facing historical and ongoing place-based discrimination, Black Philadelphians contributed significantly to the city's cultural, economic, and political life as workers, activists, artists, musicians, and politicians. However, these communities still experience significant health disparities (Brawner et al., 2015; Gripper et al., 2022), which can be attributed to various factors, including place-based discrimination and lack of access to quality healthcare.

Kensington is home to a significant low-income Hispanic-American community, predominantly comprising Puerto Ricans and Dominicans, alongside African Americans (Ribeiro, 2013). According to data from the American Community Survey (2012–2016), Kensington's average income per capita was $12,669 per year, roughly half of the average income in Philadelphia. This significant income disparity is often associated with higher disease rates and unhealthy behaviors. Lower income levels have been linked to an increased prevalence of chronic conditions and a higher likelihood of engaging in risky behaviors such as smoking and lack of physical activity. Additionally, the violent crime rate in Kensington is approximately 30% higher, with 328 violent crimes per 10,000 residents, than that of Philadelphia overall, which has a rate of 242 violent crimes per 10,000 people (Urban Health Collaborative, 2020). Such high levels of violence can have a detrimental impact on the overall health and well-being of individuals living in these communities.

Notably, Kensington also exhibits higher rates of chronic health conditions than the city. For instance, the prevalence of coronary heart disease in Kensington was 6.9%, which is slightly higher than the city-wide rate of 6.6%. Similarly, the prevalence of diabetes in Kensington was 14.4%, while the city-wide rate was 12.5% (Urban Health Collaborative, 2020). These health disparities further contribute to the lower predicted LE in the area. Regarding risky behaviors, Kensington demonstrates a higher prevalence of smoking, with a rate of 30.8% compared to the city-wide rate of 23.6%. These unhealthy behaviors can have long-term consequences for individual health and contribute to the overall disease burden within the community.

These factors, to a moderate degree, demonstrate how existing socioeconomic disparities, higher rates of violent crime, the prevalence of chronic health conditions, and risky behaviors are associated with lower predicted LE in areas such as Kensington and Stenton (as shown in Fig. 4.4).

Case Study 2: Baltimore, Maryland

Historical segregation and discrimination in Baltimore's minority communities can be traced back to the nineteenth century, when Maryland had the largest free Black population in the US, comprising 40% of its total Black population. However, despite their significant numbers, the free Black population in Baltimore was largely

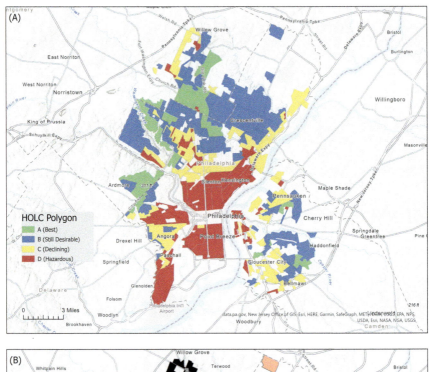

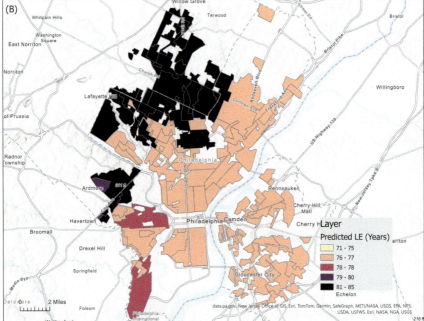

Fig. 4.2 (**a**) Redlining and (**b**) predicted life expectancy, Philadelphia, Pennsylvania

unable to join the classes of small business owners due to institutionalized policies and social practices that perpetuated spatial segregation and exclusion. Even as Baltimore's economy grew and diversified with the shift from tobacco to grain and its emergence as a milling and commerce center, the free Black population remained marginalized and segregated (See Phillips, 1997), leading to significant spatial disparities in health outcomes that persist even today. The historical context of Baltimore suggests that Eastern and Western parts of Baltimore are predominantly Black. As shown in Fig. 4.3a, b, areas that were historically labeled as "hazardous" and "declining" correspond with low predicted LE. Notably, neighborhoods identified with low LE in Baltimore, including Cherry Hill, Brooklyn, Frankford, Waltherson, and Hamilton, still have between 65% and 85% of the Black population.

Interestingly, the suburbs south of Baltimore, previously categorized as declining, show a higher predicted life expectancy. This phenomenon could be attributed to urban planning remediation and rejuvenation strategies implemented in the area, potentially leading to gentrification. Gentrification here is described as the process of renovating and improving neighborhoods, often accompanied by an influx of higher-income residents into previously disadvantaged areas. The implications of these urban planning strategies might have played a role in attracting higher-income earners back to the Southern neighborhoods. As a result, the overall socioeconomic conditions of these areas may have improved, leading to an increase in predicted LE rates. Gentrification often brings about various changes, such as infrastructure improvements, increased access to amenities and services, and revitalization of local economies (Iyanda & Lu, 2022a, b). These factors can positively influence residents' quality of life and contribute to improved health outcomes, thereby raising LE rates in the region. While gentrification may bring about some positive changes, it can also lead to the displacement of residents and widening socioeconomic disparities. Therefore, any relationship analysis between gentrification, urban planning strategies, and LE should consider the broader social and economic implications for the affected communities.

Case Study 3: Trenton, New Jersey

Based on six community-level variables, the ML algorithm predicted a higher life expectancy (79–80 years) in Trenton (Fig. 4.4a, b). The data presented in Fig. 4.4 highlight the impact of historical discrimination in Trenton, which is evident in the significant disparity in LE between minority Black/African Americans (72.2 years) and Whites (78.2 years) or Asians (86.1 years) in 2020, as reported by the New Jersey Department of Health in 2022. This disparity serves as a reflection of the long-lasting implications (or residual) of discrimination in marginalized communities.

While discriminatory practices and environmental injustices have historically contributed to the health disparities experienced by minority communities, ongoing reinvestment and revitalization efforts in these neighborhoods may play a role in

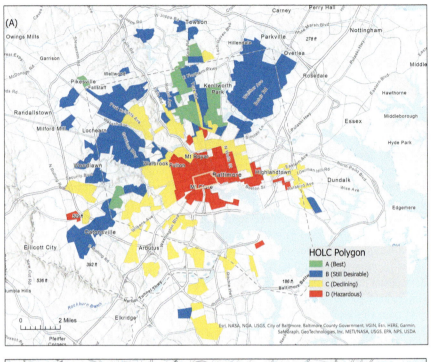

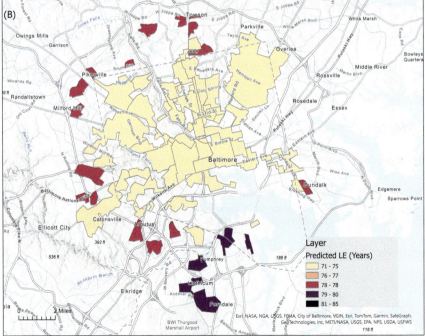

Fig. 4.3 (**a**) Redlining and (**b**) predicted life expectancy, Baltimore, Maryland

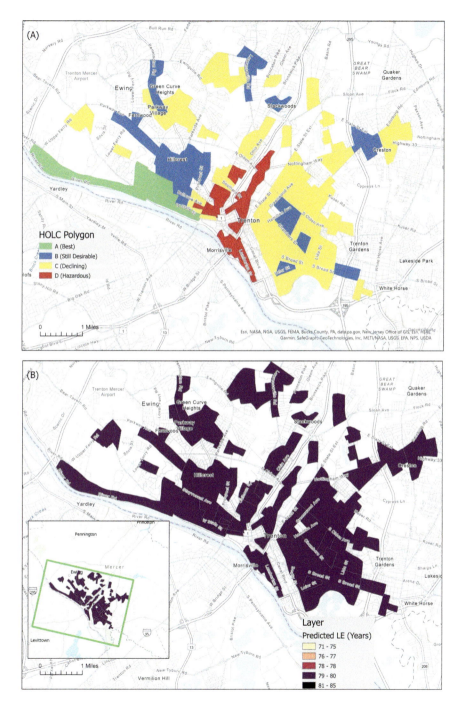

Fig. 4.4 (**a**) Redlining and (**b**) predicted life expectancy in Trenton, New Jersey

improving health outcomes. Supposedly, improved living conditions, access to better healthcare facilities, and reduced exposure to harmful environmental factors could contribute to the predicted increase in LE (Chetty et al., 2016; Mahalik et al., 2022). The potential displacement of existing residents, changes in social dynamics, and the overall complexity of the gentrification process should also be considered in a comprehensive analysis of the topic, as discussed by others (see Schnake-Mahl et al., 2020; Smith et al., 2020).

Conclusion

In this chapter, I examined the impacts of H-PBD on health outcomes, emphasizing power dynamics, social structures, and spatial inequalities. I presented evidence from developed and developing regions to understand how discrimination influences health disparities. To reveal the enduring consequences of historical injustices, I used geospatial techniques to identify spatial determinants of life expectancy. Forest-based regression predicted low life expectancy in Philadelphia and Baltimore but not Trenton. Based on current findings, policymakers must recognize these interconnected factors to develop targeted interventions and promote health equity for overburdened communities impacted by the H-PBD in the US. The study contributes to existing research by highlighting historical discrimination as a fundamental spatial-social determinant of health. Therefore, a comprehensive approach addressing structural factors and institutional inequalities is necessary, alongside community-based interventions and recognition of intersectionality. Promoting a just and equitable society requires understanding and addressing the impacts of historical place-based discrimination to achieve optimal health and well-being.

References

Alkon, A. H., Cadji, Y. J., & Moore, F. (2020). *"You Can't Evict Community Power": Food justice and eviction defense in Oakland. Food justice and eviction defense in Oakland* (pp. 223–242). New York University Press. https://doi.org/10.18574/nyu/9781479834433.003.0011

Ambade, P. N., Pakhale, S., & Rahman, T. (2022). Explaining caste-based disparities in enrollment for National Health Insurance Program in India: A decomposition analysis. *Journal of Racial and Ethnic Health Disparities*. https://doi.org/10.1007/s40615-022-01374-8

Bastos, J. L., Barros, A. J., Celeste, R. K., Paradies, Y., & Faerstein, E. (2014). Age, class and race discrimination: Their interactions and associations with mental health among Brazilian university students. *Cadernos de Saude Publica, 30*, 175–186.

Bell, G. J., Ncayiyana, J., Sholomon, A., Goel, V., Zuma, K., & Emch, M. (2022). Race, place, and HIV: The legacies of apartheid and racist policy in South Africa. *Social Science & Medicine, 296*, 114755. https://doi.org/10.1016/j.socscimed.2022.114755

Bharathi, N., Malghan, D. V., & Rahman, A. (2018). Isolated by caste: Neighborhood-scale residential segregation in Indian metros (SSRN Scholarly Paper No. 3195672). https://doi.org/10.2139/ssrn.3195672

Bowen, W. M., Salling, M. J., Haynes, K. E., & Cyran, E. J. (1995). Toward environmental justice: Spatial equity in ohio and cleveland. *Annals of the Association of American Geographers, 85*(4), 641–663. https://doi.org/10.1111/j.1467-8306.1995.tb01818.x

Braveman, P. A., Kumanyika, S., Fielding, J., LaVeist, T., Borrell, L. N., Manderscheid, R., & Troutman, A. (2011). Health disparities and health equity: The issue is justice. *American Journal of Public Health, 101*(S1), S149–S155.

Brawner, B. M., Reason, J. L., Goodman, B. A., Schensul, J. J., & Guthrie, B. (2015). Multilevel drivers of HIV/AIDS among Black Philadelphians: Exploration using community ethnography and geographic information systems. *Nursing Research, 64*(2), 100–110. https://doi.org/10.1097/NNR.0000000000000076

Bullard, R. D. (1993). Environmental racism and invisible communities. *West Virginia Law Review, 96,* 1037.

Burgos, G., Rivera, F., & Garcia, M. (2017). *Contextualizing the relationship between culture and Puerto Rican Health: Toward a place-based framework of minority health disparities.* Sociology Department, Faculty Publications. Retrieved from https://digitalcommons.unl.edu/sociologyfacpub/622

Chetty, R., Stepner, M., Abraham, S., Lin, S., Scuderi, B., Turner, N., et al. (2016). The association between income and life expectancy in the United States, 2001–2014. *JAMA, 315*(16), 1750–1766. https://doi.org/10.1001/jama.2016.4226

Chor, D., & Lima, C. R. D. A. (2005). Epidemiologic aspects of racial inequalities in health in Brazil. *Cadernos de Saúde Pública, 21,* 1586–1594.

Cobbinah, P. B., Black, R., & Thwaites, R. (2013). Tourism planning in developing countries: Review of concepts and sustainability issues. *Sustainable Development, 10,* 12.

Cooper, H. L. F., Des Jarlais, D. C., Tempalski, B., Bossak, B. H., Ross, Z., & Friedman, S. R. (2012). Drug-related arrest rates and spatial access to syringe exchange programs in New York City health districts: Combined effects on the risk of injection-related infections among injectors. *Health & Place, 18*(2), 218–228. https://doi.org/10.1016/j.healthplace.2011.09.005

Cutter, S. L. (1995). Race, class and environmental justice. *Progress in Human Geography, 19*(1), 111–122. https://doi.org/10.1177/030913259501900111

Das-Munshi, J., Lund, C., Mathews, C., Clark, C., Rothon, C., & Stansfeld, S. (2016). Mental health inequalities in adolescents growing up in post-apartheid South Africa: Cross-sectional survey, SHaW study. *PLoS ONE, 11*(5), e0154478. https://doi.org/10.1371/journal.pone.0154478

Do, D. P., Locklar, L. R. B., & Florsheim, P. (2019). Triple jeopardy: The joint impact of racial segregation and neighborhood poverty on the mental health of black Americans. *Social Psychiatry and Psychiatric Epidemiology, 54*(5), 533–541. https://doi.org/10.1007/s00127-019-01654-5

Dubowitz, T., Ncube, C., Leuschner, K., & Tharp-Gilliam, S. (2015). A Natural experiment opportunity in two low-income urban food desert communities: Research design, community engagement methods, and baseline results. *Health Education & Behavior, 42*(Suppl. 1), 87S–96S. https://doi.org/10.1177/1090198115570048

Egan, M., Abba, K., Barnes, A., Collins, M., McGowan, V., Ponsford, R., Scott, C., Halliday, E., Whitehead, M., & Popay, J. (2021). Building collective control and improving health through a place-based community empowerment initiative: Qualitative evidence from communities seeking agency over their built environment. *Critical Public Health, 31*(3), 268–279. https://doi.org/10.1080/09581596.2020.1851654

English, D., Hickson, D. A., Callander, D., Goodman, M. S., & Duncan, D. T. (2020). Racial discrimination, sexual partner race/ethnicity, and depressive symptoms among black sexual minority men. *Archives of Sexual Behavior, 49*(5), 1799–1809. https://doi.org/10.1007/s10508-020-01647-5

Gripper, A. B., Nethery, R., Cowger, T. L., White, M., Kawachi, I., & Adamkiewicz, G. (2022). Community solutions to food apartheid: A spatial analysis of community food-growing spaces and neighborhood demographics in Philadelphia. *Social Science & Medicine, 310*, 115221. https://doi.org/10.1016/j.socscimed.2022.115221

Hicken, M. T., Payne-Sturges, D., & McCoy, E. (2023). Evaluating race in air pollution and health research: Race, PM2.5 air pollution exposure, and mortality as a case study. Current Environmental. *Health Reports, 10*(1), 1–11. https://doi.org/10.1007/s40572-023-00390-y

Huang, S. J., & Sehgal, N. J. (2022). Association of historic redlining and present-day health in Baltimore. *PLoS ONE, 17*(1), e0261028. https://doi.org/10.1371/journal.pone.0261028

Huang, L., Said, R., Goh, H. C., & Cao, Y. (2023). The residential environment and health and well-being of Chinese migrant populations: A systematic review. *International Journal of Environmental Research and Public Health, 20*(4), Article 4. https://doi.org/10.3390/ijerph20042968

Huffman, M. L., & Cohen, P. N. (2004). Racial wage inequality: Job segregation and devaluation across U.S. labor markets. *American Journal of Sociology, 109*(4), 902–936. https://doi.org/10.1086/378928

Ikawa, D., & Mattar, L. (2008). Racial discrimination in access to health: The Brazilian experience. *University of Kansas Law Review, 57*, 949.

Israel, B. A., Parker, E. A., Rowe, Z., Salvatore, A., Minkler, M. L., Ópez, J., Butz, A., Mosley, A., Coates, L., Lambert, G., Potito, P. A., Brenner, B., Rivera, M., Romero, H., Thompson, B., Coronado, G., & Halstead, S. (2005). Community-based participatory research: Lessons learned from the centers for Children's Environmental Health and Disease Prevention Research. *Environmental Health Perspectives, 113*(10), 1463–1471. https://doi.org/10.1289/ehp.7675

Iyanda, A. E., & Lu, Y. (2022a). 'Gentrification is not improving my health': A mixed-method investigation of chronic health conditions in rapidly changing urban neighborhoods in Austin, Texas. *Journal of Housing and the Built Environment, 37*(1), 77–100.

Iyanda, A. E., & Lu, Y. (2022b). Social and structural determinants of self-rated health in gentrifying neighborhoods in Austin Texas: A cross-sectional quantitative analysis. *International Journal of Community Well-Being*, 1–26.

Jennings, V., Browning, M. H. E. M., & Rigolon, A. (2019). Planning urban green spaces in their communities: Intersectional approaches for health equity and sustainability. In V. Jennings, M. H. E. M. Browning, & A. Rigolon (Eds.), *Urban Green spaces: Public health and sustainability in the United States* (pp. 71–99). Springer International Publishing. https://doi.org/10.1007/978-3-030-10469-6_5

Koski, E., Scheufele, E. L., Karunakaram, H., Foreman, M. A., Felix, W., & Dankwa-Mullan, I. (2022). Understanding disparities in healthcare: Implications for health systems and AI applications. In J. M. Kiel, G. R. Kim, & M. J. Ball (Eds.), *Healthcare information management systems: Cases, strategies, and solutions* (pp. 375–387). Springer International Publishing. https://doi.org/10.1007/978-3-031-07912-2_25

Kramer, M. R., Black, N. C., Matthews, S. A., & James, S. A. (2017). The legacy of slavery and contemporary declines in heart disease mortality in the U.S. South. *SSM—Population Health, 3*, 609–617. https://doi.org/10.1016/j.ssmph.2017.07.004

Krieger, N. (2012). Methods for the scientific study of discrimination and health: An ecosocial approach. *American Journal of Public Health, 102*(5), 936–944.

Lillie-Blanton, M., & Laveist, T. (1996). Race/ethnicity, the social environment, and health. *Social Science & Medicine, 43*(1), 83–91. https://doi.org/10.1016/0277-9536(95)00337-1

Lotfata, A., Moosazadeh, M., Helbich, M., & Hoseini, B. (2023). Socioeconomic and environmental determinants of asthma prevalence: a cross-sectional study at the US County level using geographically weighted random forests. *International Journal of Health Geographics, 22*(1), 18

Mahalik, M. K., Le, T. H., Le, H. C., & Mallick, H. (2022). How do sources of carbon dioxide emissions affect life expectancy? Insights from 68 developing and emerging economies. *World Development Sustainability, 1*, 100003. https://doi.org/10.1016/j.wds.2022.100003

Maity, B. (2017). Comparing health outcomes across scheduled tribes and castes in India. *World Development, 96*, 163–181. https://doi.org/10.1016/j.worlddev.2017.03.005

Massey, D. S. (2020). Still the Linchpin: Segregation and stratification in the USA. *Race and Social Problems, 12*(1), 1–12. https://doi.org/10.1007/s12552-019-09280-1

Massey, D. S., & Denton, N. A. (1993). *American apartheid: Segregation and the making of the underclass.* Harvard University Press.

Mitchell-Walthour, G. (2017). Economic pessimism and racial discrimination in Brazil. *Journal of Black Studies, 48*(7), 675–697.

Mosse, D. (2010). A relational approach to durable poverty, inequality and power. *The Journal of Development Studies, 46*(7), 1156–1178. https://doi.org/10.1080/00220388.2010.487095

Nardone, A., Casey, J. A., Morello-Frosch, R., Mujahid, M., Balmes, J. R., & Thakur, N. (2020). Associations between historical residential redlining and current age-adjusted rates of emergency department visits due to asthma across eight cities in California: An ecological study. *The Lancet Planetary Health, 4*(1), e24–e31. https://doi.org/10.1016/S2542-5196(19)30241-4

Ndinda, C., Ndhlovu, T. P., Juma, P., Asiki, G., & Kyobutungi, C. (2018). The evolution of non-communicable diseases policies in postapartheid South Africa. *BMC Public Health, 18*(1), 956. https://doi.org/10.1186/s12889-018-5832-8

Nigra, A. E., Chen, Q., Chillrud, S. N., Wang, L., Harvey, D., Mailloux, B., Factor-Litvak, P., & Navas-Acien, A. (2020). Inequalities in public water arsenic concentrations in counties and community water systems across the United States, 2006–2011. *Environmental Health Perspectives, 128*(12), 127001. https://doi.org/10.1289/EHP7313

Northridge, M. E., & Freeman, L. (2011). Urban planning and health equity. *Journal of Urban Health, 88*(3), 582–597. https://doi.org/10.1007/s11524-011-9558-5

Paradies, Y. (2016). Colonization, racism and indigenous health. *Journal of Population Research, 33*(1), 83–96. https://doi.org/10.1007/s12546-016-9159-y

Parr, H. (2004). Medical geography: Critical medical and health geography? *Progress in Human Geography, 28*(2), 246–257.

Pavao, A. L. B., Ploubidis, G. B., Werneck, G., & Campos, M. R. (2012). Racial discrimination and health in Brazil. *Ethnicity & Disease, 22*(3), 353–359.

Phillips, C. (1997). *Freedom's port: The African American Community of Baltimore, 1790–1860.* University of Illinois Press.

Pillay-van Wyk, V., Msemburi, W., Laubscher, R., Dorrington, R. E., Groenewald, P., Glass, T., Nojilana, B., Joubert, J. D., Matzopoulos, R., Prinsloo, M., Nannan, N., Gwebushe, N., Vos, T., Somdyala, N., Sithole, N., Neethling, I., Nicol, E., Rossouw, A., & Bradshaw, D. (2016). Mortality trends and differentials in South Africa from 1997 to 2012: Second National Burden of Disease Study. *The Lancet Global Health, 4*(9), e642–e653. https://doi.org/10.1016/S2214-109X(16)30113-9

Probst, J. C., Glover, S., & Kirksey, V. (2019). Strange harvest: A cross-sectional ecological analysis of the association between historic lynching events and 2010–2014 county mortality rates. *Journal of Racial and Ethnic Health Disparities, 6*(1), 143–152. https://doi.org/10.1007/s40615-018-0509-7

Ribeiro, A. M. (2013). *"The battle for harmony": Intergroup relations between blacks and Latinos in Philadelphia, 1950s to 1980s (Doctoral dissertation).* University of Pittsburgh.

Sandoval, M. F. L., Robertsdotter, A., & Paredes, M. (2017). Space, power, and locality: The contemporary use of territorio in Latin American Geography. *Journal of Latin American Geography, 16*(1), 43–67.

Schnake-Mahl, A. S., Jahn, J. L., Subramanian, S. V., Waters, M. C., & Arcaya, M. (2020). Gentrification, neighborhood change, and population health: A systematic review. *Journal of Urban Health, 97*, 1–25. https://doi.org/10.1007/s11524-019-00400-1

Schuyler, A. J., & Wenzel, S. E. (2022). Historical redlining impacts contemporary environmental and asthma-related outcomes in Black adults. *American Journal of Respiratory and Critical Care Medicine, 206*(7), 824–837. https://doi.org/10.1164/rccm.202112-2707OC

Seitles, M. (1998). The perpetuation of residential racial segregation in America: Historical discrimination, modern forms of exclusion, and inclusionary remedies. *Journal of Land Use & Environmental Law, 14*, 89.

Sly, P. D., Carpenter, D. O., Van den Berg, M., Stein, R. T., Landrigan, P. J., Brune-Drisse, M.-N., & Suk, W. (2016). Health consequences of environmental exposures: Causal thinking in global environmental epidemiology. *Annals of Global Health, 82*(1), 3–9. https://doi.org/10.1016/j.aogh.2016.01.004

Smith, G. S., Breakstone, H., Dean, L. T., & Thorpe, R. J., Jr. (2020). Impacts of gentrification on health in the US: A systematic review of the literature. *Journal of Urban Health, 97*(6), 845–856. https://doi.org/10.1007/s11524-020-00448-4

Soja, E. W. (2013). *Seeking spatial justice* (Vol. 16). University of Minnesota Press.

Stull, V., Bell, M. M., & Ncwadi, M. (2016). Environmental apartheid: Eco-health and rural marginalization in South Africa. *Journal of Rural Studies, 47*, 369–380. https://doi.org/10.1016/j.jrurstud.2016.04.004

Suglia, S. F., Shelton, R. C., Hsiao, A., Wang, Y. C., Rundle, A., & Link, B. G. (2016). Why the neighborhood social environment is critical in obesity prevention. *Journal of Urban Health, 93*(1), 206–212. https://doi.org/10.1007/s11524-015-0017-6

Sugrue, T. J. (1995). Crabgrass-roots politics: Race, rights, and the reaction against liberalism in the urban north, 1940–1964. *The Journal of American History, 82*(2), 551–578.

Suk, W. A., Ahanchian, H., Asante, K. A., Carpenter, D. O., Diaz-Barriga, F., Ha, E. H., Huo, X., King, M., Ruchirawat, M., da Silva, E. R., & Sly, L. (2016). Environmental pollution: An underrecognized threat to children's health, especially in low- and middle-income countries. *Environmental Health Perspectives, 124*(3), A41–A45. https://doi.org/10.1289/ehp.1510517

Tan, M., Hatef, E., Taghipour, D., Vyas, K., Kharrazi, H., Gottlieb, L., & Weiner, J. (2020). Including social and behavioral determinants in predictive models: Trends, challenges, and opportunities. *JMIR Medical Informatics, 8*(9), e18084. https://doi.org/10.2196/18084

Thapa, R., van Teijlingen, E., Regmi, P. R., & Heaslip, V. (2021). Caste exclusion and health discrimination in South Asia: A systematic review. *Asia Pacific Journal of Public Health, 33*(8), 828–838.

Thornton, R. L., Glover, C. M., Cené, C. W., Glik, D. C., Henderson, J. A., & Williams, D. R. (2016). Evaluating strategies for reducing health disparities by addressing the social determinants of health. *Health affairs, 35*(8), 1416–1423.

Tung, E. L., Peek, M. E., Rivas, M. A., Yang, J. P., & Volerman, A. (2021). Association of Neighborhood Disadvantage with Racial Disparities. In COVID-19 Positivity In Chicago: Study examines the association of neighborhood disadvantage with racial disparities in COVID-19 positivity in Chicago. *Health Affairs, 40*(11), 1784–1791. https://doi.org/10.1377/hlthaff.2021.00695

Urban Health Collaborative. (2020, May 26). *Community health profile: Kensington, Philadelphia.* Urban Health Collaborative. Retrieved from https://drexel.edu/uhc/resources/briefs/kensington-brief/

van Ham, M., Tammaru, T., Ubarevičienė, R., & Janssen, H. (Eds.). (2021). *Urban socioeconomic segregation and income inequality: A global perspective.* Springer Nature. https://doi.org/10.1007/978-3-030-64569-4

Vargas, E. A., Chirinos, D. A., Mahalingam, R., Marshall, R. A., Wong, M., & Kershaw, K. N. (2021). Discrimination, perceived control, and psychological health among African Americans with hypertension. *Journal of Health Psychology, 26*(14), 2841–2850.

Venter, Z. S., Shackleton, C. M., Van Staden, F., Selomane, O., & Masterson, V. A. (2020). Green Apartheid: Urban green infrastructure remains unequally distributed across income and race geographies in South Africa. *Landscape and Urban Planning, 203*, 103889. https://doi.org/10.1016/j.landurbplan.2020.103889

Ziegler, T. B., Coombe, C. M., Rowe, Z. E., Clark, S. J., Gronlund, C. J., Lee, M., Palacios, A., Larsen, L. S., Reames, T. G., Schott, J., Williams, G. O., & O'Neill, M. S. (2019). Shifting from "Community-Placed" to "Community-Based" research to advance health equity: A case study of the heatwaves, housing, and health: Increasing Climate Resiliency in Detroit (HHH) Partnership. *International Journal of Environmental Research and Public Health, 16*(18), Article 18. https://doi.org/10.3390/ijerph16183310

Chapter 5
Resources for Health Within Latinx Communities: A Social-Spatial Determinants of Health Perspective

Elizabeth Ackert, Sigrid Van Den Abbeele, and Hannah Malak

Introduction

Latinx population health in the US is marked by a "paradox" whereby Latino/a/x individuals exhibit better-than-expected health and mortality outcomes despite their disadvantaged socioeconomic backgrounds and external barriers such as discrimination (Acevedo-Garcia & Bates, 2008; Fernandez et al., 2023).[1] Health access, however, represents one domain of Latinx population health that is an exception to this paradox. Compared to non-Latinx Whites, the US Latinx population is less likely to have health insurance or a usual source of care and is more likely to delay care due to cost (Guzman et al., 2020; Office of Disease Prevention and Health Promotion, 2023; Ortega et al., 2022; Perreira et al., 2021; Whitener & Corcoran, 2021).

These Latinx health access patterns vary spatially, however, such as by state and local areas (ASPE Office of Health Policy, 2021; Perreira et al., 2021; Whitener & Corcoran, 2021). Health uninsurance rates are highest among Latinos/as living in most of the states in the US South and Utah and Wyoming (Kaiser Family Foundation, 2023). Latinx health access is also lower in areas with smaller numbers of Latinx immigrants (at the zip code level), and in fast-growing "new" Latinx destination areas of residence (defined at the metropolitan and county levels) versus more established Latinx communities (Gresenz et al., 2009, 2012; Monnat, 2017).

---

[1] We use the terms Latino/a/x, Latinx, and Hispanic interchangeably throughout this chapter.

E. Ackert (✉) · S. Van Den Abbeele · H. Malak
Department of Geography, University of California at Santa Barbara,
Santa Barbara, CA, USA
e-mail: ackert@ucsb.edu

© The Author(s) 2026
M. A. Kolak, I. K. Moise (eds.), *Place and the Social-Spatial Determinants of Health*, Global Perspectives on Health Geography,
https://doi.org/10.1007/978-3-031-88463-4_5

Spatial variation in Latinx health access is shaped by spatial heterogeneity in "demand" factors, such as the need for and affordability of care, and by "supply" factors, such as the local availability of health providers. To be sure, the availability of specific types of health services varies according to both place-based (e.g., urban/rural) and population-based (e.g., percent in poverty, racial/ethnic composition) factors (Ackert et al., 2021; Naylor et al., 2019; Parker, 2021; Rural Health Information Hub, 2021). The availability of and access to generalized and specialized health services has been linked to the utilization of care, such as the use of mental health services by psychiatrists, as well as to health outcomes, such as late-stage breast cancer (Cook et al., 2013; Dai, 2010; Dinwiddie et al., 2013). Therefore, addressing inequalities in the "geography of opportunity" in access to health providers could reduce both place-based and ethno-racial health disparities for the Latinx population and other groups (Acevedo-Garcia et al., 2008; Derose et al., 2007; Osypuk & Acevedo-Garcia, 2010).

In this chapter, we investigate the issue of spatial variability in access to health providers among the Latinx population. We focus on the interplay of Latinx community attributes and the availability of federally qualified health centers (FQHCs), an important type of "safety net" health care. Our work is guided by two frameworks that we combine into one social-spatial determinants of health (SSDoH) framework: (a) Schulz et al.'s (2002) framework of racial segregation as a fundamental determinant of racial disparities in health, and (b) Bronfenbrenner and Morris' (2006) ecological model of human development. The Schulz et al. framework highlights the "fundamental," "intermediate," and "proximate" determinants of racial health disparities, while the Bronfenbrenner and Morris framework emphasizes the concept of ecologically nested mechanisms affecting health and human development. We demonstrate how these combined frameworks can be used both to guide the operationalization of SSDoH and to attend to issues of measurement at different scales (i.e., across nested ecological contexts).

We focus on the relationship between two potential "fundamental" determinants of health disparities—Latinx residential segregation and Latinx community histories as immigrant destinations—and one "intermediate" determinant of health disparities, the presence of FQHCs. We consider the ecologically nested contexts of the census tract (the "neighborhood") and the county (the "community"). We draw from multiple data sources, including the decennial censuses (1990 and 2010), the American Community Survey (ACS; 2015–2019), and the Health Resources and Services Administration (HRSA) (FQHC data through 2019). Using quantitative analyses, we show how the interplay of Latinx residential segregation and Latinx population change relates to the availability of FQHCs in Latinx neighborhoods and communities.

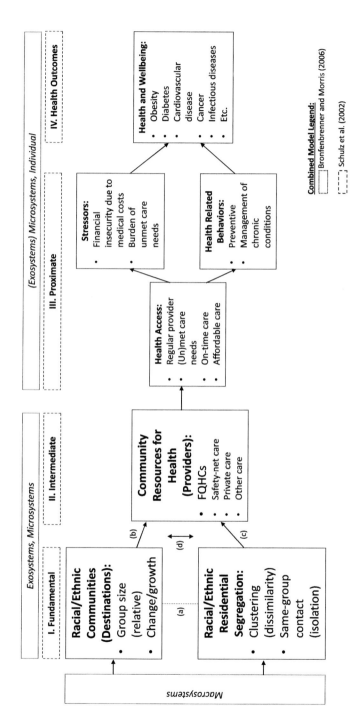

Fig. 5.1 A Combined SSDoH Framework

Social-Spatial Determinants of Health Framework

To address the issue of access to FQHCs across Latinx communities, we develop an SSDoH framework (Fig. 5.1). We draw first from Schulz et al.'s (2002) framework of racial segregation as a fundamental determinant of racial disparities in health. The Schulz et al. framework was motivated by an interest in examining how racial residential segregation could ultimately affect racial health disparities in the Detroit Metropolitan Area and Detroit City. This framework highlights the "fundamental," "intermediate," and "proximate" determinants of racial health disparities. Fundamental causes are those that shape access to intermediate health-promoting resources (e.g., supply of health providers). These fundamental causes are the outcomes of macrosocial political, social, and economic conditions, which are also considered fundamental determinants of racial health disparities in this framework. In the Schulz et al. framework, two key fundamental causes of interest are racial/ethnic residential segregation and economic inequalities, which are both interrelated.

Intermediate factors link fundamental determinants of health with proximate determinants of health outcomes and health disparities. Schulz et al. (2002) note that while fundamental determinants can be difficult to remedy using policy levers, intermediate determinants have more potential to change via key interventions. For example, in their framework, intermediate determinants include the physical environment (such as land use and industrial pollution), community infrastructure (such as health services, grocery stores, and housing), and the social environment (such as relations between racial/ethnic groups). They note that health disparities could be reduced by tackling intermediate determinants stemming from fundamental determinants, such as limiting industrial pollutants or providing quality housing.

Proximate determinants of health are direct determinants of health that are well established in the literature and policy on health outcomes and disparities (Lucyk & McLaren, 2017; U.S. Department of Health and Human Services, 2023). In the Schulz et al. (2002) framework, these proximate determinants include exposure to stressors, health-related behaviors, and social integration and social support. Finally, the framework includes health outcomes of interest, which can be several health and development outcomes throughout the life course.

We also incorporate Bronfenbrenner and Morris' (2006) ecological model of human development into our SSDoH framework. The Bronfenbrenner and Morris framework proposes the idea of ecologically nested mechanisms affecting human health and development. This framework is often applied to understanding how ecologically-nested contexts affect children and adolescents and health and development, but it is also broadly applicable to health and development across the life course. The Bronfenbrenner and Morris framework is similar to the Schulz et al. (2002) framework in that it connects factors and processes that may be distal to the individual to those that are more proximate determinants of health and development.

In the Bronfenbrenner and Morris framework, the "macrosystem" is the most distal ecological context that encompasses broad institutional systems of a society, including economic, social, educational, legal, and political systems (Rosa & Tudge, 2013). The macrosystem influences lower ecological levels, especially the next lowest ecological level in the framework, the "exosystem," by determining the values and qualities of each ecological level. For example, a neoliberal government within a capitalist economic system may emphasize both private insurance markets and private health services, both of which could hinder health access.

The defining feature of the exosystem is its indirect influence on human development and health (Bronfenbrenner & Morris, 2006; Rosa & Tudge, 2013). The exosystem includes (but is not limited to) environmental contexts such as neighbors in a neighborhood, social welfare services, health services, mass media, and industries. Importantly, many policy contexts are part of the exosystem (Rosa & Tudge, 2013). Like the intermediate determinants in the Schulz et al. (2002) framework, the exosystem can be an important site to intervene in individual health and development outcomes through policy reforms.

The exosystem context can shape lower-level contexts that are more proximal to the individual, and thus more directly influence individual health and development outcomes. These proximal ecological inputs make up the "microsystem." The microsystem includes any place, person, or group that may exert a direct influence on individual health and development and includes (but is not limited to) the family, peers, direct neighbors, managers and coworkers, schoolteachers, and health care professionals. Individuals may also be embedded in multiple microsystems, and this system of microsystems is referred to as the "mesosystem." Finally, at the center of the nested ecological context is the individual, who may possess certain traits or characteristics that further influence health and well-being or that could moderate the influences of higher-level ecological contexts on their outcomes, such as their age, gender identity, and health status.

Figure 5.1 presents our combined SSDoH framework. One aspect of this framework that differs from the Schulz et al. (2002) framework is that we consider macrosystems (referred to as "macrosocial factors" in the Schulz et al. framework) to be a separate set of determinants of fundamental determinants of health, and we thus display the macrosystem outside of the fundamental determinants as the most distal determinants of health. When combined with the Schulz et al. framework, the Bronfenbrenner and Morris (2006) framework also allows for a consideration of the scale of measurement—the level of measurement and/or aggregation of fundamental, intermediate, and proximate determinants of health outcomes (e.g., individual, family, neighborhood, etc.). For example, residential segregation can be considered a fundamental determinant of health and health disparities, but aspects of segregation could be measured at the exosystem level (the metro area or county level, such as the index of dissimilarity or isolation index) and the microsystem level (the racial/ethnic composition of the immediate block group area where a person lives). Similarly, it is possible to consider the measurement of proximate determinants of health outcomes, as well as the outcomes themselves, at different scales and/or levels of aggregation, such as the exosystem level (population health), the microsystem level (parental health status), or the individual level (child health status).

Operationalizing Key Framework Concepts: The Case of FQHCs in Latinx Communities

The framework in Fig. 5.1 guides our operationalization of key attributes of Latinx communities and safety-net care (FQHCs) and hypothesized linkages between them (lines/arrows a–d in Fig. 5.1). Our study measures two fundamental attributes of Latinx communities: (a) Racial/ethnic communities (Latinx "destinations"), and (b) Racial/ethnic residential segregation. Latinx destinations are categorized based on Latinx population size and change between 1990 and 2010. Since the 1990s, the Latinx population has increasingly resided outside of the established "Big 5" areas of settlement (NY, CA, TX, FL, IL) and the American Southwest (Lichter & Johnson, 2009; Massey & Capoferro, 2008). The emergence of "new" Latinx destinations has been shaped by macrosystem factors such as the changing geography of low-skilled labor opportunities and stringent immigration laws and border policies that have reduced circular migration and increased settlement in nontraditional areas of immigration (Massey, 2008; Zúñiga & Hernández-León, 2005). Destinations are theoretically meaningful because the history and growth of areas of Latinx settlement can shape group size, intergroup relations, and institutional resources, which are all factors relevant to the incorporation of Latinx immigrants and their children and descendants (Waters & Jiménez, 2005). Destinations are also directly related to healthcare access for Latinos/as, with lower health access in "newer" versus more established Latinx destinations of residence, defined at the metropolitan and county levels (Gresenz et al., 2012; Monnat, 2017).

Racial/ethnic residential segregation is also an important fundamental determinant of health and health disparities. In fact, residential segregation is a key fundamental determinant of health disparities in the Schulz et al. (2002) framework. In our study, we measure residential segregation according to dissimilarity—evenness versus clustering of groups within a geographic area, or isolation or exposure—the potential for same-group versus different group contact among the average group member in the average neighborhood (Massey & Denton, 1988). Residential segregation by race/ethnicity is meaningful because, like destinations, segregation shapes intergroup relations and access to and quality of institutional resources (Leventhal & Brooks-Gunn, 2000; Morenoff & Lynch, 2004; Schulz et al., 2002). Residential segregation is also intertwined with destinations, with higher Latinx-White segregation in newer areas of Latinx residence (Hall, 2013; Lichter et al., 2010).

Finally, we focus on FQHCs as important community resources for health that are intermediate determinants of health outcomes and disparities. FQHCs are healthcare facilities funded under the federal Health Center program, and they serve as an important point of "safety-net" care for vulnerable populations, such as public housing residents, individuals who are homeless, and migrant workers (Health Resources and Services Administration, 2023b). These centers are able to provide low-cost or no-cost care because they are reimbursed for services by the Bureau of Primary Health Care and Centers for Medicare and Medicaid Services. The Latinx population is overrepresented among FQHC patients and is the second-largest racial/ethnic group among FQHC patients after non-Latinx Whites (Health Resources and Services Administration, 2023a). Importantly, the supply of FQHCs varies spatially across the USA, in part because of geographic variation in the demographic, health outcome, and health care supply criteria used to measure medical underservice.

The prior literature demonstrates that Latinx destinations and racial/ethnic segregation are systematically related to one another (Fig. 5.1, line a) and that both are linked to the supply of FQHCs (Fig. 5.1, arrows b and c). Several studies find that Latinx-White dissimilarity is higher or increasing in newer areas of Latinx residence (Hall, 2013; Lichter et al., 2010; Park & Iceland, 2011). Both Latinx destinations and racial/ethnic residential segregation are also associated with FQHC availability. The supply of FQHCs is lower in newer versus established Latinx destination counties (Ackert et al., 2021; Parker, 2021). However, FQHCs are more prevalent in more racially/ethnically segregated areas than in areas with lower levels of residential segregation (Ko & Ponce, 2013).

In our study, we examine the interplay of Latinx destinations and racial/ethnic residential segregation as predictors of FQHC supply (Fig. 5.1, lines/arrows a–d). We anticipate that more FQHCs will be available in Latinx communities that are established Latinx destinations with high levels of Latinx segregation (Fig. 5.1, arrows b–d). Conversely, we expect that fewer FQHCs will be found in "new" or "other" Latinx destination areas with lower levels of Latinx segregation (Fig. 5.1, arrows b–d). Based on our SSDoH framework (Fig. 5.1), we also examine this issue at two geographic scales: the county and the census tract. The county is used to represent the broader community (exosystem) that likely exerts an indirect influence on health outcomes and disparities. For the county-level analysis, we measure Latinx destinations, Latinx segregation (both isolation and dissimilarity), and FQHC supply at the county level. It is also possible, however, that segregation and the supply of FQHCs are related to neighborhood conditions and processes that exert indirect (exosystem) or direct (microsystem) influences on health outcomes. Though definitions of neighborhoods vary widely in the literature, here we are referring to the administrative boundaries defined by the Census Bureau meant to encapsulate the proximate residential area in which someone lives (Diez Roux, 2001). For this reason, we also measure aspects of Latinx segregation and the supply of FQHCs at the census tract level.

A Study of Latinx Destinations, Racial/Ethnic Residential Segregation, and FQHCs

Our SSDoH framework guides our conceptualization of how Latinx destinations and racial/ethnic residential segregation could shape the supply of FQHCs and informs our measurement of these key concepts and the scale of measurement. To execute our study, we create two datasets, a county-level dataset and a census tract-level dataset. We use the 2010 county and census tract boundaries. We begin with data on the key outcome of interest—FQHCs. We obtained publicly-available data on FQHCs from HRSA (Health Resources and Services Administration, 2022). We restrict the analysis to FQHCs and look-a-like sites (LALs). We focus on FQHCs established by 2019 because the COVID-19 pandemic changed the landscape of FQHCs, as the federal government sought to use FQHCs to address the pandemic, such as for testing and vaccine distribution (Biden, 2021; National Association of Community Health Centers, 2022). Each FQHC in the data has a county identified, which allowed us to determine how many FQHCs were in each county. Each FQHC in the data also has an address. In ArcGIS Pro, we uploaded an attribute table with these addresses to create one data layer with points for FQHCs, and obtained a 2019 census tract GIS data layer (based on 2010 census tract boundaries/shapefiles). We used the clip tool in ArcGIS Pro to assign each FQHC to its corresponding census tract. We then exported the tract identifiers for each FQHC in the data.

To measure Latinx destinations, we use data from the 1990 and 2010 decennial censuses. Following our previous work in this area (Ackert et al., 2021), we define Latinx destinations at the county level (with county boundaries harmonized to 2010), and we divide counties into three groups: established, new, and other destinations. Established destinations are counties that were 9% or more Latinx in 1990 (the national average Latinx percent for that year). New destinations are non-established counties that experienced higher-than-median Latinx growth among all non-established counties (272% growth or higher) between 1990 and 2010 and were at least 5% Latinx by 2010, or that had lower-than-median growth and were at least 16% Latinx by 2010 (the national average Latinx percent for that year). Other destinations were all other counties that did not meet those criteria to be considered an established or new destination.

We measure aspects of Latinx residential segregation at both the county and census tract levels. Using data from the 2015–2019 ACS (5-year survey), we calculate the Latinx-White index of dissimilarity and the Latinx index of isolation, which are two commonly used indicators of segregation (Massey & Denton, 1988). The index of dissimilarity is calculated based on counts of Latinx and non-Latinx-White residents in census tracts and counties and is a measure of evenness versus clustering of Latinx and White populations within a county (Forest, 2005; U.S. Census Bureau, 2021). The index of dissimilarity values ranges from 0.0 to 1.0, with 0.0 indicating no segregation (i.e., two groups equally distributed across tracts within the county) and 1.0 indicating total segregation (i.e., two groups occupy separate tracts within the county). Conceptually, the index of dissimilarity values represents the proportion of one of the groups that would need to move tracts to ensure an equal distribution of the two groups across all tracts.

The index of isolation is a measure of average exposure to a member of the same racial/ethnic group in census tracts in a county and is also based on counts of Latinx and non-Latinx-White residents in tracts, tract total populations, and the total Latinx population in the county (Forest, 2005; U.S. Census Bureau, 2021). The index of isolation also ranges from 0.0 to 1.0 and conceptually represents the probability that a Latinx person shares a tract with another Latinx person. At the tract level, we also measure the percentage of Latinx residents in the tract (also from the 2015–2019

ACS). The tract percent Latinx is indicative of Latinx population size within a county, Latinx residential clustering within a county, and the likelihood of living near Latinx neighbors.

In multivariate models, we also include several variables from the ACS (2015–2019) that may be correlated with Latinx destinations, Latinx residential segregation, and/or FQHC supply. In the county-level models, these variables include the percent Hispanic (in dissimilarity models), percent Black, median income, the GINI index (a measure of income inequality), proportion uninsured, poverty rate, proportion over age 65, and number of tracts. In tract-level models, these variables include percent Black, median income, the GINI index, the poverty rate, the proportion over age 65, and the total tract population size.

Our analytic approach is to first show descriptive differences in the supply of FQHCs and Latinx residential segregation across destinations. Our county-level dataset for descriptive statistics includes 2,902 counties (approximately 92.3% of all counties, based on 2010 boundaries), and our tract-level dataset includes 71,458 census tracts (approximately 96.4% of all census tracts, based on 2010 boundaries). We do not have full coverage of all counties and tracts in the USA because we exclude Puerto Rico, drop some counties that were not fully accounted for in our county boundary harmonization (for 1990 to 2010 destination data), and/or had missing values in the HRSA data on FQHCs. After conducting a descriptive analysis, we use mixed effects (multilevel) logistic regression models with varying (random) intercepts to estimate the likelihood that a county or tract has at least one FQHC versus no FQHCs. For county-level models, counties are nested within states, and for tract-level models, tracts are nested within counties. We further drop 370 tracts for the mixed effects logistic regression model analysis because of missing values on key covariates, resulting in a sample size of 71,088 tracts for the tract-level mixed effects logistic regression models.

Study Results

Table 5.1 shows descriptive statistics for all counties (panel a) and all tracts (panel b) and then stratifies these results by destination counties (established, new, or other). The results in Table 5.1 confirm that FQHC supply, Latinx-White segregation, and other population characteristics vary across destinations. New destination counties have a lower supply of FQHCs than established destinations, consistent with previous studies (Ackert et al., 2021; Parker, 2021). New destination counties have approximately 59% of the number of FQHCs per 10,000 residents as established destination counties. New destinations have a higher index of dissimilarity values than established destinations, which has also been found in the prior literature (Hall, 2013; Lichter et al., 2010). However, because the Latinx population is a smaller proportion of the total population in new versus established destination counties (12.5% versus 40.6%, respectively), the Latinx population in new destinations tends to be less isolated in these areas, with a lower index of isolation value at

Table 5.1 Descriptive statistics for all counties (panel a) and all tracts (panel b) and then stratified results by destination counties (established, new, or other)

Variable	All counties (N = 2902)	Established destination counties (N = 299)	New destination counties (N = 510)	Other counties (N = 2093)
(a) County-level descriptive statistics				
FQHCs (raw #)	4.1	13.3	4.1	2.8
FQHCs per capita (per 10 k)	0.714	0.730	0.427	0.782
Hispanic-White Dissimilarity Index	0.323	0.307	0.347	0.319
Hispanic-White Isolation Index	0.124	0.469	0.189	0.059
Percent Hispanic	9.1%	40.6%	12.5%	3.8%
Percent Black	9.6%	5.2%	11.4%	9.8%
Total population	111,426	335,991	163,571	66,639
Median income	$27,666	$27,757	$29,219	$27,274
GINI coefficient	0.446	0.455	0.446	0.445
Percent uninsured	9.47%	13.55%	10.89%	8.54%
Poverty rate	10.45%	11.56%	9.29%	10.57%
Proportion over age 65	18.51%	16.35%	17.34%	19.10%

Variable	All tracts in all counties (N = 71,458)	Tracts in established destination counties (N = 20,404)	Tracts in new destination counties (N = 17,280)	Tracts in other counties (N = 33,774)
(b) Tract-level descriptive statistics				
FQHCs (raw #)	0.163	0.184	0.122	0.172
FQHCs per capita (per 10 k)	0.047	0.052	0.034	0.052
Percent Hispanic	16.4%	36.4%	13.3%	5.8%
Percent Black	13.4%	11.3%	14.7%	14.1%
Total population	4486	4791	4828	4128
Median income (*370 tracts NA*)	$67,113	$72,316	$71,311	$61,818
GINI coefficient (*256 tracts NA*)	0.425	0.428	0.419	0.427
Poverty rate	14.1%	14.9%	12.6%	14.3%
Proportion over age 65	16.4%	14.5%	16.4%	17.6%

the county level (a difference of 0.280) and lower percent Hispanic within the tract at the tract level (a difference of 23.1 percentage points).

Several population attributes may help to explain why the supply of FQHCs is higher in established destination counties and in tracts in established destination counties. At the county level, established destinations have lower median incomes, higher poverty rates, and higher noninsurance rates than new destinations. Tracts in established destinations also have higher poverty rates but slightly higher median incomes than those in new destinations. However, new destinations have several attributes that could make them viable candidates for FQHCs relative to established destinations, including higher percentages of Black populations (because many of these new destinations are in the South) and higher percentages of older residents (aged 65 and over).

Table 5.2 displays the results (log-odds) from mixed effects logistic regression models predicting whether a county or a tract has any FQHCs versus no FQHCs, adjusting for population characteristics that could be related to destinations, segregation, and FQHC supply. At the county level, there is no significant adjusted association between destinations and FQHC supply, but both measures of Latinx residential segregation are significantly related to having any FQHCs in the county. Specifically, the log-odds of having at least one FQHC increase as Latinx-White dissimilarity increases and as Latinx-White isolation increases.

This pattern is shown in Fig. 5.2, which displays the predicted probability of having any FQHC in the county by levels of Latinx-White dissimilarity (left panel), Latinx-White isolation (right panel), and destination type (both panels), with all covariates held at their observed values. An established destination county with a Latinx-White dissimilarity index value of 0.25 (low segregation) would have a predicted probability of having at least one FQHC of 0.7 (70%), whereas an established destination county with a Latinx-White dissimilarity index value of approximately 0.875 (very high segregation) would have a predicted probability of having at least one FQHC of 0.8 (80%). Similarly, an established destination with a Latinx-White

Table 5.2 Results (log-odds) from mixed effects logistic regression models

County-level variables	County-level variable fixed effects estimates (log-odds)		Tract-level variables	Tract-level variable fixed effects estimates (log-odds)
	Dissimilarity models	Isolation models		
New destination county (ref. established)	−0.25	−0.16	New destination county (ref. established)	−0.3027***
Other destination county (ref. established)	−0.17	−0.03	Other destination county (ref. established)	0.1443***
Hispanic-White dissimilarity index	1.137**	–	Percent hispanic	0.0115***
Hispanic-White isolation index	–	1.99*		

Note: Results from mixed effects logistic regression models with counties nested within states (varying intercept models; all covariates fixed). County-level models control for: Percent Hispanic (dissimilarity only), Percent Black, Median Income, GINI Index, Proportion Uninsured, Poverty Rate, Proportion Over Age 65, and Number of Tracts

Note: Results from mixed effects logistic regression models with tracts nested within counties (varying intercept models; all covariates fixed). Tract-level models control for: Percent Black, Median Income, GINI Index, Poverty Rate, Total Population, and Proportion Over Age 65

Signif. codes: ***$p < 0.001$, **$p < 0.01$, *$p < 0.05$

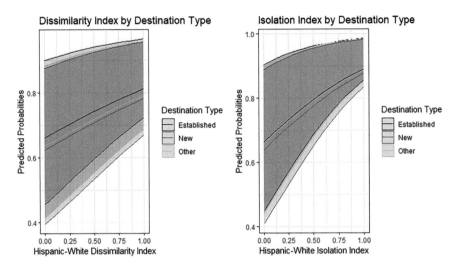

Fig. 5.2 Predicted probability of having any FQHC in the county by levels of Latinx-White dissimilarity (left panel), Latinx-White isolation (right panel), and destination type (both panels)

isolation value of 0.125 (lower value) would have a predicted probability of having at least one FQHC of 0.70 (70%), but an established destination with a Latinx-White isolation value of 0.75 (high value) would have a predicted probability of having at least one FQHC of 0.84 (85%). The results displayed in Fig. 5.2 also indicate that for each level of dissimilarity or isolation, established destinations have a higher predicted probability of having at least one FQHC relative to new and other destinations, but these differences are not statistically significant.

For the tract level, both destinations and segregation (measured by percent Latinx per tract) are significantly associated with having any FQHC in the tract. In these tract-level results, a tract in a new destination county has lower log-odds of having any FQHC, whereas a tract in another destination county has higher log-odds of having any FQHC, relative to tracts in established destination counties. Additionally, similar to the results for county Latinx-White isolation, the log-odds of having any FQHC in the tract increase as the percentage of Latinx in the tract increases.

Discussion

These results confirm prior findings that unadjusted levels of FQHC supply are lower in new versus established destination counties (Ackert et al., 2021; Parker, 2021). They also confirm that segregation is higher in new versus established destinations when measuring dissimilarity—clustering versus evenness (Hall, 2013; Lichter et al., 2010)—but lower in new versus established destinations when measuring Latinx-White isolation (at the county level) and percent Latinx in the neighborhood (at the tract level). The adjusted association between destinations and FQHC supply, net of measures of Latinx-White residential segregation and other population characteristics, depends on the geographic scale of measurement. The relationship between destinations and having any FQHCs is not significant at the county level, but tracts in new destination counties are less likely to have any FQHCs than those in established counties in adjusted models. In contrast, indicators of Latinx-White residential segregation are significantly related to FQHC supply at both the county and tract levels. Net of destinations and population compositional characteristics, the likelihood of having any FQHCs increases as Latinx-White dissimilarity increases (at the county level), as Latinx-White isolation increases (at the county level), and as the percent Latinx increases (at the tract level).

Conclusion

We demonstrate that a framework of SSDoH and health disparities (Bronfenbrenner & Morris, 2006; Schulz et al., 2002) can guide a conceptualization of how racial/ethnic communities and racial/ethnic residential segregation (two "fundamental" determinants of health and health disparities) are related to community resources for health ("intermediate" determinants of health and health disparities). This framework also informs the measurement of these key constructs, including issues of measurement level or scale.

In this study, measures of racial/ethnic segregation at both the county ("exosystem") and the census tract ("exosystem," "microsystem") levels emerge as consistent, salient fundamental health determinants that are connected to the intermediate health determinant of FQHC supply. Both counties and census tracts are more likely to have FQHCs within them when segregation is higher. For Latinx destinations, the results are less consistent across geographic levels (scales). At the county level, the lower supply of FQHCs observed among new versus established destinations is attenuated in adjusted models. That is, the association between Latinx destinations and FQHC supply can be explained by a combination of factors related to Latinx-White residential segregation (a concurrent fundamental determinant) and population composition. At the tract level, being in a new versus an established destination is consequential; tracts in new destinations are less likely to have any FQHCs than those located in established destinations. This finding suggests that scale matters when trying to understand destinations and FQHC supply—differences in neighborhood-level processes rather than community-level processes may better help to explain differences in FQHC supply across destinations.

This framework and our measurement approach can serve as a model for future work seeking to conceptualize and measure SSDoH and health disparities. The framework, operationalization of key concepts, and consideration of issues of scale can help to expand an assessment of social-spatial relationships between communities, segregation, community resources, and health for the Latinx population. This framework can also be used to scrutinize the proximate and distal factors that contribute to Latinx healthcare access disparities. Finally, this approach can be used to investigate other constructs and social-spatial associations that inhibit or promote health for other racial/ethnic, socioeconomic, or demographic groups.

References

Acevedo-Garcia, D., & Bates, L. M. (2008). Latino health paradoxes: Empirical evidence, explanations, future research, and implications. In P. H. Rodríguez, P. R. Sáenz, & A. P. C. Menjívar (Eds.), *Latinas/os in the United States: Changing the Face of América* (pp. 101–113). Springer US. https://doi.org/10.1007/978-0-387-71943-6_7

Acevedo-Garcia, D., Osypuk, T. L., McArdle, N., & Williams, D. R. (2008). Toward a policy-relevant analysis of geographic and racial/ethnic disparities in child health. *Health Affairs, 27*(2), 321–333.

Ackert, E., Hong, S. H., Martinez, J., Van Praag, G., Aristizabal, P., & Crosnoe, R. (2021). Understanding the health landscapes where Latinx immigrants establish residence in the US. *Health Affairs, 40*(7), 1108–1116. https://doi.org/10.1377/hlthaff.2021.00176

ASPE Office of Health Policy. (2021). *Health insurance coverage and access to care among Latinos: Recent trends and key challenges.* Retrieved from https://aspe.hhs.gov/sites/default/files/documents/68c78e2fb15209dd191cf9b0b1380fb8/ASPE_Latino_Health_Coverage_IB.pdf.

Biden, J. (2021). *National strategy for the COVID-19 response and pandemic preparedness.* Retrieved from https://www.whitehouse.gov/wp-content/uploads/2021/01/National-Strategy-for-the-COVID-19-Response-and-Pandemic-Preparedness.pdf.

Bronfenbrenner, U., & Morris, P. A. (2006). *The bioecological model of human development.* Handbook of Child Psychology.

Cook, B. L., Doksum, T., Chen, C., Carle, A., & Alegría, M. (2013). The role of provider supply and organization in reducing racial/ethnic disparities in mental health care in the U.S. *Social Science & Medicine, 84*, 102–109. https://doi.org/10.1016/j.socscimed.2013.02.006

Dai, D. (2010). Black residential segregation, disparities in spatial access to health care facilities, and late-stage breast cancer diagnosis in metropolitan Detroit. *Health & Place, 16*(5), 1038–1052.

Derose, K. P., Escarce, J. J., & Lurie, N. (2007). Immigrants and health care: Sources of vulnerability. *Health Affairs, 26*(5), 1258–1268. https://doi.org/10.1377/hlthaff.26.5.1258

Diez Roux, A. V. (2001). Investigating neighborhood and area effects on health. *American Journal of Public Health, 91*(11), 1783–1789.

Dinwiddie, G. Y., Gaskin, D. J., Chan, K. S., Norrington, J., & McCleary, R. (2013). Residential segregation, geographic proximity and type of services used: Evidence for racial/ethnic disparities in mental health. *Social Science & Medicine, 80*, 67–75. https://doi.org/10.1016/j.socscimed.2012.11.024

Fernandez, J., García-Pérez, M., & Orozco-Aleman, S. (2023). Unraveling the hispanic health paradox. *Journal of Economic Perspectives, 37*(1), 145–168. https://doi.org/10.1257/jep.37.1.145

Forest, B. (2005). *Measures of segregation and isolation.* Retrieved from https://www.dartmouth.edu/~segregation/IndicesofSegregation.pdf.

Gresenz, C. R., Rogowski, J., & Escarce, J. J. (2009). Community demographics and access to health care among U.S. Hispanics. *Health Services Research, 44*(5p1), 1542–1562. https://doi.org/10.1111/j.1475-6773.2009.00997.x

Gresenz, C. R., Derose, K. P., Ruder, T., & Escarce, J. J. (2012). Health care experiences of Hispanics in new and traditional US destinations. *Medical Care Research and Review, 69*(6), 663–678.

Guzman, L., Chen, Y., & Thomson, D. (2020). *The rate of children without health insurance is rising, particularly among Latino children of immigrant parents and white children (No. 2020–05).* National Research Center on Hispanic Children and Families. Retrieved from https://www.childtrends.org/publications/rate-children-without-health-insurance-rising-particularlyamong-Latino-children-immigrant-parents-white-children

Hall, M. (2013). Residential integration on the new frontier: Immigrant segregation in established and new destinations. *Demography, 50*(5), 1873–1896. https://doi.org/10.1007/s13524-012-0177-x

Health Resources and Services Administration. (2022). *FQHCs and LALs by state.* Retrieved from https://data.hrsa.gov/data/reports/datagrid?gridName=FQHCs.

Health Resources and Services Administration. (2023a). *National Health Center Program Uniform Data System (UDS) Awardee Data.* Retrieved from https://data.hrsa.gov/tools/data-reporting/program-data/national.

Health Resources and Services Administration. (2023b, May). *What is a health center?* Retrieved from https://bphc.hrsa.gov/about-health-centers/what-health-center.

Kaiser Family Foundation. (2023). *Uninsured rates for the nonelderly by race/ethnicity.* KFF. Retrieved from https://www.kff.org/uninsured/state-indicator/nonelderly-uninsured-rate-by-raceethnicity/

Ko, M., & Ponce, N. A. (2013). Community residential segregation and the local supply of federally qualified health centers. *Health Services Research, 48*(1), 253–270. https://doi.org/10.1111/j.1475-6773.2012.01444.x

Leventhal, T., & Brooks-Gunn, J. (2000). The neighborhoods they live. In: The effects of neighborhood residence on child and adolescent outcomes. *Psychological Bulletin, 126*(2), 309–337. https://doi.org/10.1037/0033-2909.126.2.309

Lichter, D. T., & Johnson, K. M. (2009). Immigrant gateways and hispanic migration to new destinations. *International Migration Review, 43*(3), 496–518. https://doi.org/10.1111/j.1747-7379.2009.00775.x

Lichter, D. T., Parisi, D., Taquino, M. C., & Grice, S. M. (2010). Residential segregation in new Hispanic destinations: Cities, suburbs, and rural communities compared. *Social Science Research, 39*(2), 215–230. https://doi.org/10.1016/j.ssresearch.2009.08.006

Lucyk, K., & McLaren, L. (2017). Taking stock of the social determinants of health: A scoping review. *PLoS One, 12*(5), e0177306. https://doi.org/10.1371/journal.pone.0177306

Massey, D. S. (2008). *New faces in new places: The changing geography of American immigration.* Russell Sage Foundation.

Massey, D. S., & Capoferro, C. (2008). *The geographic diversification of American immigration* (pp. 25–50). New Faces in New Places.

Massey, D. S., & Denton, N. A. (1988). The dimensions of residential segregation. *Social Forces, 67*(2), 281–315. https://doi.org/10.1093/sf/67.2.281

Monnat, S. M. (2017). The new destination disadvantage: Disparities in Hispanic health insurance coverage rates in metropolitan and nonmetropolitan new and established destinations. *Rural Sociology, 82*(1), 3–43.

Morenoff, J. D., & Lynch, J. W. (2004). What makes a place healthy? Neighborhood influences on racial/ethnic disparities in health over the life course. *Critical Perspectives on Racial and Ethnic Differences in Health in Late Life*, 406–449.

National Association of Community Health Centers. (2022, August). *Research fact sheets and infographics.* NACHC. Retrieved from https://www.nachc.org/research-and-data/research-fact-sheets-and-infographics/.

Naylor, K. B., Tootoo, J., Yakusheva, O., Shipman, S. A., Bynum, J. P. W., & Davis, M. A. (2019). Geographic variation in spatial accessibility of U.S. Healthcare providers. *PLoS One, 14*(4), e0215016. https://doi.org/10.1371/journal.pone.0215016

Office of Disease Prevention and Health Promotion. (2023). *Healthy people 2030: Increase the proportion of people with health insurance- AHS-01, Data.* Retrieved from https://health.gov/healthypeople/objectives-and-data/browse-objectives/health-care-access-and-quality/increase-proportion-people-health-insurance-ahs-01/data?group=Race%2FEthnicity&state=United+States&from=2019&to=2021&populations=#edit-submit.

Ortega, A. N., Chen, J., Roby, D. H., Mortensen, K., Rivera-González, A. C., & Bustamante, A. V. (2022). Changes in coverage and cost-related delays in care for Latino individuals after elimination of the affordable care act's individual mandate. *JAMA Network Open, 5*(3), e221476. https://doi.org/10.1001/jamanetworkopen.2022.1476

Osypuk, T. L., & Acevedo-Garcia, D. (2010). Beyond individual neighborhoods: A geography of opportunity perspective for understanding racial/ethnic health disparities. *Health & Place, 16*(6), 1113–1123. https://doi.org/10.1016/j.healthplace.2010.07.002

Park, J., & Iceland, J. (2011). Residential segregation in metropolitan established immigrant gateways and new destinations, 1990–2000. *Social Science Research, 40*(3), 811–821.

Parker, E. (2021). Spatial variation in access to the health care safety net for Hispanic immigrants, 1970–2017. *Social Science & Medicine, 273*, 113750. https://doi.org/10.1016/j.socscimed.2021.113750

Perreira, K. M., Allen, C. D., & Oberlander, J. (2021). Access to health insurance and health care for hispanic children in the United States. *The Annals of the American Academy of Political and Social Science, 696*(1), 223–244. https://doi.org/10.1177/00027162211050007

Rosa, E. M., & Tudge, J. (2013). Urie Bronfenbrenner's theory of human development: Its evolution from ecology to bioecology. *Journal of Family Theory & Review, 5*(4), 243–258.

Rural Health Information Hub. (2021, July). *Map of health professional shortage areas: Primary care, by County, 2021*. Retrieved from https://www.ruralhealthinfo.org/charts/5.

Schulz, A. J., Williams, D. R., Israel, B. A., & Lempert, L. B. (2002). Racial and spatial relations as fundamental determinants of health in Detroit. *The Milbank Quarterly, 80*(4), 677–707. https://doi.org/10.1111/1468-0009.00028

U.S. Census Bureau. (2021, November 21). *Housing patterns: Appendix B: Measures of residential segregation*. Census.Gov. Retrieved from https://www.census.gov/topics/housing/housing-patterns/guidance/appendix-b.html

U.S. Department of Health and Human Services, Office of Disease Prevention and Health Promotion. (2023). *Social determinants of health—healthy people 2030*. Retrieved from https://health.gov/healthypeople/priority-areas/social-determinants-health.

Waters, M. C., & Jiménez, T. R. (2005). Assessing immigrant assimilation: New empirical and theoretical challenges. *Annual Review of Sociology*, 105–125. https://doi.org/10.1146/annurev.soc.29.010202.100026

Whitener, K., & Corcoran, A. (2021). *Getting back on track: A detailed look at health coverage trends for Latino children—center for children and families*. Georgetown University Health Policy Institute Center for Children and Families. Retrieved from https://ccf.georgetown.edu/2021/06/08/health-coverage-trends-for-latino-children/

Zúñiga, V., & Hernández-León, R. (2005). New destinations: Mexican immigration in the United States. .

Chapter 6
Toward a Transdisciplinary Theory of Access: Medical Geography, Intersectionality, and the Human Right to Health

Avery R. Everhart

Introduction

"Everything is related to everything else but near things are more related than distant things," (Tobler, 1970). These words written as "the first law of geography" by the late Waldo Tobler in Economic Geography as part of the International Geographic Union's Commission on Quantitative Methods in 1970 have since informed the very foundation of our modern means and methods for conducting spatial analysis. More than 50 years later, this law remains a vital contribution to the field of geography; however, it is not as often taken up by those working in, or concerned with, public health. In this volume that seeks to bridge these two fields, geography and public health, and synthesize across our methods of social and spatial analysis, my contribution is to elaborate upon Tobler's theory such that we might draw from seemingly disparate fields to construct a new, transdisciplinary theory. From my perspective, the first law suggests a strong local interdependence, with which subfields of public health like those concerned with infectious diseases are quite familiar. Tobler's law also suggests a significant, though weaker, global interdependence, and one that we resist or ignore at our own peril. Indeed, the COVID-19 pandemic has, hopefully, taught us with a new kind of devastation how sincerely and inescapably interconnected people are globally. Therefore, I take the implications of Tobler's first law quite seriously, even if the implications may be far more widespread than his intentions. In this chapter, I propose a transdisciplinary theory of access to care that also aims to be transnationally adaptable. In doing so, I focus my intervention on one social, spatial determinant of health, healthcare access and quality, and demonstrate how geographical thinking and spatial analysis, coupled with theoretical and practical insights from other fields, can help us better understand how "access" influences

A. R. Everhart (✉)
Department of Geography, University of British Columbia, Vancouver, BC, Canada
e-mail: avery.everhart@ubc.ca

© The Author(s) 2026
M. A. Kolak, I. K. Moise (eds.), *Place and the Social-Spatial Determinants of Health*, Global Perspectives on Health Geography,
https://doi.org/10.1007/978-3-031-88463-4_6

and also is influenced by the other determinants of health. Though the other four major determinants of health—economic stability, education access and quality, social and community context, and neighborhood and built environment—are not my focus, the fields to which I turn to craft this theory of access should prove instructive for understanding how these determinants are not only social, but also spatial and structural.

By drawing upon methodological contributions from medical geography and spatial epidemiology, we can see how data can be wielded not only to document or model disparities in health outcomes or access to healthcare, but also to predict potential impacts of interventions designed to alleviate those disparities at multiple scales. Even where there is a lack of robust data infrastructure, such as with the "small numbers issue" faced by many of the subpopulations who also experience systemic barriers to care, the field of medical geography can be instructive. As I will demonstrate in this section, qualitative methods, creative adaptation of existing methods, or even advanced quantitative techniques like spatial microsimulation can supplement an incomplete data infrastructure. Where medical geography excels at methodological innovation, it may lack a strong theoretical backing or the means to adopt robust measures of geographic phenomena, such as law and policy, that are not easily quantified. To augment medical geography's contributions, I turn to two other fields: intersectionality, and health and human rights.

Intersectionality offers a means of understanding the work racialized gender in particular does to privilege some at the expense of others. And since its coinage in the late 1980s, intersectionality has expanded to account for other complex social and structural processes that advantage and disadvantage everyone along lines of social differences simultaneously in addition to race and gender. How it pertains to health, and especially healthcare access, will be the bulk of what I unpack in the section focused on the theory. However, this chapter does not provide ample space to fully unpack its origins, trace its contemporary applications, or capture the depth and breadth of its potential impacts. Perhaps no single piece of writing can. Nevertheless, attuning us to what the theory offers for those of us interested in the determinants of health, and especially how intersectional analysis can be infused with geographic thinking, can expand our existing understanding and application of the socio-spatial determinants of health. In fact, Kimberle Crenshaw (1989) first posited intersectionality as a deeply spatial metaphor, and this fact shaped how intersectional thinking has evolved from its origins in Black feminist legal scholarship to its arguably mainstream status today (Rodó-Zárate & Jorba, 2022). The later section on intersectionality unpacks what it does, in my view, to advance our understanding of access to healthcare and the socio-spatial determinants of health more broadly through an explicitly geographic lens.

The field of health and human rights offers a legal framework and theoretical spine for applying the methods of medical geography and the critical sociolegal analysis of intersectionality. While a comprehensive history of the origins of international human rights law on the *right to health* is beyond this chapter's scope, acknowledging the theoretical roots of health and human rights frameworks is essential to developing a transdisciplinary theory of access that foregrounds the human right to health. What I hope to accomplish by introducing it is to highlight how it harmonizes with intersectionality and how its inherent geographic thinking lends itself well to thinking about the socio-spatial determinants of health at scale. Health and human rights often emphasize the move from global to local and vice versa, and here I will elaborate on this move to suggest means of conducting rights-based, intersectional analyses using both quantitative and qualitative geographic methods to capture a more holistic portrait of healthcare access. Moreover, international human rights law, from drafting to passage and from adoption by member states to ratification in the country, is inherently geographic. While there are indeed universal standards of what constitutes certain foundational rights, not all countries are signatory parties to even some of the most foundational International Covenants, and the process of ratification, meaning the process of incorporating new human rights standards into national law and policy by member states, is a difficult process to oversee. In these ways, health and human rights offer those of us invested in the socio-spatial determinants of health an empirical means of adapting geographic thinking and intersectional analysis to different geopolitical and sociolegal contexts.

This chapter is meant as both a theoretical intervention and a means of speculating on methodological innovation that transdisciplinary thinking can enable. While there is no empirical analysis here, the goal is to take seriously what Lisa Bowleg calls "intersectional praxis" (Bowleg, 2021). It is not enough to advance our understanding of a phenomenon or conceptual framework—in our case, the socio-spatial determinants of health and healthcare access. Rather, we must advance our understanding insofar as it empowers us to propose meaningful solutions to the inequities faced by those who experience poor health outcomes driven by the socio-spatial determinants. My contribution in this chapter is to sketch out a transdisciplinary theory of access to healthcare that pieces together fields that are not often in conversation so that we can perhaps, in an intersectional fashion, create something that exceeds the sum of its parts. I began with Tobler's first law to ground us in geographical thinking and spatial analysis, and each section will speak to how we can use both as a bridge to create pathways of understanding between disciplines with overlapping concerns, yet radically different intellectual genealogies, methodologies, and epistemologies. The result will necessarily be imperfect. My hope is that it will prove useful even if only as a starting point.

Medical Geography and Spatial Epidemiology

Those working at the intersections of public health and geography know the difficulties in communicating the utility of geographic thinking and spatial analysis for an audience who is chiefly concerned with exposures and outcomes, relative risks, and tabular, that is aspatial, statistics. What's been established is that geography, whether proxied by state, county, or region, is not a variable for which one can simply control in a larger regression model. A major intervention into the field of public health from geographers has been that space and place do not co-vary; they are instead the very grounds upon and the context in which everything else happens. Since the "spatial turn" in health research (Richardson et al., 2013), there has been a proliferation of work that incorporates geographic thinking and spatial analysis into work on health access and outcomes at varying levels of rigor. What I find most instructive for articulating a transdisciplinary theory of access in this context of socio-spatial determinants of health is to turn to theories of geographic context and a suite of methods that have inspired quantitative work on geographic access to healthcare for at least two decades.

In terms of theorizing the role of geographic context, Mei-Po Kwan has been a prolific figure in medical geography whose contributions are crucial for understanding the spatial turn in health and the methodological conundrums of using spatial analysis to unearth geographic patterns of health access and outcomes. She articulated the "uncertain geographic context problem" as an elaboration of the modifiable areal unit problem (MAUP) wherein studies that measure effects of area-based attributes, such as neighborhood-level indicators of social deprivation, can be influenced by how the contextual units, like neighborhoods, are defined and whether they deviate from ground truth (Kwan, 2012a). Given how frequently social determinants of health rely on neighborhood-level measures, it stands to reason that if data availability dictates our definitions of neighborhoods and where a given neighborhood's boundaries lie, then we may not derive the kind of meaningful insights needed to intervene on those determinants and improve health access and outcomes. Thankfully, Kwan has since proposed multiple solutions, including using GIS and GPS for individual-level space-time data, construction of individual or collective activity spaces, and even the use of qualitative data to understand how things we distill into geographic units, like activity spaces and neighborhoods, are experienced (Kwan, 2012b, 2018). Kwan especially recommends the integration of temporal data and methods as a means of enriching empirical understanding of geographic context (Kwan, 2010, 2013). If we take Kwan's contributions seriously, she insists that we can and should try to capture the richness of geographic and temporal context with empirical means, but that we must also accept the inherent limitations of trying to turn people's lived experiences of where they live, and when, into something as discrete as data.

Kwan's work demonstrates the difficulties and the promise of translating people's sense of space and place into data that can be used to better understand the larger geographic context in which they live their lives, including how and where they access healthcare. To better understand how the broader field of medical geography conceptualizes access, we can turn to the suite of methods perhaps most frequently used for quantifying spatial access to healthcare: the floating catchment area series. While the method was originally used for measuring employment access and availability of jobs (Wang, 2000; Peng, 1997), Wei Luo and Fahui Wang adapted the floating catchment area method to measure access to primary care in a study highlighting the limitations of the US Health Resources and Services Administration's method for identifying areas of care shortages (2003). Luo and Wang's method first calculates a physician-to-population ratio for a series of defined threshold travel times from the location of that physician's practice. The second step is then to find the sum of all physician-to-population ratios for each population location, summarized as the centroid of a census tract or other administrative geographic unit, and for each drive-time threshold. The logic behind this is that (1) accessibility can be defined as the relative ease by which a population can access healthcare at a given location, and (2) patient populations are more likely to access a nearby physician than a more remote one. In this way, it's clear that Tobler's first law influences the method because it quantifies access based on travel time and weighs accessibility by distance. Since then, multiple researchers have adapted the floating catchment area method (Luo & Qi, 2009; Wan et al., 2012; McGrail, 2012), but always with a focus on access to primary care for the general population. Others have adapted measures like this toward more computational means of understanding real-time access, such as the Rational Agent Access Model, which accounts for congestion and multiple types of travel modes (Saxon & Snow, 2020). Still, others have synthesized the various means of using methods like the floating catchment area for measuring spatial access to health services (Chen & Jia, 2019; Wang, 2012). However, the methods are data intensive and thus difficult to apply to either subpopulations or to specialty care that the general population may not need or access. I have previously used the three-step floating catchment area method to measure access to transgender-specific healthcare (Everhart et al., 2023) but had to rely upon a dataset of my own construction to populate the clinician and facility data needed for the analysis (Everhart et al., 2022). Indeed, more work can and should be done to incorporate the geographic context, following Kwan's body of research and advances in floating catchment area methods.

Ultimately, these floating catchment area methods implicitly define access almost exclusively as where healthcare facilities are located and measure spatial access using either Euclidean, geodesic or road network-based distances and travel times. In this way, the floating catchment area series of methods are limited to what Abdullah Khan (1992) calls "potential spatial access," or a geographically explicit means of understanding the supply side of health services within a given health system, without real information on utilization, or "realized access," of services, or the barriers and facilitators influencing such realized access. Moreover, the broader geographic context in which the socio-spatial determinants of health are experienced and lived out could be captured empirically and adopted methodologically within the floating catchment area methods, and accounted for by how we as researchers interested in the determinants of health and access to care interpret and contextualize results. To that end, what I propose is that we can learn much more about access by turning to other fields that have theorized access differently. While there have been significant advancements in our methodological toolkit in medical geography, much less attention has been paid to aspects beyond the spatial and temporal dimensions of where and when care is available. To understand not only the potential or realized spatial access, but also the interplay between them, we must broaden our disciplinary horizons to other fields where geographic thinking and spatial analysis may be less present or rigorous, but theories of access have outshone our own. Our existing methods are useful for capturing with increasing granularity the spatial and temporal patterns of potential spatial access to healthcare, but the volume of data and information required to conduct said analyses leaves us with a significant gap in our capacities to measure and understand additional dimensions of access. I propose that the methods themselves, like floating catchment area analyses, can be applied in and through other theoretical frameworks, like intersectionality and human rights, to expand our notion of access, better our understanding of facilitators and barriers to healthcare, and promote geographical thinking, without sacrificing our empirical or scientific methods and rigor. In the next sections, I outline why intersectionality and health and human rights are fields uniquely positioned to aid in expanding our thinking about access and how we can adopt their theories and applications to our spatial analysis.

Intersectionality

Intersectionality is a theoretical framework for conceptualizing the ways in which complex, interlocking systems oppress marginalized groups in order to privilege dominant ones. Since the term itself first appears in two landmark papers by renowned Black feminist scholar Kimberle Crenshaw (Crenshaw, 1989, 1991), there has been an explosion of interest in the theory. In fact, as Jennifer C. Nash has argued, intersectionality has become synonymous with Black feminism and has even been made to do the work of field formation for women's studies programs and departments around the world (Nash, 2018). There has been work calling for the formation of a field of "intersectionality studies" (Cho et al., 2013), and indeed the work of theorizing the complex interplay of different forms of structural power in and through the lives of those experiencing the intersections thereof has proliferated in the interdisciplinary humanities and social sciences. However, an adaptation of intersectionality for empirical work in public health has followed a different trajectory. McCall's (2005) foundational piece elucidated the existing approaches to intersectionality as inter-categorical, intra-categorical, and anti-categorical. And this distinction in understandings and applications of intersectionality has proven paramount for unpacking the explosion of interest in the theoretical framework within the empirical health and social sciences. As one key example, Lisa Bowleg's (2008, 2012) pioneering work in bringing intersectionality to bear on empirical public health scholarship has inspired numerous endeavors, with varying degrees of theoretical rigor as Elle Lett notes (Lett, 2022), to apply intersectional thinking on a field that may be at odds with it, at least in part due to its reliance on methods that do not easily lend themselves to intersectional analysis (Bauer et al., 2021). Importantly, Bowleg (2021) has argued for an intersectional praxis that seeks to move past mere documentation of disparities, for fear of calcifying them or essentializing the differences between people based on categories that often overlap, toward work that proposes solutions to them. Yet there is one aspect of intersectionality's origins that, arguably, has been deemphasized over time: the spatial nature of the metaphor used to first propose the theory. As I illustrate below, much of the empirical, quantitative applications of intersectionality rely on categorical comparisons, the adaptation of traditional statistical methods, and the use of demographic categories as proxies for exposure to systemic oppression or privilege. However, this chapter overall suggests that intersectional thinking can be incorporated into spatially explicit methods and facilitate rich theoretical insights into the geographic context in which individuals and populations access, or do not access, healthcare.

As Abichahine and Veenstra (2016) emphasize, three principles of intersectionality have emerged as the concept has been developed since its inception. They are *simultaneity*, the notion that multiple axes of inequality privileging some at the expense of others operate at the same time and cannot be isolated; *multiplicativity*, which suggests that these axes interact to create identities and experiences that exceed the mere sum of their parts; and *multiple jeopardy*, the theory that those at the intersection of multiple axes of oppression experience multiplicative amounts of disadvantage (Abichahine & Veenstra, 2016, p. 693). This study in particular focuses on inter-categorical analysis to measure potentially outsized, that is multiplicative, effects of structural racism, sexism, and homophobia on leisure activity. Greta Bauer and Ayden Scheim advanced both our theoretical and methodological understanding by producing an index of intersectionality for inter-categorical analysis (Scheim & Bauer, 2019), demonstrating how discrimination mediates health inequalities (Bauer & Scheim, 2019), and comparing intra- and inter-categorical analysis through quantitative population health data (Bauer & Scheim, 2019). The methodological innovation continued with the application of multilevel modeling and latent class analysis (Garnett et al., 2014; Earnshaw et al., 2018; Taggart et al., 2019), and these advanced methods have been lauded for their capacity to derive cohesive social groupings within large datasets beyond mere singular categorical demographic variables, especially for those axes of inequality that may be multidimensional (Bauer et al., 2021, 2022; Agénor, 2020). However, in nearly every example of these studies and articles that advance our capacities to adopt intersectional thinking in quantitative health sciences research, there is no mention of geography, space, or place. An affinity for the socio-ecological model, and especially Nancy Krieger's work in extending the concept to incorporate historical injustices, has sparked some interest in context, but that is almost always conceptualized as spatially with placeholders like school, work, home, or neighborhood standing in for real geographic thinking.

What is missing in the adaptation of intersectional theory into quantitative health inequities research is attention to space and place (Bambra, 2022). More specifically, I suggest that a transdisciplinary theory of access must take seriously not only space and place, but also scale. To that end, much of health inequalities research has focused either on qualitative work to provide rich narrative insights into intersectional experiences of health and its systems or on quantitative, population-level hypothesis testing that derives macro-level insights from neighborhood-level comparisons and (often) ecological data. I ask how we might incorporate both to derive micro-, mezzo-, and macro-level insights and what kinds of data would be needed to do so. For example, how might we mobilize data from electronic health records (EHRs) to understand the patient-clinician interactions that are necessarily inflected with structural racism, sexism, ableism, homophobia, and transphobia? Can we measure implicit bias among clinicians through their notes about patients in EHRs? Could patient-level information, especially geographic information like addresses, become a means to derive novel insights about anything from travel burden to access care to the impact that the lived experience of socio-spatial determinants of poor health outcomes has on experiences with clinic staff and providers? Perhaps most importantly, what are the ethical stakes of conducting such research, and what might be done to value patient privacy, consent, and refusal while still conducting research?

Beyond these interpersonal experiences in clinical settings, there are a myriad of other factors that structurally determine healthcare for anyone within a given health system. Intersectionality as a framework might inspire us in medical geography to pathologize the power structures that construct the environments rather than the individuals or communities experiencing poor access to care or poor health outcomes. It might also push us to consider how historical geographic context influences access to care and other socio-spatial determinants of health. Those living at different intersections of structural social power experience the same health system differently, and intersectionality offers not only a theoretical means of understanding how social power shapes those differences, but also an ethical imperative for combating those inequalities.

Health and Human Rights

To better understand how health systems are both *structured*, that is constructed by those with decision-making power with varying motivations and goals in providing care, and *structuring*, that is how health systems scaffold those groups with differing relationships to social power within the matrix of domination, we can turn to the field of health and human rights. Rights-based approaches to health emphasize a holistic understanding of health and well-being that emanates not only from practical understandings of how people experience varying degrees of good or poor health, but also from international law. Sofia Gruskin and Laura Ferguson's work, both separately and in their collaborative endeavors, prove instructive in understanding the far-reaching implications for health systems the right to health has, if taken seriously, and for our understanding of access to care as a key aspect of socio-spatial determinants of health. More specifically, their elaboration on what is called the Availability, Accessibility, Acceptability, and Quality (AAAQ) framework as a crucial means of applying the right to health is most useful for my proposal of a transdisciplinary theory of access (Gruskin et al., 2010). However, to unpack this, we need to understand the international legal covenants that begat the framework and how these covenants have been differently taken up, or in some cases not at all, in different geographical contexts.

Building off of the United Nations (UN) Charter, which was signed in July 1945 and took effect in October of the same year, the multi-national collaborative effort established both the World Health Organization and later the Universal Declaration of Human Rights (UDHR) in 1948. While the right to health is first enshrined in Article 25 of the UDHR (Universal Declaration of Human Rights, 217 A (III), 1948), the International Covenant on Economic, Social and Cultural Rights (ICESCR) further enshrined the right to the highest attainable standard of physical and mental health in its 12 articles in 1966 (UN Committee on Economic, Social, and Cultural Rights, 2000). Notably, though the ICESCR took effect in 1976, some countries like the USA have signed the agreement but have not ratified it, meaning it does not exist within their domestic legal system. Since its enactment decades ago, there has been a Committee on Economic, Social, and Cultural Rights tasked with ongoing review of how the covenant is applied within member states' legal systems and with reviewing violations thereof for signatory parties within the UN. What interests me most here is the fact that this international legal precedent for health as a human right exists and how it has been refined and extended since the beginning of the twenty-first century.

The AAAQ framework emanates from General Comment No. 14 on the right to the highest attainable standard of health (UN Committee on Economic, Social, and Cultural Rights, 2000). In said comment, it defines availability as "sufficient quantity [of] functioning healthcare facilities, goods, services," which the committee notes are "varying by context," but should "address the underlying determinants of health" (UN Committee on Economic, Social, and Cultural Rights, 2000). According to the committee, accessibility has four pillars: non-discrimination, physical accessibility, affordability, and access to information (UN Committee on Economic, Social, and Cultural Rights, 2000). Much of the medical geographic research measuring access to healthcare has only gone as far as investigating physical accessibility, about which the committee says "health facilities, goods and services must be within safe physical reach for all sections of the population, especially vulnerable or marginalized groups," (UN Committee on Economic, Social, and Cultural Rights, 2000). Quantitative geography has yet to fully adopt a multidimensional view of access that evaluates acceptability and quality of services along with availability and spatial accessibility, but with noteworthy exceptions to this rule in the areas of food access research (Shannon et al., 2021) and access to medication for opioid use disorder (Joudrey et al., 2022; Bommersbach et al., 2023). This gap, even as it begins to close, however, represents an opportunity. As Gruskin et al. (2010) argue, any human rights-based approach to crafting health policy or programming should be prepared to address a series of questions posed by proper engagement with AAAQ. Among myriad other things, they query whether disaggregated data is collected to monitor for discrimination; affected communities' perspectives are incorporated into the design and implementation of policies and programs; decision-making processes are transparent with built-in mechanisms for accountability between affected communities and all stakeholders; and if multi-sectoral responses to specific health issue concerns are facilitated at each stage (Gruskin et al., 2010, p. 138–40).

Importantly, their suggestions for ensuring that these rights-based approaches engage with the principles of AAAQ are limited to qualitative data collection such as interviews and focus groups with affected communities and vulnerable or hard-to-reach populations. I suggest that these qualitative approaches can be supplemented with quantitative methods for monitoring the protection and promotion of the right to health. This could include a review of health information materials available at healthcare facilities, inter-categorical and intersectional analyses of disaggregated data collected in clinics and by health surveillance systems, comprehensive analysis of formal complaints filed, and other means of quantifying the *right to health* in a geographic context. Indeed, each of these elements that make up a rights-informed understanding of true access to health could be quantified and incorporated into existing spatial methods like the floating catchment area series. For example, instead of only weighing distance by drive time between population centers as available healthcare facilities, other factors such as availability of interpretation and translated materials could also be used as weights. This expansion on an existing methodological framework to integrate the *human right to health* would enable us to understand the broader picture of access to healthcare through the AAQ framework in the specific geographic contexts in which care is or is not provided. The AAAQ framework is particularly advantageous because it is already designed as a universal legal standard and therefore can be adapted not only to different geopolitical contexts and in different health and legal systems, but also at different subnational or even transnational scales. Inflecting the human right to health with geographic thinking and applying it with our existing tools for spatial analysis could enable better comparisons across geopolitical and sociolegal contexts and advance our understanding of how the law affects health, and vice versa.

A Transdisciplinary Theory

Accessing healthcare is a complex process. How potential patients source information about health conditions, find providers and clinicians, navigate health systems, and adhere to medications and treatments have been understudied as part of the social determinants of health. In addition, how clinicians provide care, the quality of the care they offer, their level of expertise and experience in working with specific conditions or marginalized populations, and how they embed themselves in public versus private health systems necessarily affect health outcomes for their patients. While healthcare access and quality are one of the core five social determinants of health, how care is delivered and how it is accessed vary significantly over space and time. Medical geographers have made great strides in quantifying the availability of services, especially primary care, in the US, but less attention has been paid to other aspects of healthcare access. Intersectionality's origins in Black feminist analysis of law in a US American context elucidates the importance of legal systems in structuring the distribution of life chances that people experience along lines of racialized gender, as well as class, ability, and myriad other axes of privilege and oppression. For those of us concerned with expanding our understanding of sociospatial determinants of health, it is necessary to also consider how the legal and health systems interact, overlap, and mutually influence one another. Where law within intersectionality is often treated as something punitive, a mechanism for doling out punishment, health, and human rights as a field can extend our application of intersectionality to conceptualize human rights law as a means of protecting and promoting rights.

Intersectionality can be incorporated into this "rights-based floating catchment area method" in myriad ways. One could be that the simplistic demographic categories we use as proxy measures for exposure to systemic oppression, (e.g., self-identification as Black becomes exposure to systemic anti-Blackness), instead become multidimensional variables where data is available or is viable to collect. Gender might look like less a self-selected categorical variable and more like a series of variables captured by nuanced questions incorporating legal sex, gender modality, assigned sex at birth when relevant, gender expression and presentation, and even levels of conformity to gendered social roles. Beyond this expansion of our limited capacities to capture complex social identities, we may also incorporate a historically grounded and equity-focused perspective to give data a richer, more nuanced context. Combating the uncertainty of geographic context that Kwan cautions against with an attunement to intersectionality could influence not only data collection or analysis, but also interpretation of results. Taking seriously Bowleg's (2021) call for "intersectional praxis" in tandem with the international legal standard of the human right to the highest attainable standard of health and well-being enables us to take intersectionality globally and adapt this rights-based quantitative method to different geopolitical and sociolegal contexts.

The transdisciplinary theory of access that I've proposed here need not remain untested. An empirical adaptation of this theory could look like a floating catchment area analysis that incorporates not only the acceptability and quality of AAAQ, but also renames medical geography's understanding of access as availability and takes seriously the four pillars of accessibility outlined in the framework. In fact it would be possible to build on some recent examples of quantitative work that take an expansive definition of access and highlight that what geography often calls spatial access is actually just availability (Bommersbach et al., 2023). A simple adaptation might look like incorporating more than mere location data about the healthcare facilities included in our studies such that even complex concepts could be recast as dichotomous variables. Indeed, as a theory of access, this framework could be adopted to research other socio-spatial determinants of health, such as access to quality education, access to green space, access to water, sanitation and hygiene, access to safe and affordable housing, and other central concerns that structurally influence health and well-being. Each of these domains could also be understood through the lens of intersectionality, articulated through a human rights framework, and be adapted to different sociolegal and geopolitical contexts at different scales.

A significant reimagination of how we understand access to healthcare as a core socio-spatial determinant of health is needed to better investigate how access is affected by structural power, as well as how it influences other socio-spatial determinants, at different geographic scales and in different sociolegal and geopolitical contexts. A thorough review of fields that have often been totally distinct from one another is essential to bridge research gaps and drive new understanding of the complexities of health. The transdisciplinary approach I have put forth foregrounds how the socio-spatial determinants of health, especially access to quality healthcare, take place in the lives of individuals and can be investigated and empirically observed at multiple scales. I have also demonstrated that, in keeping with Tobler's First Law of Geography, the intersecting systems and structures of power are necessarily inter-related, but that these powerful structures and systems take place differently at different spatial and temporal scales. Finally, the theory I propose here emanates from the human right to the highest attainable standard of health and well-being and insists upon a spatially explicit, intersectional, and mixed methods approach to reimagine how availability, accessibility, acceptability, and quality of healthcare can not only be studied more critically and accurately, but also improved for populations in all geopolitical and sociolegal contexts.

References

Abichahine, H., & Veenstra, G. (2016). Inter-categorical intersectionality and leisure-based physical activity in Canada. *Health Promotion International.* https://doi.org/10.1093/heapro/daw009

Agénor, M. (2020). Future directions for incorporating intersectionality into quantitative population health research. *American Journal of Public Health, 110*(6), 803–806. https://doi.org/10.2105/AJPH.2020.305610

Bambra, C. (2022). Placing intersectional inequalities in health. *Health & Place, 75,* 102761. https://doi.org/10.1016/j.healthplace.2022.102761

Bauer, G. R., & Scheim, A. I. (2019). Methods for analytic intercategorical intersectionality in quantitative research: Discrimination as a mediator of health inequalities. *Social Science & Medicine, 226,* 236–245. https://doi.org/10.1016/j.socscimed.2018.12.015

Bauer, G. R., Churchill, S. M., Mahendran, M., Walwyn, C., Lizotte, D., & Villa-Rueda, A. A. (2021). Intersectionality in quantitative research: A systematic review of its emergence and applications of theory and methods. *SSM—Population Health, 14,* 100798. https://doi.org/10.1016/j.ssmph.2021.100798

Bauer, Greta R., Mahendran, Mayuri, Walwyn, Chantel, and Shokoohi, Mostafa. (2022). Latent variable and clustering methods in intersectionality research: systematic review of methods applications. *Social Psychiatry & Psychiatric Epidemiology 57,* no. 2, 221–237. https://doi.org/10.1007/s00127-021-02195-6

Bommersbach, T., Justen, M., Bunting, A. M., Funaro, M. C., Winstanley, E. L., & Joudrey, P. J. (2023). Multidimensional assessment of access to medications for opioid use disorder across urban and rural communities: A scoping review. *International Journal of Drug Policy, 112,* 103931. https://doi.org/10.1016/j.drugpo.2022.103931

Bowleg, L. (2008). When black + lesbian + woman ≠ black lesbian woman: The methodological challenges of qualitative and quantitative intersectionality research. *Sex Roles, 59*(5–6), 312–325. https://doi.org/10.1007/s11199-008-9400-z

Bowleg, L. (2012). The problem with the phrase women and minorities: Intersectionality—an important theoretical framework for public health. *American Journal of Public Health, 102*(7), 1267–1273. https://doi.org/10.2105/AJPH.2012.300750

Bowleg, L. (2021). Evolving intersectionality within public health: From analysis to action. *American Journal of Public Health, 111*(1), 88–90. https://doi.org/10.2105/AJPH.2020.306031

Chen, X., & Jia, P. (2019). A comparative analysis of accessibility measures by the two-step floating catchment area (2SFCA) method. *International Journal of Geographical Information Science, 33*(9), 1739–1758. https://doi.org/10.1080/13658816.2019.1591415

Cho, S., Crenshaw, K. W., & McCall, L. (2013). Toward a field of intersectionality studies: Theory, applications, and praxis. *Signs: Journal of Women in Culture and Society, 38*(4), 785–810. https://doi.org/10.1086/669608

Crenshaw, K. (1989). Demarginalizing the intersection of race and sex: A black feminist critique of antidiscrimination doctrine, feminist theory and antiracist policies. *University of Chicago Legal Forum, 1989*(1), 139–167.

Crenshaw, K. (1991). Mapping the margins: Intersectionality, identity politics, and violence against women of color. *Stanford Law Review, 43*(6), 1241. https://doi.org/10.2307/1229039

Earnshaw, V. A., Rosenthal, L., Gilstad-Hayden, K., Carroll-Scott, A., Kershaw, T. S., Santilli, A., & Ickovics, J. R. (2018). Intersectional experiences of discrimination in a low-resource urban community: An exploratory latent class analysis. *Journal of Community & Applied Social Psychology, 28*(2), 80–93. https://doi.org/10.1002/casp.2342

Everhart, A. R., Ferguson, L., & Wilson, J. P. (2022). Construction and validation of a spatial database of providers of transgender hormone therapy in the US. *Social Science & Medicine, 303*, 115014. https://doi.org/10.1016/j.socscimed.2022.115014

Everhart, A. R., Ferguson, L., & Wilson, J. P. (2023). Measuring geographic access to transgender hormone therapy in Texas: A three-step floating catchment area analysis. *Spatial and Spatio-Temporal Epidemiology, 24*, 100585. https://doi.org/10.1016/j.sste.2023.100585

Garnett, B. R., Masyn, K. E., Austin, S. B., Miller, M., Williams, D. R., & Viswanath, K. (2014). The intersectionality of discrimination attributes and bullying among youth: An applied latent class analysis. *Journal of Youth and Adolescence, 43*(8), 1225–1239. https://doi.org/10.1007/s10964-013-0073-8

Gruskin, S., Bogecho, D., & Ferguson, L. (2010). 'Rights-based approaches' to health policies and programs: Articulations, ambiguities, and assessment. *Journal of Public Health Policy, 31*(2), 129–145. https://doi.org/10.1057/jphp.2010.7

Joudrey, P. J., Kolak, M., Lin, Q., Paykin, S., Anguiano, V., & Wang, E. A. (2022). Assessment of community-level vulnerability and access to medications for opioid use disorder. *JAMA Network Open, 5*(4), e227028. https://doi.org/10.1001/jamanetworkopen.2022.7028

Khan, A. A. (1992). An integrated approach to measuring potential spatial access to health care services. *Socio-Economic Planning Sciences, 26*(4), 275–287. https://doi.org/10.1016/0038-0121(92)90004-O

Kwan, M.-P. (2010). Space-time and integral measures of individual accessibility: A comparative analysis using a point-based framework. *Geographical Analysis, 30*(3), 191–216. https://doi.org/10.1111/j.1538-4632.1998.tb00396.x

Kwan, M.-P. (2012a). How GIS can help address the uncertain geographic context problem in social science research. *Annals of GIS, 18*(4), 245–255. https://doi.org/10.1080/19475683.2012.727867

Kwan, M.-P. (2012b). The uncertain geographic context problem. *Annals of the Association of American Geographers, 102*(5), 958–968. https://doi.org/10.1080/00045608.2012.687349

Kwan, M.-P. (2013). Beyond space (as we knew it): Toward temporally integrated geographies of segregation, health, and accessibility: Space–time integration in geography and GIScience. *Annals of the Association of American Geographers, 103*(5), 1078–1086. https://doi.org/10.1080/00045608.2013.792177

Kwan, M.-P. (2018). The limits of the neighborhood effect: Contextual uncertainties in geographic, environmental health, and social science research. *Annals of the American Association of Geographers, 108*(6), 1482–1490. https://doi.org/10.1080/24694452.2018.1453777

Lett, E. (2022). Crossing lines does not equal intersectionality. *Journal of Behavioral Medicine, 45*(6), 983–984. https://doi.org/10.1007/s10865-022-00375-6

Luo, W., & Qi, Y. (2009). An enhanced two-step floating catchment area (E2SFCA) method for measuring spatial accessibility to primary care physicians. *Health & Place, 15*(4), 1100–1107. https://doi.org/10.1016/j.healthplace.2009.06.002

Luo, W., & Wang, F. (2003). Measures of spatial accessibility to health care in a GIS environment: Synthesis and a case study in the Chicago Region. *Environment and Planning B: Planning and Design, 30*(6), 865–884. https://doi.org/10.1068/b29120

McCall, L. (2005). The complexity of intersectionality. *Signs: Journal of Women in Culture and Society, 30*(3), 1771–1800. https://doi.org/10.1086/426800

McGrail, M. R. (2012). Spatial accessibility of primary health care utilising the two step floating catchment area method: An assessment of recent improvements. *International Journal of Health Geographics, 11*(1), 50. https://doi.org/10.1186/1476-072X-11-50

Nash, Jennifer. 2018. Black Feminist Reimagined: After Intersectionality. Durham: Duke University Press. ISBN 978-1-4780-0059-4. https://doi.org/10.1215/9781478002253

Peng, Z.-R. (1997). The jobs-housing balance and urban commuting. *Urban Studies, 34*(8), 1215–1235. https://doi.org/10.1080/0042098975600

Richardson, D. B., Volkow, N. D., Kwan, M.-P., Kaplan, R. M., Goodchild, M. F., & Croyle, R. T. (2013). Spatial turn in health research. *Science, 339*(6126), 1390–1392. https://doi.org/10.1126/science.1232257

Rodó-Zárate, M., & Jorba, M. (2022). Metaphors of intersectionality: Reframing the debate with a new proposal. *European Journal of Women's Studies, 29*(1), 23–38. https://doi.org/10.1177/1350506820930734

Saxon, J., & Snow, D. (2020). A rational agent model for the spatial accessibility of primary health care. *Annals of the American Association of Geographers, 110*(1), 205–222. https://doi.org/10.1080/24694452.2019.1629870

Scheim, A. I., & Bauer, G. R. (2019). The intersectional discrimination index: Development and validation of measures of self-reported enacted and anticipated discrimination for inter-categorical analysis. *Social Science & Medicine, 226*, 225–235. https://doi.org/10.1016/j.socscimed.2018.12.016

Shannon, J., Reese, A. M., Ghosh, D., Widener, M. J., & Block, D. R. (2021). More than mapping: Improving methods for studying the geographies of food access. *American Journal of Public Health, 111*(8), 1418–1422. https://doi.org/10.2105/AJPH.2021.306339

Taggart, T., Powell, W., Gottfredson, N., Ennett, S., Eng, E., & Chatters, L. M. (2019). A Person-centered approach to the study of black adolescent religiosity, racial identity, and sexual initiation. *Journal of Research on Adolescence, 29*(2), 402–413. https://doi.org/10.1111/jora.12445

Tobler, W. R. (1970). A computer movie simulating urban growth in the Detroit region. *Economic Geography, 46*, 234. https://doi.org/10.2307/143141

UN Committee on Economic, Social, and Cultural Rights. (2000). *General comment no. 14 (2000), The right to the highest attainable standard of health (article 12 of the International Covenant on Economic, Social and Cultural Rights)*. United Nations.

Universal Declaration of Human Rights, 217 A (III) (1948).

Wan, N., Zou, B., & Sternberg, T. (2012). A three-step floating catchment area method for analyzing spatial access to health services. *International Journal of Geographical Information Science, 26*(6), 1073–1089. https://doi.org/10.1080/13658816.2011.624987

Wang, F. (2000). Modeling commuting patterns in Chicago in a GIS environment: A job accessibility perspective. *The Professional Geographer, 52*(1), 120–133. https://doi.org/10.1111/0033-0124.00210

Wang, F. (2012). Measurement, optimization, and impact of health care accessibility: A methodological review. annals of the association of american geographers. *Association of American Geographers, 102*(5), 1104–1112. https://doi.org/10.1080/00045608.2012.657146

Part III
Integrating Social-Spatial Determinants of Health in Practice

Chapter 7
Operationalization of Social-Spatial Determinants of Health

Annemarie G. Hirsch

Introduction

The passage of the Affordable Care Act in 2010 sought to transform US healthcare from a fee-for-service model to a value-based model. This value-based model involves a provider payment system based on patient health outcomes rather than the volume of procedures performed (Porter, 2009). This focus on value-based care models has sparked widespread interest among US health systems to integrate social-spatial determinants of health measures into their health care delivery. Concurrently, there has been growth in electronic health record (EHR) adoption and geocoded datasets, expanding the potential for automated approaches for incorporating these measures into care. In 2014, the National Academy of Medicine (NAM, formerly the Institute of Medicine) recommended the inclusion of standard social-spatial determinants of health measures in EHR systems (Institute of Medicine [IOM], 2014). Nearly 10 years later, there is still no standardization among health systems in their measurement and use of social-spatial determinants of health data. A minimal number of health systems have started to implement and evaluate the use of these data in point-of-care decision-making, population health strategies, and reimbursement models; however, widespread usage is far from the reality among health systems and healthcare providers. This chapter provides a scoping review of how organizations in the US health system are currently using social-spatial determinants of health measures.

The chapter is presented in five sections that each describe a way that social-spatial determinants of health measures are being used by the US. health system: EHR records, risk prediction, resource allocation, reimbursement models, and

A. G. Hirsch (✉)
Department of Population Health Sciences, Center for Community Environment and Health, Geisinger, Danville, PA, USA
e-mail: aghirsch@geisinger.edu

© The Author(s) 2026   105
M. A. Kolak, I. K. Moise (eds.), *Place and the Social-Spatial Determinants of Health*, Global Perspectives on Health Geography,
https://doi.org/10.1007/978-3-031-88463-4_7

surrogates for individual-level social determinants of health. Each section begins with an overview of how the measures are being used, followed by specific examples of their use. For the specific examples, a description of the social-spatial determinants of health measures used (e.g., specific indices), the rationale for their selection, data sources, and the impact of their use are presented when these data are available in peer-reviewed or gray literature. Examples were selected based on timeliness (i.e., recent publications), completeness of information available, and relevance to the section topic.

Integrating Social-Spatial Determinants into the Electronic Health Record

As of 2019, approximately three-quarters of office-based physicians and nearly all nonfederal acute care hospitals in the USA had adopted a certified EHR system (Office of the National Coordinator for Health Information Technology, 2022). EHRs generally capture demographic (e.g., age, sex, race, ethnicity), behavioral (e.g., tobacco, alcohol, illicit drug use), and clinical (e.g., vital signs, diagnoses, medication orders, procedures, and outpatient, inpatient, emergency department encounters) information in both discrete data fields and free-text clinical notes. Social determinants of health, at the individual or area level, are not typically captured in the EHR. However, over the last decade, there has been increasing pressure on health systems to collect and record these data.

In 2014, the NAM's Committee on the Recommended Social and Behavioral Domains and Measures for Electronic Health Records recommended that The Office of the National Coordinator for Health Information Technology (ONC) and the Centers for Medicare and Medicaid Services (CMS) include the addition of 12 measures to capture 11 social and behavioral domains in the meaningful use regulations for EHRs based on the following criteria: strength of evidence of the association of the domain with health; usefulness of the domain to clinical care decisions, population health policies, and research; availability of a reliable and valid measure; burden of obtaining and/or storing the data; sensitivity to patient discomfort; and accessibility of the data. One of these domains was "neighborhoods and communities compositional characteristics." In this domain, the report suggested that EHRs capture residential addresses and neighborhood median household incomes (IOM, 2014). The committee recommended measuring median household income at the census tract level using American Community Survey (ACS) data. The report justified the use of census tract measures based on the homogeneity of socioeconomic characteristics at this scale and their common use as proxies for neighborhoods. Because the ACS measures only a sample of households each year, the report suggested pooling the data across multiple years to obtain the estimates for income, with the number of years serving as a function of the sample size per year in any given tract (IOM, 2014).

The committee considered several other compositional characteristics for standard inclusion in EHRs: air pollution, allergens, other hazardous exposures, land use, access to nutritious food options, public transportation, parks, health care and social services, education and job opportunities, safety and violence, social cohesion, and social organization. While the committee acknowledged the potential clinical utility of these measures, they noted that the data required to create them are not routinely available in a standardized format; the processes used to create the measures can be complex; a number of different approaches to measurement exist; and the validity and usefulness of different types of measures remain a topic of active research (IOM, 2014). Thus, these domains were not included in the 2014 recommendations (IOM, 2014).

Residential addresses are regularly captured in EHRs for clinical communications and billing purposes. However, nearly a decade after the NAM report, median household income is not routinely incorporated in EHRs. This slow adoption may be the result of a variety of factors including limited resources, lack of incentives or requirements by payors or regulators, and minimal clinician training on how to respond to these data (Wark et al., 2021). Rather, health systems are only recently starting to build and pilot the integration of a variety of social-spatial determinants of health into the EHR. Here, we describe three examples of these efforts.

Cincinnati Children's Hospital Medical Center EHR Alert

In 2015, Cincinnati Children's Hospital Medical Center (CCHMC) built automated alerts in the EHR that triggered a multidisciplinary intervention based on the characteristics of the community in which patients resided (Beck et al., 2019). CCHMC selected two neighborhoods in their service area based on disproportionately high rates of all-cause morbidity and the following social-spatial determinants of health: median household income, housing instability, food insecurity, and poor transportation access. When a child from a high-risk neighborhood is hospitalized at CCHMC, an automatic alert from the EHR is sent to a multidisciplinary team of physicians, nurses, social workers, and community engagement consultants. The EHR alert prompted an in-depth chart review and a bedside huddle focused on the potential preventability of hospitalization, identifiable care gaps (e.g., need for vaccination), and transition needs. When appropriate, patients were connected with additional support during inpatient hospitalization (e.g., social work consultation, school services to help with schoolwork completion in the hospital) and, if needed, transition-related services such as post-discharge nurse visits. In these same communities, CCHMC started to proactively ensure outreach to children with asthma had a ready supply of medications in advance of the fall exacerbation season.

The CCHMC evaluated the impact of these interventions on the neighborhood inpatient bed-day rate (the number of days neighborhood children spent hospitalized divided by the number of children living in the neighborhood). Compared to before the implementation of these community-targeted programs, CCHMC observed an 18% decrease in the inpatient bed-day rate and a 20% decrease in hospitalizations (Beck et al., 2019). No such reductions were observed in demographically similar neighborhoods. As of 2023, only the combined impact of the bedside huddle and proactive outreach had been evaluated; the impact of these interventions alone has not been published.

The Envirome Web Service

In 2020, the Children's Mercy Kansas City (CMKC) hospital system integrated a clinical decision support tool, the Envirome Web Service (EWS), into care delivery (Kane et al., 2021). The EWS geocoded patient addresses in real time and presented key indicators called envirome indicators to clinicians. The envirome patient profile contained the following indicators, measured at the census tract level: socioeconomic (e.g., median family income, median household income, percent of population in college); demographics (e.g., median age, percent of population under 18, percent of population over 65, population by race); and food access and context (e.g., food desert, urbanicity, low income census tract) (Kane et al., 2021). All these elements were pulled from the Census Bureau and the US Department of Agriculture. A pilot study of the EWS found that 6.1% of clinicians with access to the tool were regular users who opted to maintain the EWS in their custom workflows, logging more than 100 EWS sessions per year (Kane et al., 2021). During the development of the EWS, stakeholders expressed concern regarding a lack of actionable content, potentially explaining this limited adoption. Information on how the EWS was used in care and the impact of the EWS on health outcomes was not available.

Community Vital Signs

With funding from the Patient-Centered Outcomes Research Institute, the Accelerating Data Value Across a National Community Health Center Network (ADVANCE) Clinical Data Research Network (funded by the Patient-Centered Outcomes Research Institute), the Robert Graham Center (RGC), OCHIN (originally called the Oregon Community Health Information Network), and Health Landscape partnered with US federally qualified health centers to develop a program to incorporate a core set of community measures into the EHR, described as Community Vital Signs (Bazemore et al., 2016). This program integrates a novel interface, the Community Vital Signs Geocoding Application Interface (API), which would use residential addresses in the EHR to generate geographic identifiers (e.g., county, census tract). The API links these geographic identifiers to join a core set of Community Vital Signs and returns the linked data to the EHR. The program included six categories of indicators: built environment (e.g., fast food restaurants and liquor stores per 100,000 population, population density); environmental exposures (e.g., median housing structure age, percent of occupied housing units without complete plumbing facilities); neighborhood economic conditions (e.g., percent of vacant addresses, overall percentile ranking for the Centers for Disease Control and Prevention Social Vulnerability Index (SVI)); neighborhood race/ethnic composition; neighborhood resources (e.g., Modified Retail Food Environment Index, percent of people living more than 1 mile from a supermarket); neighborhood socioeconomic composition (e.g., median household income, percent below 100% the Federal Poverty Level (FPL)); and a social deprivation index (Butler et al., 2013). These indicators were derived from publicly available data sources, such as the US Census Bureau, Environmental Public Health Tracking Network, and the Environmental Protection Agency (Bazemore et al., 2016).

The Community Vital Signs API, like the aforementioned EWS, demonstrated proof of concept that it is technically feasible to geocode patient addresses in the EHR, assign patients a broad range of social-spatial determinants of health measures, and display these measures in the EHR. However, unlike the EHR alert evaluated at CCHMC, the data presented to clinicians by these programs were not designed to trigger a specific clinical intervention. In fact, the creators provide little guidance as to how clinicians and health systems should use the collected social-spatial determinants of health data in their clinical care, limiting the utility of these data in the EHR.

Risk Prediction

One of the potential values of integrating social-spatial determinants of health into the EHR has been the potential for improved risk prediction. Researchers have been exploring whether incorporating social-spatial determinants of health in the predictive modeling of health outcomes can improve model performance for the prediction of a range of outcomes, including hospital readmission, adherence to preventive care, social work referrals, and mortality. Chen et al., (2022) conducted a review of these studies and found that, in most cases, models evaluated the addition of a publicly available deprivation index (e.g., SVI, Social Deprivation Index (SDI), Townsend Index, and Area Deprivation Index (ADI)), measured at the zip code or census tract level, to models containing individual-level demographic information as well as relevant clinical data (e.g., comorbid disease). In most cases, the review found that the addition of social-spatial determinants of health to predictive models did not enhance model performance. However, more recent studies reported that social-spatial determinants of health measures enhanced the performance of predictive models in certain sub-populations. Zhang et al. (2020), for example, evaluated whether the addition of social-spatial determinants of health measures could improve the Simplified HOSPITAL score model prediction of potentially avoidable

30-day hospital readmission or death. The Simplified HOSPITAL score included the following individual-level variables: frequency of prior hospital admissions; urgency admission; last available hemoglobin and sodium levels; discharge from an oncology division; and the index hospital length of stay. The team added the following measures at the census tract level: socioeconomic status (e.g., median income, unemployment rate, percent with high school or high school-equivalent diploma, percent foreign-born, percent without insurance, and percent dual-eligible for Medicaid and Medicare); felony rate; walkability score; Gini income equality coefficient; a composite score reflecting household composition and disability; and a composite score for minority status and language. The study found that the addition of these measures did not improve prediction in the general population but did improve the performance of the tool among individuals on Medicaid, patients 65 years of age or older, and obese patients.

Two studies found that the impact of adding social-spatial determinants of health measures differed by race. Hammond et al. (2020) evaluated how the addition of individual- and area-level social determinants of health impacted risk-adjustment models for annual costs of care, all-cause hospitalization, cardiovascular hospitalization, and death. Models initially included age, sex, the original reason for Medicare eligibility, dual Medicaid enrollment, institutionalization in long-term care, and 83 clinical conditions. Of the seven domains of social determinants of health added to this model, six were individual level (alcohol abuse, access to care, economic status, financial strain, marital status, education), and the seventh was the rural versus urban status of the patient's residence. Without the social determinants of health, these models were sufficient predictors among White patients but underpredicted all-cause hospitalization, cardiovascular hospitalization, and costs among Black and Hispanic patients. Adding the social determinants of health measures improved the performance of the risk-adjustment models in the Black and Hispanic populations but not among the White population. Similarly, Segar et al. (2022) evaluated the impact of adding the Distressed Communities Index, the SVI, and the Graham Center 2011–2015 SDI at the zip code level to a model predicting in-hospital mortality in heart failure. Adding these measures enhanced model performance only among the Black population.

As EHRs begin to bring social-spatial determinants of health data into EHRs through programs such as the EWS and Community Vital Signs, such models could incorporate a broader set of social-spatial determinants that may improve risk prediction, particularly in vulnerable sub-populations of patients. Further evaluation of the health impact of using these predictive models to guide care is needed. This evaluation should continue to explore potential differences in the value of risk prediction by sociodemographic features, as early evidence suggests that the impact of incorporating social-spatial determinants into risk prediction models may vary by patient characteristics.

Health Care Resource Allocation

While the ACA advanced interest in the integration of social-spatial determinants of health into health care, the concept is not new. For decades, the federal government has considered social-spatial determinants of health in resource allocation to support population health. Below, we describe two examples of federal programs that use social-spatial determinants of health measures for resource allocation. The first case involved the 1970 creation of the National Health Service Corps (NHSC) that was created to address shortages in health personnel (Health Resources & Services Administration [HRSA], 2023). The second describes the more recent guidelines for the allocation of COVID-19 vaccines.

Health Professional Shortage Areas

The Health Professional Shortage Area (HPSA) and Medically Underserved Areas (MUAs) or Medically Underserved Populations (MUPs) are designations established by the federal government to identify geographic areas, population groups, or facilities that are experiencing a shortage of healthcare professionals (e.g., primary care, dental care, mental health) (U.S. Department of Health & Human Services [HHS], 2019). The HSPA designation is used by over 30 federal programs to identify areas and populations in need of assistance, including the NHSC, Nurse Corps, Indian Health Services loan repayment program, the CMS HPSA bonus payment program, the CMS rural health clinic program, and the J-1 Visa waiver. Three scoring criteria are applied to classify areas as HPSA: population-to-provider ratio, percent of population below 100% of the FPL, and travel time to the nearest source of care outside of the HPSA (HHS, 2019). MUA and MUP are designated according to an Index of Medical Underservice. This index involves four variables: the ratio of primary care physicians per 1000 population, infant mortality, percentage of the population with incomes below the poverty level, and percentage of the population over 65 years of age or older (HHS, 2019). MUA designations inform grants for community health centers, certification of Rural Health Clinics, and Public Health Service Grant Programs. To be certified as a Federally Qualified Health Center, a clinic must serve an area designated as an MUA or MUP (CMS, 2023).

Vaccine Allocation

In 2020, the National Academies of Science, Engineering, and Medicine (NASEM) was tasked with assisting the Centers for Disease Control and Prevention's (CDC) Advisory Committee on Immunization Practice in developing an equitable allocation framework for COVID-19 vaccines. NASEM recommended that vaccine access should be prioritized based on the SVI at the US Census tract level or, when data are available, SVI-derived indices that include COVID-19-specific measures (e.g., indicators of comorbidities and health system resource availability) (NASEM, 2020). Schmidt et al. (2021) reviewed the vaccination plans of the 64 jurisdictions to which the CDC dispenses vaccines and found that 43 jurisdictions used a geographic measure of disadvantage, and 37 specifically used a disadvantage index to organize their vaccination distribution process. Of the 37 jurisdictions that used a disadvantage index, 29 used the SVI. Among participating jurisdictions, the reason for the usage of these indices differed. Almost half of the jurisdictions report using the disadvantage indices to allocate more vaccines and vaccine appointments (46%); others used it to identify priority areas and populations (46% of jurisdictions), plan outreach and communications (32% of jurisdictions), plan dispensing sites (22% of jurisdictions), and monitor vaccine receipt (11% of jurisdictions). Additional work is needed to determine whether these methods were successful in an equitable distribution.

Social-Spatial Determinants in Reimbursement Models

Unlike the usage of social-spatial determinants of health in resource allocation in the USA, the use of these determinants in healthcare reimbursement models is relatively new. Many value-based programs do not account for the role that social risk factors have in influencing health outcomes, thus potentially disadvantaging plans or providers that care for patients who, because of social inequities, face disproportionate barriers to health and health care (Breslau et al., 2022). In this section, we will summarize recent efforts in the USA to integrate social-spatial determinants of health measures into healthcare reimbursement models.

In the last decade, there have been efforts at the state and federal levels to adjust reimbursement systems to account for measures of social-spatial determinants of health. As of 2023, there are two state-based Medicaid programs that include measures of social-spatial determinants of health. Since 2017, the Massachusetts Medicaid Agency, MassHealth, has used the Neighborhood Stress Score (NSS7) in their calculation to adjust their system's payments to providers (Breslau et al., 2022). The NSS7 is derived from addresses geocoded at the census block group level using seven census variables that were identified in a principal components analysis of 2013 Massachusetts Medicaid data. These components include the percentage of families with incomes below FPL; percent of families with incomes <200% of the FPL; percent of adults who are unemployed; percent of households receiving public assistance; percent of households with no car; percent of households with children and a single parent; and percent of people aged 25 or older who have no high school degree. While weights were used in the original principal component analysis, the program calculates the NSS7 using an unweighted sum, as the weights varied little across the seven variables (Breslau et al., 2022). The second state-based Medicaid program to incorporate measures of social-spatial determinants of health is Arizona. In 2021, Arizona's Medicaid Agency, AHCCCS, started adjusting capitation payments to managed care organizations participating in the AHCCCS Complete Care Program. The adjustment incorporates four individual-level social risk factors and a modified version of the SVI at the 5-digit zip code level (Breslau et al., 2022). In 2023, risk adjustment for social-spatial determinants of health will begin at the federal level. CMS implemented a new model called the Accountable Care Organization (ACO) Realizing Equity, Access, and Community Health (REACH) (CMS, 2022). Among the new health equity components of REACH is a health equity benchmark that shifts programming dollars from organizations serving fewer underserved patients to those caring for more, using a risk-adjustment score based on ADIs.

To date, little is known about the impact of these relatively new risk-adjustment programs in the USA. Conversely, such risk adjustments have existed for decades in Europe. Since the 1980s, New Zealand has used an area-level deprivation index to identify geographic areas of need and reallocate funds to the most deprived areas. Initially, New Zealand used a large area estimate of need, the Health and Equity Index of Deprivation, for healthcare fund allocation. However, the scale of the tool overlooked small pockets of need. In 1997, New Zealand's Ministry of Health started to implement the New Zealand Index of Deprivation (NZDep), an index that measured need on the Mesh Block scale (90 people on average) (Huffstetler & Phillips, 2019). The NZDep was created from factors collected in the five-year census of 1991 and comprised nine deprivation characteristics. Deprivation was scaled from 1 (low) to 10 (high). Two main types of deprivation are included in the NZDep: material deprivation (e.g., resources, services, physical environment) and social deprivation (e.g., relationships, roles, and responsibilities of members of society). Areas may have one dominant trait of deprivation or may share a combination of social and material deprivation. The reimbursement adjustment based on the NZDep can result in as much as a threefold increase in healthcare funding (Huffstetler & Phillips, 2019). In the United Kingdom, National Health Services has been adjusting payments using the Carr-Hill formula since 2004. This formula, known as the global sum allocation, adjusts reimbursement on clinic-level characteristics (e.g., age and sex structure) as well as rurality, as determined by population density (Huffstetler & Phillips, 2019). Both the New Zealand and United Kingdom approaches have been found to improve access to care (e.g., shorter wait times); however, data on the impact on health outcomes and health disparities have not been published (Foley, 2018; Huffstetler & Phillips, 2019).

Individual Versus Social-Spatial Determinants of Health

Social determinants of health at the individual level are being integrated into EHR systems, evaluated in predictive models, and used in reimbursement adjustment. In 2015, the Department of Health and Human Services required that health systems use the tenth revision of the International Classification of Diseases codes (ICD-10) for EHR documentation. This version of ICD codes includes a growing list of codes designed to capture a patient's individual-level social characteristics, referred to as Z-codes (e.g., unemployed, homelessness, and victim of crime) (Truong et al., 2020). The CMS recently instituted a requirement that the health risk assessments used by all Medicare Advantage Special Needs Plans include at least one question from a list of specified screening instruments in each of three domains (food security, housing stability, and transportation access). In its Inpatient Quality Reporting Program, the CMS requires reporting of process measures for the collection of social determinants of health (e.g., percent of patients screened; percent of patients who screen positive). Furthermore, the Joint Commission has issued new standards that require hospitals to screen patients for health-related social needs and provide information about community resources and support services to patients who screen positive.

The collection of individual-level social determinants is often more logistically challenging than obtaining social-spatial determinants of health. Unlike the many spatial measures available from secondary data sources, capturing individual-level data often requires primary data collection that can be disruptive to clinical workflows, perceived as intrusive by patients, and have greater issues with missing data. Truong et al. (2020), for example, reported that the uptake of the ICD-10 Z-codes has been slow and poorly reflects the actual burden of social needs experienced by patients. Thus, social-spatial determinants of health measures are often viewed as a proxy for individual-level social determinants. In some cases, social-spatial measures are viewed as an interim step until individual-level social determinants can be more systematically captured. For example, a Society of General Internal Medicine's Position Statement noted, "Area-level indices using census data to evaluate the social vulnerability of communities are immediately available…such indices can be used until individual social risk data can be collected." A report commissioned by the Department of HHS on area-level social determinants of health stated,

> *Individual-level health-related social needs information is not widely available, and thus, developing measures to directly target funds based on these needs is not currently feasible. As an interim step, area-level measures of social needs or deprivation could be used.*

Viewing spatial-social determinants of health measures as proxies for individual-level measures is potentially problematic. First, using these measures as proxies undervalues the role that community context has in health, independent of individual-level factors. Neighborhoods possess social attributes that could plausibly affect the health of individuals, independent of individual-level factors (Diez Roux & Mair, 2010). Second, the integration of social-spatial determinants of health is important for informing place-level health policies and interventions. Third, although individual- and spatial-level measures of social determinants overlap, such measures are not equivalent (Cottrell et al., 2020). Using social-spatial determinants of health measures in place of individual-level measures may mean that patients who could benefit from targeted interventions are missed (Cottrell et al., 2020).

Conclusions

Healthcare policymakers, payors, and providers are starting to explore the integration of social-spatial determinants of health measures into health care. These determinants are being introduced in EHR documentation, risk prediction, reimbursement models, and allocation of resources for health. Pilot work has demonstrated the technical feasibility of integrating these data; however, adoption has been slow. The delay in more widespread adoption is likely due to be a combination of factors, including limited resources; lack of incentives; and a dearth of guidance on how providers and health systems can respond to this data.

Efforts to bring social-spatial determinants of health measures to health care have been dominated using existing socioeconomic indices (e.g., ADI, SVI) that are based on publicly available data (e.g., U.S. Census Bureau data) and applied to predefined administrative spatial boundaries (e.g., census tract, zip code). Selection of measures is based on logistical demands (e.g., data availability) and evidence of the relation of these measures to health outcomes. However, even the integration of publicly available and well-established indices requires technical, financial, and personnel resources that may present barriers to adoption (Wood et al., 2021). Of concern is that health systems serving populations that would most benefit from using social-spatial determinants of health measures may be less equipped to integrate these measures. Financial support or incentives for capturing these data may help health systems overcome these barriers to adoption.

Early evidence suggests the value of integrating social-spatial determinants of health in risk models differs by sociodemographic and clinical characteristics of the population (e.g., age, race, body mass index). Consistent with these findings, studies have demonstrated a synergistic effect of social-spatial determinants of health, race, and health outcomes, such that stronger associations between these determinants and health outcomes have been observed in Black versus non-Hispanic White populations (Mode et al., 2016). Similarly, stronger associations have been observed in older versus younger populations (Wang et al., 2020). Moreover, there is good evidence that different indices used to measure social-spatial determinants of health are not equally effective in uncovering health inequalities for different population subgroups (Allik et al., 2020). Thus, the measures selected to capture social-spatial determinants of health and the application of these measures in health care should be evaluated in subgroups of the population, as the value of integrating these determinants may differ across groups.

Though theory points to the immense benefit of knowledge of the community context that can have on health care delivery and health outcomes, there is little guidance for providers regarding how to use these data in their practice. Moreover, there is limited research on how responding to these data impacts health outcomes and mitigates disparities. Thus, providers express reservations about using social-spatial determinants of health in the clinical care setting, as they do not see these data as actionable. Funding should be allocated to support such research and curriculum for physicians and allied health professionals should educate students on how to care for patients in the context of their social-spatial health determinants.

References

Allik, M., Leyland, A., Ichihara, M. Y. T., & Dundas, R. (2020). Creating small-area deprivation indices: A guide for stages and options. *Journal of Epidemiology and Community Health, 75*, 20–25.

Bazemore, A. W., Cottrell, E. K., Gold, R., Hughes, L. S., Phillips, R. L., Angier, H., Burdick, T. E., Carrozza, M. A., & DeVoe, J. E. (2016). "Community vital signs": Incorporating geocoded social determinants into electronic records to promote patient and population health. *Journal of American Medical Informatics Association, 23*, 407–412.

Beck, A. F., Anderson, K., Rich, K., Taylor, S. C., Iyer, S. B., Kotagal, U. R., & Kahn, R. S. (2019). Cooling the hot spots where child hospitalization rates are high: A neighborhood approach to population health. *Health Affairs, 38*(9), 1433–1441.

Breslau, J., Martin, L., Timbit, J., Qureshi, N., & Zajdman, D. (2022). *Landscape at the area level deprivation measures and other approaches to account for social risk and social determinants of health in health care payments.* RAND Health Care.

Butler, D. C., Petterson, S., Phillips, R. L., & Bazemore, A. W. (2013). Measures of social deprivation that predict health care access and need within a rational area of primary care service delivery. *Health Services Research, 28*(2 Pt 1), 539–551.

Centers for Medicare & Medicaid Services (CMS). (2022, February 24). *Accountable Care Organization (ACO) Realizing Equity, Access, and Community Health (REACH) model.* Retrieved from https://www.cms.gov/newsroom/fact-sheets/accountable-care-organization-aco-realizing-equity-access-and-community-health-reach-model.

Centers for Medicare & Medicaid Services (CMS). (2023, August). *Federally Qualified Health Center.* Medical Learning Network. Retrieved from https://www.cms.gov/files/document/mln006397-federally-qualified-health-center.pdf.

Chen, A., Ghosh, A., Gwynn, KB., Newby, C., Henry, T.L., Pearce, J., Fleurant, M., Schmidt, S., Bracey, J., & Jacobs, E.A. (2022) Society of General Internal Medicine Position Statement on Social Risk and Equity in Medicare's Mandatory Value-Based Payment Programs. *JGIM.* 37(12):3178–87.

Cottrell, E. K., Hendricks, M., Dambrun, K., Cowburn, S., Pantell, M., Gold, R., & Gottlieb, L. (2020). Comparison of community-level and patient-level social risk data in a network of community health centers. *JAMA Network Open, 3*(10), e2016852.

Diez Roux, A. V., & Mair, C. (2010). Neighborhoods and health. *Annals of the New York Academy of Sciences, 1186*, 125–145.

Foley, J. (2018). Social equity and primary healthcare financing: Lessons from New Zealand. *Australian Journal of Primary Health, 24*(4), 299–303.

Hammond, G., Johnston, K., Huang, K., & Joynt Maddox, K. E. (2020). Social determinants of health improve the predictive accuracy of clinical risk models for cardiovascular hospitalization, annual cost, and death. *Circulation: Cardiovascular Quality and Outcomes, 13*, e006752. https://pubmed.ncbi.nlm.nih.gov/32412300/

Health Resources & Services Administration (HRSA). (2023). Mission, work, and impact. *National Health Service Corps, 57*(6S1), S82–S88. Retrieved from https://nhsc.hrsa.gov/about-us

Huffstetler, A. N., & Phillips, R. L. (2019). Payment structures that support social care integration with clinical care: Social deprivation indices and novel payment models. *American Journal of Preventive Medicine, 57*(6S1), S81–S88.

Institute of Medicine (IOM). (2014). *Capturing social and behavioral domains and measures in electronic health records: Phase 2.* The National Academies Press.

Kane, N. J., Gerkovich, M. M., Breitkreutz, M., Rivera, B., Kunchithapatham, H., & Hoffman, M. A. (2021). The Envirome web service: Patient context at the point of care. *Journal of Bioinformatics, 119*, 103817.

Mode, N. A., Evans, M. K., & Zonderman, A. B. (2016). Race, neighborhood economic status, income inequality and mortality. *PLoS One*, (11, 5), e0154535.

National Academies of Sciences, Engineering, and Medicine (NASEM). (2020). *Framework for equitable allocation of COVID-19 vaccine*. The National Academies Press. https://doi.org/10.17226/25917

Office of the National Coordinator for Health Information Technology. (2022). *Secondary quick stats. 2022*. HealthIt. Retrieved from https://www.healthit.gov/data/quickstats/adoption-electronic-health-records-hospital-service-type-2019-2021.

Porter, M. E. (2009). A strategy for health care reform—Toward a value-based system. *NEJM, 361*(2), 109–112.

Schmidt, H., Weintraub, R., Williams, M. A., Miller, K., Buttenheim, A., Sadecki, E., Wu, H., Doiphode, A., Nagpal, N., Gostin, L. O., & Shen, A. A. (2021). Nature medicine. Equitable allocation of COVID-19 vaccines in the United States. *Nature Medicine, 27*, 1298–1307.

Segar, M. W., Hall, J. L., Jhund, P. S., Powell-Wiley, T. M., Morris, A. A., Kao, D., Fonarow, G. C., Hernandez, R., Ibrahim, N. E., Rutan, C., Navar, A., Stevens, L. M., & Pandey, A. (2022). Machine learning-based models incorporating social determinants of health vs. traditional models for predicting in-hospital mortality in patients with heard failure. *JAMA Cardiology, 7*(8), 844–854.

Truong, H. P., Luke, A. A., Hammond, G., Wadhera, R. K., Reidhead, M., & Joynt Maddox, K. E. (2020). Utilization of social determinants of health ICD-10Z codes among hospitalized patients in the United States. *Medical Care, 58*(12), 1037–1043.

U.S. Department of Health & Human Services (HHS). (2019, August 1). *Health Professional Shortage Areas (HPSAs) and Medically Underserved Areas/Populations (MUA/P) shortage designation types*. Health Professional. Retrieved from https://www.hhs.gov/guidance/document/hpsa-and-muap-shortage-designation-types.

Wang, B., Eum, K. D., Kazemiparkouhi, F., Li, C., Manjourides, J., Pavlu, V., & Suh, H. (2020). The impact. of long-term PM 2.5 exposure on specific causes of death: Exposure-response curves and effect modification among 53 million US Medicare beneficiaries. *Environmental Health, 19*, 1–2.

Wark, K., Cheung, K., Wolter, E., & Avey, J. P. (2021). Engaging stakeholders in integrating social determinants of health into electronic health records: A scoping review. *International Journal of Circumpolar Health, 80*(1), 1943983. Retrieved from https://www.ncbi.nlm.nih.gov/pmc/articles/PMC8276667/

Wood, E., Sanders, M., Frazier, T (2021). The practical use of social vulnerability indicators in disaster management. *International Journal of Disaster Risk Reduction*. 63: 102464.

Zhang, Y., Zhang, Y., Sholle, E., Abedian, S., Sharko, M., Turchioe, M. R., Wu, Y., & Ancker, J. S. (2020). Assessing the impact of social determinants of health on predictive models for potentially avoidable 30-day readmission or death. *PLoS ONE, 15*(6), e0235064. https://pubmed.ncbi.nlm.nih.gov/32584879/

Chapter 8
Spatial Practices That Reshape the Social Determinants of Health for Families with Young Children Affected by Disadvantage

Yvonne Young, Nick Hopwood, and Donna Rooney

Introduction

Early childhood experiences are a key social determinant of health, according to the World Health Organization (WHO, 2008). These experiences can impact a child's development and are compounded for children in families experiencing socioeconomic disadvantage (Hertzman, 2010; Moore et al., 2015). A nurturing family environment and social connection are crucial for these families. Taking these determinants into account, therefore, is essential for services that support families with young children affected by socioeconomic disadvantage. This chapter focuses on social connections between families and services, families and children, families and communities, and between families themselves. Such connections have been identified as important for services seeking to improve the social determinants of health of young children in families affected by disadvantage (Moore, 2021b).

Currently, Australian children living in socioeconomically disadvantaged areas have a higher representation among children who are behind developmentally in more than one domain when starting school (Australian Early Development Census, 2021). This serves to hinder their future life chances (Marmot, 2012). Goldfield et al. (2018) argue that understanding child disadvantage from a social determinants' perspective enables a better understanding of the "complex and multifaceted ways in which disadvantage can manifest" (p. 223). Developmental, health, education, and social issues need to be identified early and responded to. Siloed solutions are inadequate given the nature of the disadvantage experienced by these families (Logan et al., 2018; Moore & Fry, 2011). Delivering services in an integrated, place-based way, therefore, has been considered best practice for some time in Australia (Press et al., 2010), Europe (Glass, 1999), and the United States (Hines, 2017). Integrated service delivery (ISD) is characterized as a holistic approach "joining up

Y. Young (✉) · N. Hopwood · D. Rooney
Faculty of Arts and Social Sciences, University of Technology Sydney, Sydney, Australia

© The Author(s) 2026
M. A. Kolak, I. K. Moise (eds.), *Place and the Social-Spatial Determinants of Health*, Global Perspectives on Health Geography,
https://doi.org/10.1007/978-3-031-88463-4_8

social services to provide a better service to service users" (OECD, 2012, p. 3). More specifically in relation to families, ISD is "the process of building connections between services in order to work together as one to deliver services that are more comprehensive and cohesive and more responsive to the needs of families" (Prichard et al., 2010, p. 5). Although long established as a desirable approach, the question of how ISD can reshape social determinants of health (SDoH) for families with young children remains a key priority in addressing social disadvantage in the early years around the world (H. M. Government, 2021; Marmot, 2020; Moore, 2021a). To address the lacuna in the practices of ISD, the emphasis in this chapter is on social connections as a contributing factor to SDoH (Moore, 2021b).

Place-based child and family learning centers (CFLCs) are informal spaces that operate with the explicit ethos of ISD within Tasmania, Australia. CFLCs offer a variety of services including early education, health, legal, play, and other support for families in some of the country's most disadvantaged communities. CFLCs vary in design, but all have an informal public space. Everyone visiting a center either passes through them or spends time within them. Spaces can be accessed by parents with or without an appointment. Within these spaces are features designed for children's play (e.g., sand pits), adult social interaction, open offices, and kitchens. Staff call these spaces "the floor" and parents often call them "play spaces"; however, in this chapter, we refer to them as *shared public spaces*.

Spatial theory can conceptualize the way these spaces enable connections that become the beginnings of ISD. Spaces that produce connections in services for families hold the potential to shed light on how social-spatial determinants of health can be reshaped where children's health and development are at risk due to socioeconomic disadvantage in their community. In this way, shared public spaces can be understood not merely as containers or stages for certain things to happen (play, waiting, making coffee), but as important contributors to improved childhoods and life chances. Accordingly, this chapter integrates a spatial perspective informed by Massey (2005), who views space as dynamic, socially constructed, ongoing, and a coming together of trajectories or stories thus far. The focus is not on formal aspects of services that happen behind closed doors, but within shared public spaces where crucial connections are being produced. This perspective reveals how ISD is accomplished, enriching the idea of place-based services with insights at a granular level pertaining to key practices that created intersections between multiple trajectories, charging connections with what we term "depth in the moment" (see Fig. 8.1).

Fig. 8.1 Conceptualizing practices, space, and connection

In this view, practices are defined broadly as activities engaged by people in shared public spaces. Trajectory is defined as movement from one thing to another, and an intersection of trajectories is the coming together of multiple or different trajectories. Moments with depth are defined as having:

1. Sustained and iterative engagement
2. A redirection of attention from problems to be fixed to production of spaces rich with intersecting trajectories that create depth in the moment
3. Significance in small, mundane acts.

As such, moments of depth are directly linked to addressing the SDoH, especially social connection, and timely access to services that provide support for both the child and the family.

We argue that a spatial reading of these shared public spaces and the deformalized services delivered helps understand how spatial-social determinants of health are established, operate, and can be reshaped, by linking key considerations that are particularly important for families living in disadvantaged circumstances. These considerations include access to safe, comfortable family spaces bursting with opportunities for social connections and pathways to multiple services. We identify the significant practices that produce shared public spaces as places where these imperatives are addressed holistically, as part of ISD, and in so doing, reshape the social-spatial determinants of health at a local, often family-specific level.

The chapter begins with a discussion of current literature on ISD. Next, the study and the sites of the study are described in detail. Following, three key practices of *hanging out*, *consuming food*, and *negotiating* are outlined before looking at how these produce connections with depth in the moment in three different parts of shared public spaces (sandpits, kitchens, and open office areas). The importance of such depth in reshaping SDoH is considered before drawing conclusions about the value of spatial perspectives.

Defining Integrated Service Delivery

Integrated Service Delivery (ISD) is not a new concept. It has been referred to by many different names, including wrap-around, place-based, one stop shop, and joined up services (OECD, 2012), and most recently, family hubs (Honisett et al., 2023). These different terms, however, refer to approaches with key principles in common: families can access more than one service for child care, and there is some kind of connection between these services. There is an extensive body of research literature alongside policy documents focusing on ISD. Yet there is little diversity in approach within this work, which is overwhelmingly evaluative in nature and predominantly employs quantitative analyses (Roberts et al., 2014). Such evaluation is typically concerned with formal structural factors such as colocation and outcomes such as school readiness (Byron, 2010; H. M. Government, 2021; Melhuish, 2016; UK Government, 2013). While important, these foci overlook unfolding, localized features and outcomes of ISD that are accomplished in the moment, and often in less formal aspects of practice.

Although place-based initiatives have been around for a long time and there is a large body of literature to draw on, there is still much to learn about how to make place-based approaches best work for families (Harris et al., 2023). A new line of work has emerged that takes a broader view of what ISD can mean and its value. The idea of social connection or social cohesion and their relationship to well-being has come into play (Balenzano, 2020; Moore, 2021b). In one Australian study based on interviews and focus groups in eight communities identified as disadvantaged across six states, participants identified the opportunity to connect within their community as a priority (Tanton et al., 2021). A participant in their research expressed it in this way:

> *I think we need some sort of family and community services here, a connecting space, whether it's events or activities, but also helping the family that needs to be networked with something else. A linking place, a bump-in place* (Tanton et al., 2021, p. 194).

What this participant is valuing—social connection—is also recognized in research that shows that social connectedness can lead to longer life, better health, and improved well-being (Holt-Lundstad and Steptoe, 2022; Martino et al., 2015). Holt-Lundstad and Steptoe (2022) argue that social connection is an underappreciated determinant of physical health and that preventing social isolation can improve health. This participant is also identifying a link between providing social connection and providing services, which is what place-based integrated child and family services aim to do.

Recent studies, while mostly evaluative, are focusing on different things, such as the mental health of children and families (Honisett et al., 2022). Positive childhood experiences are now recognized as a counter to the risk of adverse childhood experiences leading to adult mental health issues (Bethell et al., 2019). At the heart of early childhood experiences are connections made in warm and caring environments. Understanding how parents experience service delivery by considering not just the services, but how they are delivered, is crucial (Bulling & Berg, 2018; Butler et al., 2020). This study contributes to understandings of how services are delivered by taking connection as a starting point. Traditionally, ISD has been seen as something that is formally delivered in offices, where colocation is regarded as the spatial key. In contrast, this study focuses on how shared public spaces and practices within them generate connections that can underpin ISD. The ISD created in this way reshapes the social determinants of health for these families.

Positioning the Study

In Australia, and elsewhere, there have been moves toward developing integrated services in place-based centers to address the needs of families with young children that are impacted through socioeconomic disadvantage (H. M. Government, 2021;

Honisett et al., 2023). Disadvantage in families is when families are at risk of adverse impacts from being exposed to multiple social and economic stressors. A prior study of persistent and multilayered disadvantage across Australian states found that disadvantage in Tasmania is concentrated in a small number of geographic locations (Tanton et al., 2021). Six of these locations account for 36% of the most under-resourced positions across all indicators. There are 37 indicators grouped into domains such as social distress, health, community safety, housing, education, lifetime disadvantage, and the environment. Significantly, there is evidence of multilayered disadvantage, as two of the six locations have 19 indicators in the top 5% of disadvantage.

Tasmanian CFLCs exemplify a widespread approach, offering integrated services for families with children under 5 years of age. From the outset CFLCs committed to an ethos of ISD (Department of Education (DoE), 2011). Services available include child health nurses, early education teachers, psychologists, and speech pathologists. However, three factors set these centers apart and make them distinctive:

1. Investment in establishing community support prior to opening. Typically, this involved enabling a group drawn from the local community to make decisions about the building and 12 months of staff involvement with community activities and running playgroups in existing places (Moore, 2021b; Prichard et al., 2015).
2. Innovative practices that facilitate needs-based situational responses are encouraged, rather than the usual known approaches (Hopwood, 2018).
3. Ongoing high level investment by successive governments. In 2009, the Tasmanian Government announced the establishment of child and family centers (now CFLCs). Twelve CFLCs opened around the State between 2011 and 2014. There are now 15 operational CFLCs around Tasmania, with three CFLCs due to open in 2024.

Research Approach

This study was conducted in three Tasmanian CFLCs, each in locations identified as highly disadvantaged. CFLC study sites were determined/enrolled in discussion with the Education Department who manage them.

To create new, finely detailed understandings of how ISD is accomplished in the shared public spaces within CFLCs, design decisions prioritized the generation of granular data linked to specific instances, for which an ethnographic approach was appropriate. Data were generated through 120 hours of participant observations in shared public spaces and 40 semi-structured interviews with staff, volunteers, and parents. All participation was subject to informed consent. In some instances, specific details could not be reported so as to protect participants' confidentiality, hence our focus here is on key patterns that were found across all three CFLC study sites. To ground the study in theory, we integrate a spatialized conceptual model that upholds a granular, as-it-happens approach, drawing specifically on Massey (2005). Massey (2005) argues that space is an "intersection of a multiplicity of trajectories" (p. 113). The notion of intersecting trajectories underpins the examination of how connections were made and extends to the analysis of their importance and relevance to ISD. Massey refers to space as *a coming together of stories thus far*, meaning that space is always open-ended and never finished, but also that connections have histories and backgrounds in their movement from what was to what is coming to be. This is highly relevant to the ongoing, never-finished work of supporting young children affected by disadvantage. It also disrupts dominant ways of thinking about ISD, focusing less on formal structures and outcomes and more on sometimes fleeting but nonetheless significant interactions that emerge in the vagaries and contingencies of day-to-day practice.

Shared Public Spaces as Produced

A spatial practice perspective understands any space, including shared public spaces in integrated service centers, as produced rather than given. Their spatial-social characteristics reflect what people do rather than being fixed by the container in which practices happen. Prior research has identified that these public shared spaces are perceived as safe and comfortable by parents (Jose et al., 2019; Prichard et al., 2015). In interviews, parents expressed their willingness to come to these spaces because they felt comfortable in them. Parents were enthusiastic about being able to "drop in" to these spaces. As one parent explained:

> [W]e can be having a bad day, it doesn't matter what state we are in, I don't have to change my clothes, I can just rock up with tired grumpy kids. The kids can play and I can take breath, sit down and have a cup of coffee.

This was a typical comment from parents who valued the drop-in nature of the informal space and felt that they could turn up at their worst. Some parents whose children had specific needs and who were nervous about how their children's behavior would be perceived appreciated the relaxed nature of the space. One mother, who came with three young children, spent several hours three times a week in the shared informal space expressed her appreciation for the space by responding in this way to a question about what she would like to improve:

> [T]here's nothing different that I need to change for me, and that's why I continue to come. If I go somewhere and I struggle with the kids because there might be something that's going to set them off, I don't continue to go because I can't cope in that situation cause I struggle myself with anxiety. However, here I am just 100% comfortable with knowing that everything's safe and everything's fun for my kids and they love coming and there's not a thing I would change not a thing.

Another parent disclosed that she had mental health issues that rendered her dysfunctional, and so her house was messy, and she enjoyed coming to a clean, comfortable space. Other parents mentioned that they were living in temporary accommodations unsuitable for children and wanted their children to be able to play freely. These findings align with increasing recognition that for many families, housing conditions, which have long been recognized as a major social determinant of health, have become more problematic. In inadequate housing conditions, access to attractive, free spaces within the community is crucial for family well-being (Joseph et al., 2023).

In these comments from parents, the qualities to which they refer should not be seen as innate, but rather produced through staff engaging in practices that have place-based effects. This perspective invites questions about the professional practices involved in these spaces, what is distinctive about them, how they become entangled, and how their enactment unfolds.

Three Key Practices

Three practices were identified as commonly enacted in shared public spaces in the CFLCs to produce safe spaces of warm connection: hanging out, consuming, and negotiating. Each is considered in turn below, followed by an explanation of how these practices led to depth in the moment.

Hanging out

Hanging out can be characterized as a casual presence with purpose. In our study findings, noticing, conversations, and modeling (especially interactions with children) were all part of the hanging out practices. Skillful approaches to conversation produced space as safe and rich with connection. Usually, staff members began chatting informally with families about neutral things such as the weather and traffic. Sometimes the conversation did not progress beyond that, but at other times, it developed into more personal conversations about family matters. These were spontaneous, informal interactions, yet they had a purpose.

In the extract below, a staff member explains her approach to chatting as part of *hanging out* practices in these spaces. She points to the importance of acknowledging the presence of families:

> *Therefore, it's about being available. Therefore, it truly is, just, just acknowledging our community, so whoever's here at the centre with their kids, it's saying hello. It's greeting them. Again, you're trying to remember last conversations.*

She then describes the strategy of engaging families in neutral conversations:

> *It's asking them what they had for breakfast that morning. Therefore, there are big things that you obviously want to discuss and check in with the families, but I think it's also important to do the smaller... [J]ust that everyday conversation is how I would talk to my colleague, how I would talk to a friend. How would I talk to another mum?*

Then she explains the importance of not broaching tricky topics too soon:

> *Therefore, I think when you have that balance of tricky but also, you know, you can keep things at that… you know, because, sometimes as well, I think, when our families, when they're in that tricky spot, I think sometimes you can actually escalate a problem by going straight in. Therefore, it might not need that kind of attention. Therefore, you can start at a lower, always start at a lower base, and build on that, I truly do try and keep things simple.*

The kind of conversations and connections resulting from hanging out in shared public spaces were different from those that might happen in an office setting. Hanging out had a distinctive purpose and value in the production of shared public spaces as key sites at which ISD was accomplished.

Consuming

Consuming practices refer not just to the physical act of eating but more broadly to the preparation, consumption, and sharing of food and drink. Furthermore, practices around food were often combined with other practices, such as story-reading sessions. One center leader described consuming practices as "a hook" to encourage families to come to the center. Observations and interviews confirmed this, as well as other significant contributions consuming practices made to the production of shared public spaces and to ISD as accomplished in the CFLCs. For example, routine weekly food bank deliveries enabled the development of connections over time. In one center, a family's involvement with the center began with the father visiting weekly to pick up bread. Initially, only eye contact was made with staff. Gradually, over weeks and months, conversations started around the food bank but then moved elsewhere, developing from neutral ones to discussion of more personal matters and eventually disclosure of problems that the family wanted help with. Ultimately, the rest of the family came and spent time in the center.

In another center, staff noticed that a father who only ever came to pick up food and did not engage with staff was looking for fresh bread. There was no fresh bread left, and so the center leader told the center assistant to get bread from the freezer and give it to him. The center leader explained that "we need to take the opportunity whenever we can to do something which will give us a connection, a way into a family." These are two of the many examples that show how consuming practices were pivotal in engaging families gently and effectively into the wider suite of more formal services offered through the CFLCs. The operation of the food bank through the shared public space provided a soft, gradual entry into the more formal aspects, an entry that could happen at a pace determined by the family.

Negotiating

Negotiating practices refer to how expectations of appropriate behavior were not simply enforced according to a fixed set of rules. There were rules and norms around the use of language, physical force and throwing of objects, and the consumption of food in certain places (such as sandpits). However, moments where behavior broke away from these norms triggered nuanced, emergent practices that negotiated the contingencies of situations and circumstances. This might involve anticipating things that could become problematic, making judgments to ignore minor things in order to avoid parents feeling overly watched or judged, or working with families to find ways out of challenging situations (e.g., when parents might use foul language in front of children, or a child might aggressively rip a toy out of another child's hands, or when conflict arose between parents).

Managing the shared public space in this way enabled the other two practices of hanging out and consuming to take place in a safe environment. Safety here refers not just to physical comfort or avoiding risk. In line with prior research, the safety produced by negotiating practices was about ensuring families did not feel judged, something that they can feel very acutely in shared, public spaces, especially if their children behave in challenging ways (Boag-Munroe & Evangelou, 2012; The Southern initiative and the Co-Design Lab, 2016). One mother expressed her appreciation in this way:

> Um, and I love coming here and not feeling judged in the slightest for anything. You know, I'm breastfeeding my baby right now and I never feel judged for that. Even from the girls and from all the parents, so yeah.

Practices enacted in a three-step process transformed what might be experienced as judgmental correction or "telling off" into an opportunity to connect and support families, while also producing broader safety crucial to making the space so comfortable and valuable for families. The first step involved efforts to notice possible issues early before a problem developed; the second involved offering practical help in the moment (such as assisting to calm a frustrated child or modeling calm assertive behavior management); the third involved staff staying with parents after an incident so that their attention was not defined only by an immediate problem but rather endured into calmer, positive moments where a different connection could be established.

Place-Based Practices

Following the description of the fine-grained ethnographic, spatial practice approach used in our study, we now look in detail at practices in each of the three sites within the shared public space. In each of these sites, two of the three key practices were enacted:

- **Sandpits**: hanging out and negotiating
- **Kitchens**: consuming and hanging out
- **Open offices**: negotiating and consuming

Sandpit Sites

At sandpit sites, connections resulted from intersections of trajectories that were typically spontaneous and unplanned. Connections arose between parents and workers and sometimes between parents from different families. These were sites of free play for children, and unsurprisingly, unexpected incidents could occur, which were observed and responded to through hanging out and negotiating practices. There were repeating triggers from which trajectories came together, namely a child throwing sand, conflict between children, and a child distressed or having a tantrum.

These triggers were common across sandpit sites in all three research locations. Although the resulting connections were unplanned and spontaneous, they had depth. Hanging out practices led to triggers being noticed and often informal conversations being initiated. When needed, negotiating practices (following the three steps discussed above) ensured that connections with parents were supportive and not just corrective in nature.

Negotiating practices sometimes involved strategic ambivalence, that is, when staff noticed something but chose to monitor rather than actively intervene. If the situation developed into an incident that needed a response, staff were able to move in quickly and offer support. Sometimes physical trajectories of moving toward the sandpit to model something or speak to parents came into play. When a child threw sand, the physical trajectory of staff moving within the space would intersect with parents going from struggling with a child's behavior to (supported by and connected with staff) feeling in control of the situation. Depth in the moment here depended on the experience and skill of the staff, who disrupted a trajectory of deteriorating relations between parent and child. After the situation had calmed down, they would stay and chat with their parents. It was through precisely such practices that trajectories of parents moving from not knowing staff to knowing staff and perhaps moving toward a position of trust were accomplished.

Negotiating practices could produce trajectories that led to other sites such as kitchens, transforming a need to manage behaviors into an opportunity to connect. For example, children were not allowed to eat in the sandpits. Instead of just being asked or told to stop, staff would offer to go with a parent and child to somewhere near the kitchen. This prompted not just a movement away from the sandpit but led to the sharing of food. In this way, what might have been a simple matter of reminding parents of rules produced moments of connection with depth.

Kitchen Sites

At kitchen sites, the combination of hanging out and consuming practices created intersections of trajectories, which in turn produced connections with depth. In contrast to the spontaneous connections made in the sandpit, connections made in the kitchens were typically planned and occurred routinely in three ways: cooking classes; staff taking breaks and lunch; and the routine provision of free food (i.e., food bank). For example, the routine of staff having coffee just before the center opened created connections with depth. Permanent staff chatted with visiting staff. This was part of a staff epistemic trajectory (Table 8.1), which came about not from primary contact with families but from what other staff had noticed, perhaps while hanging out. This was particularly relevant when there was a new family with a

Table 8.1 Examples of trajectories in shared public spaces of CFLCs

Kind of trajectory	From	To	Example
Physical	A site	Another site	Moving from kitchen to sandpit
Social	People on their own	People spending time with other people	People chatting together
Epistemic	Unknowing	New ways of knowing	People coming to know about X
Obstructed	Being obstructed by a problem, confidence, or negative experience	Problem, confidence, or experience addressed	People can move on or move on differently

child who was showing signs of delayed development in speech or movement. Here, the depth stems from the knowledge permanent staff gained from observing play in a naturalistic environment or chatting to parents and the professional expertise of the visiting staff member. Four kinds of trajectories were observed – physical, social, epistemic and obstructed (Table 8.1). These trajectories and the resulting depth could occur over several visits.

It is important to note how a Masseyan approach to understanding space also brings temporality into focus. From Massey's perspective, space is not what is left if we freeze time but is rather charged with temporality. This temporality is tied not to duration but rather to movement in the stories thus far that are extended as trajectories intersect. Momentary connections around kitchens in shared public spaces involved intersections of both immediate trajectories of movements through or changing activities in a center but also longer-term ones, of changing relationships, building trust, and professionals' deeper understanding of what mattered to families and the support they needed. Thus, consuming practices around the food bank one week could add depth to similar moments a week later.

Open Office Sites

Consuming and negotiating practices were predominant in the open office sites. The consumption practices here were different from those in kitchens and produced different intersections of trajectories. These intersections still created depth in the moment, often relating to the diverse perspectives that different members of staff and visitors brought to bear. In one center, staff intensified the coming together of trajectories around cooking by scheduling visiting professionals' sessions to coincide with cooking classes. Visiting staff from diverse services could come into the kitchen space, leading to chats about cooking, which often led to chats about other things. Parents who had come to the center for a cooking class would find themselves talking about things such as a concern about their child or how they were feeling stressed. The depth in these incidental conversational moments came from the fact that they were talking to people who often were familiar with these types of problems and could begin to offer pathways to help. Therefore, a parent obstructed trajectory (Table 8.1) could intersect with a staff epistemic trajectory, as parents started to find ways to address problems and staff learned more about the families' situation. Thus, planning cooking lessons around visiting professionals' schedules led to moments with depth as staff and families chatted. Negotiating practices in open offices often involved staff coming together to share information about incidents as they unfolded. When situations needed a quick response, these intersections created depth, enabling decisions to be made that considered first-hand observation and diverse professional expertise.

Intersecting Trajectories and Depth in the Moment

Hanging out, consuming, and negotiating practices created trajectories that intersected, producing moments that, although ephemeral, had significance and depth. Trajectories here refer not only to concrete movements but also, in a Masseyan spirit, to complex movements, such as those related to knowledge and emotional positions, as presented in Table 8.1.

The trajectories shown above operated on different temporalities. Longer-term trajectories, on which movement was slower and often iterative, included children's behavior, sleep and health, or families' struggles with finances or navigating government organizations such as Centrelink (Australian government organization responsible for welfare payments). Others were more acute situations requiring rapid response, such as family conflict. Some combined both immediate and longer-term movements, such as responses to reported domestic violence.

Intersections of different kinds of trajectories produced meaningful connections. Staff moving physically into the shared space for a break might intersect with a family in a social trajectory. Some intersections depended on others, such as when parents were hesitant to connect with staff but became comfortable doing so once trajectories connecting them with other parents had been established. This was particularly important for parents who mistrusted services and were reluctant to engage with staff (Prichard, 2018). The staff epistemic trajectory could involve staff connecting with parents themselves, or it could come from interaction with other staff who had gained an understanding of a family.

We identified patterns of intersection of trajectories at different sites within shared public spaces. Focusing on sandpits, kitchens, and open office areas highlights important features of these patterns, where trajectories came together to produce connections with depth. These moments with depth were not an endpoint but were part of a continuing story; advancing the work of supporting parents; brokering access across the suite of services delivered; and enabling staff to offer the holistic support that ISD seeks to accomplish. While some connections were planned and others were spontaneous, the combination of planned and emergent connections was crucial to realizing the potential of the shared public space.

Social Determinants of Health Reshaped Through Connections and Depth

Given the current high level of interest in place-based centers offering ISD, it is appropriate to offer fresh ways of understanding how integrated services can be provided. Several studies and policy documents have expressed concern about how COVID-19 exposed a failure to address the SDoH (Marmot et al., 2020) and the resulting inequity (Marmot, 2020; Marmot et al., 2021; Honisett et al., 2023). Early childhood features prominently in these concerns, and place-based approaches feature strongly in responses, including proposed family hubs in Australia and the UK. The present study offers important new understandings of how place-based approaches might address SDoH, specifically revealing how shared public spaces can be produced through informal practices (hanging out, consuming, negotiating) as spaces of connection that underpin families' access to support from more formal services.

Understanding ISD from a spatial perspective offers an alternative to the more common structural focus. This study suggests that informal spaces need to be taken seriously as places where significant work is done. These are not merely waiting spaces but can be produced as spaces where connections with depth abound, contributing to the wider accomplishment of ISD.

Moments with depth arise from the intersection of trajectories, the coming together of stories thus far. Such an understanding links more general ideas of place-based services and colocation with a more fine-grained account of how spaces are produced in particular ways. First, achieving depth in the moment requires sustained and often iterative engagement. Foregrounding the moment does not suggest isolated, fleeting encounters in which problems are solved. Rather, our analysis highlights how depth in one moment often depends on connections made through prior intersections of trajectories. This is of relevance for families who are "hard to reach." With these families, outreach can help to encourage a first visit, but the real challenge lies in sustaining engagement (Boag-Munroe & Evangelou, 2012). Depth in one moment influences subsequent connections and creates conditions for future moments of depth. Massey explains it this way:

> That tree which blows now in the wind out there beyond the train window was once an acorn on another tree, will one day hence be gone. That field of yellow oil-seed flower, product of fertilizer and European subsidy, is a moment—significant but passing—in a chain of industrialized agricultural production (Massey, 2005, p. 119).

In the context of place-based services and their role in addressing SDoH, the "chains" are those of connection between professionals and families, between families and services, and between families and others in their community. These connections are always part of something that remains under construction. Construction need not be continuous, nor unbroken. Trajectories of family engagement with centers might be disrupted, temporarily severed, and then repaired. They may deepen at different paces and along different fronts. The spatial practice perspective accommodates this complexity in the reshaping of SDoH in a way that is much more difficult if the focus is on more formal structures or bound provisions such as interventions with fixed beginnings and ends.

Second, one of the inherent problems in addressing SDoH are the tensions between the desire to improve things and wariness about disempowering people by solving problems on their behalf, or driving people away from services when they feel judged and professional agendas determining what is done rather than what matters to families. The moments of depth documented here happen through the intersection of trajectories that arise in the production of informal, shared public spaces. Practices of consuming, hanging out, and negotiating provide a healthy balance between the planned (e.g., cooking classes, play activities) and the spontaneous, which are both highly conducive to connections that develop at a pace set by parents but can equally respond to the immediacy of the moment. The tension between the urge to fix the problem and the need to avoid taking over might be alleviated by redirecting "professionals'" attention from problems needing fixing to finding ways to produce spaces in a way that promotes connections with depth in the moment.

Third, building on the prior points, the spatial perspective adopted here reveals the significance of seemingly small, mundane acts. Eye contact when a parent picks up free food; sharing a cup of coffee; the response when a child throws sand—these and other acts are far from trivial when they produce shared, public spaces as safe (nonjudgmental) but also rich with connections. This redirects evaluation away from measurables such as attendance at formal programs, frequency, and duration of visits, and instead towards the depth that can be produced within and across moments. Such moments often occur in the enactment of practices that might otherwise be regarded as low value, incidental, or even (especially in the case of hanging out), dead time that should be filled with formal appointments. Shifting from thinking about the SDoH to the social-spatial determinants of health offers a granular, in-the-moment view of how to move the needle on children's health and development outcomes: small things are revealed to have large effects on the unfolding, never-finished accomplishment of ISD. Collectively, these key points suggest ways to sharpen practices and strengthen access to and links between services without the need for expensive structural redesign but rather by recognizing the value of producing shared, public spaces in particular ways.

Conclusion

This study offers a fresh understanding of the social-spatial determinants of health by investigating how ISD can be accomplished in shared public spaces in place-based or co-located services. It identifies three practices that can produce spaces that are safe and rich in intersecting trajectories that generate connections with depth in the moment. These moments were significant in both being attuned to the moment and also underpinning the broader unfolding of ISD in responsive, patient ways that met family needs. Identifying these moments with depth can play a part in learning how to reshape the social-spatial determinants of health for families with young children.

This study illustrates that adopting a spatial approach offers new ways of thinking that can contribute to effective ISD. The focus of our study has been on services supporting families with young children. The principles we described, however, of foregrounding practices in shared spaces, rather than on formal provisions behind closed doors, and being ready and able to recognize the value that accrues from informal practices (planned and spontaneous) in creating relevant connections with depth can be taken up in diverse contexts. These contexts include disability services and aged care provision, where social connection is also an important social determinant of health. The Masseyan foundation, viewing space as the intersection of trajectories and tracing the coming together of stories thus far, brings us up close to practices that drive and reshape the social-spatial determinants of health.

Acknowledgments The authors acknowledge the Palawa and Pakana people of Lutruwita, upon whose ancestral lands the Child and Family Learning Centres now stand, and the Gadigal people of the Eora Nation, upon whose ancestral lands the University of Technology, Sydney, now stands. The authors pay respect to the elders past and present as traditional custodians of knowledge in these places. Sincere thanks are offered to the families, staff, and volunteers of the CFLCs involved.

References

Australian Early Development Census (AEDC). (2021). *Australian early development census national report 2021*. Commonwealth of Australia.

Balenzano, C. (2020). Promoting family well-being and social cohesion: The networking and relational approach of an innovative welfare system in the Italian context. *Child and Family Social Work, 26*(1), 100–110. https://doi.org/10.1111/cfs.12793

Bethell, C., Jones, J., & Gombojav, N. (2019). Positive childhood experiences and adult mental and relational health in a statewide sample: Associations across adverse childhood experience levels. *JAMA Pediatrics, 173*(11), 2–10. https://doi.org/10.1001/jamapediatrics.2019.3007

Boag-Munroe, G., & Evangelou, M. (2012). From hard to reach to how to reach: A systematic review of the literature on hard-to-reach families. *Research Papers in Education, 27*(2), 209–239. https://doi.org/10.1080/02671522.2010.509515

Bulling, I., & Berg, B. (2018). "It is our Children": Exploring intersectoral collaboration in family centres. *Child & Family Social Work, 23*, 726–734. https://doi.org/10.1111/cfs.12469

Butler, J., Gregg, L., Calam, R., & Wittkowski, A. (2020). Parents' perceptions and experiences of parenting programmes: A systematic review and meta synthesis of the qualitative literature. *Clinical Child and Family Psychology Review, 23*(2), 176–204. https://doi.org/10.1007/s10567-019-00307-y

Byron, I. (2010). Place-based approaches to addressing disadvantage: Linking science and policy. *Family Matters, 84*, 20–22.

Department of Education (DoE). (2011). *Tasmania's Child and Family Centres (CFCs) initiative: CFC statewide outcomes framework*. State of Tasmania Department of Education.

Glass, N. (1999). Sure start: The Development of an early intervention programme for young children in the United Kingdom. *Children and Society, 13*, 257–264. https://doi.org/10.1002/CHI569

Goldfield, S., O'Connor, M., Cloney, D., Gray, S., Redmond, G., Badland, H., Williams, K., Mensah, F., Woolfenden, S., Kvalsvig, A., & Kochanoff, A. (2018). Understanding child disadvantage from a social determinants perspective. *Journal of Epidemiology & Community Health, 72*(3), 223–229. https://doi.org/10.1136/jech-2017-209036

H. M. Government. (2021). *The best start for life: A vision for the 1001 critical days.* The Early Years Healthy Development Review Report.

Harris, D., Cann, R., Dakin, P., & Narayanan, S. (2023). *Place-based initiatives in Australia: An overview.* ARACY.

Hertzman, C. (2010). Framework for the social determinants of early childhood development. In R. E. Tremblay, M. Bolvin, & R. V. Peters (Eds.), *Encyclopedia on early child development.* University of Montreal.

Hines, J. (2017). An overview of head start programme studies. *Journal of Instructional Pedagogies, 18*, 1–10.

Holt-Lundstad, J. (2022). Social connection as a public health issue: The evidence and a systemic framework for prioritizing the "social" in social determinants of health. *Annual Review Public Health, 43*, 92–213. https://doi.org/10.1146/annurev-publhealth-052020-110732

Holt-Lundtsad, J., & Steptoe, A. (2022). Social isolation: An underappreciated determinant of physical health. *Current Opinion Psychology, 43*, 232–237. https://doi.org/10.1016/j.copsyc.2021.07.012

Honisett, S., Loftus, H., Hall, T., Sahle, B., Hiscock, H., & Goldfield, S. (2022). Do integrated hub models of care improve mental health outcomes for children experiencing adversity? A systematic review. *International Journal of Integrated Care, 22*(2), 24., 1-14. https://doi.org/10.5334/ijic.6425

Honisett, S., Cahill, R., Callard, I., Eapen, V., Eastwood, G. R., Graham, C., Heery, L., Hiscock, H., Hodgins, M., Hollands, A., Jose, K., Newcomb, D., O'Loughlin, Ostojic, K., Sydenham, E., Tayton, S., Woolfenden, S., & Goldfield, S. (2023). *Child and family hubs: An important front door for equitable support for families across Australia.* National Child and Family Hubs Network. https://doi.org/10.25374/MCRI.22031951

Hopwood, N. (2018). *Creating better futures: Report on Tasmania's child and family centres.* University of Technology Sydney.

Jose, K., Christensen, D., van de Lagewegen, W., & Taylor, C. (2019). Tasmanian child and family centres building parenting capability: A mixed method study. *Early Child Development and Care, 189*(14), 2360–2369. https://doi.org/10.1080/03004430.2018.1455035

Joseph, N., Burn, A. M., & Anderson, J. (2023). The impact of community engagement as a public health intervention to support the mental well-being of single mothers and children living under housing insecure conditions. *BMC Public Health*, 1–26. https://doi.org/10.1186/s12889-023-16668-7

Logan, D., Rubenstein, L., & Fry, R. (2018). *Place-based collective impact: An Australian response to childhood vulnerability.* Centre for Community Child Health.

Marmot, M. (2012). Fair society healthy lives. *Public Health, 126*(1), 4–10. https://doi.org/10.1016/j.puhe.2012.05.014

Marmot, M. (2020). Health equity in England: The Marmot review 10 years on. *British Medical Journal, 368*, m693. https://doi.org/10.1136/bmj.m693

Marmot, M., Allen, J., Goldbalt, P., Herd, E., & Morrison, J. (2020). *Build back fairer.* Retrieved from https://www.instituteofhealthequity.org/.../build-back-fairer.

Marmot, M., Marteau, T., & Rutter, H. (2021). Changing behavior: An essential component of tackling health inequalities. *British Medical Journal, 372*, n322. https://doi.org/10.1136/bmj.n332

Martino, J., Pegg, J., & Frates, E. P. (2015). The connection prescription: Using the power of social interactions and the deep desire for connectedness to empower health and wellness. *American Journal of Lifestyle Medicine, 11*(6), 466–475. https://doi.org/10.1177/1559827615608788

Massey, D. (2005). *For space.* Sage.

Melhuish, E. (2016). Longitudinal research and early years policy development in the UK. *International Journal of Childcare and Education Policy, 10*, 3. https://doi.org/10.1186/s40723-016-0019-1

Moore, T. G. (2021a). *Core care conditions for children and families: Implications for integrated child and family services*. Centre for Community Child Health. https://doi.org/10.25374/MCRI.14593878.v1

Moore, T. G. (2021b). *Developing holistic integrated early learning services for young children and families experiencing socioeconomic vulnerability*. Centrefor Community Child Health. https://doi.org/10.25374/MCRI.14593890

Moore, T. G., & Fry, R. (2011). *Place-based approaches to child and family services: A literature review*. Murdoch Children's Research Institute & The Royal Children's Hospital Centre for Community Child Health.

Moore, T. G., McDonald, M., Carlton, L., & O'Rourke, K. (2015). Early childhood development and the social determinants of health inequities. *Health Promotion International, 30*(2), ii102–ii115. https://doi.org/10.1093/heapro/dav031

OECD. (2012). *Joined up services*. Retrieved from www.oecd.org.

Press, F., Sumision, J., & Wong, S. (2010). Integrated early years provision in Australia: A research project for the professional support coordinators alliance. Retrieved from www.childautsralia.org.au

Prichard, P. (2018). *Transformations in parenting: New possibilities through peer-led interventions*. Western Sydney University.

Prichard, P., Purdon, S., & Chaplyn, J. (2010). *Moving forward together: A guide to support the integration of service delivery for children and families*. Murdoch Children's Research Institute.

Prichard, P., O'Byrne, M., & Jenkins, S. (2015). *Supporting Tasmania's Child and Family Centres: The journey of change through a learning and development strategy*. Tasmanian Early Years Foundation with the Centre for Community Child Health.

Roberts, J., Donkin, A., & Pillas, D. (2014). *Measuring what matters: A guide for children's centres*. Institute of Health Equity.

Tanton, R., Dare, L., Miranti, R., Vidyattama, Y., Yule, A., & McCabe, M. (2021). *Dropping off the edge 2021: Persistent and multilayered disadvantage in Australia*. Jesuit Social Services.

The Southern initiative and the Co-Design Lab. (2016). *Parents' experience of early years in South Auckland: Early years challenge*. Auckland Council. Retrieved from https://www.aucklandco-lab.nz/s/Parents-experience-of-early-years-in-south-auckland

UK Government. (2013). *Foundation years sure start centres: Fifth report of session 2013–2014*. Retrieved from www.publications.parliament.uk/educom.

World Health Organization. (2008). *Commission on social determinants of health. Closing the gap in a generation: Health equity through action on the social determinants of health*. World Health Organization.

Chapter 9
Design Guidelines for Safe Environment to Improve Aging in Place

Juliana Tasca Tissot, Lizandra Garcia Lupi Vergara, Widya A. Ramadhani, and Wendy A. Rogers

Introduction

Aging is a shared experience of all human beings. The process is long, progressive, and cumulative, impacting physiological aspects of human life (Kirkwood, 2017). The impact of aging is noticeable when it interferes with the functional performance and social relationships of older adults. Declining physical functions like mobility can impact older adults' everyday activities (Ramadhani & Rogers, 2022). The inability to perform activities of daily living can disrupt older adults' health and independence, which are among the important elements of aging in place (Rogers et al., 2020c).

Older adults engage in various everyday activities, categorized into three types. First, basic activities of daily living (ADLs) are essential for self-care (Katz et al., 1959). Examples of ADLs are eating, using the toilet, bathing, dressing, and ambulating. Second, instrumental activities of daily living (IADLs) include activities essential to one's independence to live in one's own home (Lawton & Brody, 1969), such as cooking, housekeeping, shopping, and managing medication. The last category is enhanced activities of daily living (EADLs). Activities in this category can contribute to happiness, well-being, and a sense of fulfillment, such as using technology, social, and cultural activities (Rogers et al., 2020b). The ability to independently conduct these everyday activities is critical for older adults to age successfully.

J. T. Tissot (✉)
Federal University of Pelotas, Pelotas, Rio Grande do Sul, Brazil

L. G. L. Vergara
Federal University of Santa Catarina, Florianópolis, Santa Catarina, Brazil

W. A. Ramadhani · W. A. Rogers
University of Illinois at Urbana-Champaign, Urbana, Illinois, USA

© The Author(s) 2026
M. A. Kolak, I. K. Moise (eds.), *Place and the Social-Spatial Determinants of Health*, Global Perspectives on Health Geography,
https://doi.org/10.1007/978-3-031-88463-4_9

137

Older adults employ various strategies to achieve successful aging, such as maintaining functional capacity, staying engaged, minimizing disabilities, and adapting to their changing needs (Rogers et al., 2020a; Rowe & Kahn, 1997, 2015). These strategies are grounded in the life course perspective of aging, which acknowledges that "change introduced at one stage of the life course may alter the needs and opportunities at other stages" (Rowe & Kahn, 2015, p. 595). Older adults' adaptive strategies to maintain independence can be behavioral, such as changing the method of doing things (Ramadhani, 2023). Strategies can also be environmental, such as modifying the home environment to minimize physical barriers (Ramadhani, 2023). Hence, it is critical to support older adults holistically so that they can be adaptive to the changes that happen across the life course, which include physical, behavioral, and social adaptations (Penney, 2013).

Holistic support for aging in place can be viewed through the lens of Social Determinants of Health (SDoH) (U.S. Department of Health and Human Services, n.d.). The five domains of SDoH—economic stability, education access and quality, healthcare access and quality, social and community context, neighborhood and built environment—are also conditions that impact the health, functioning, and quality of life of older adults who are aging in place (Fig. 9.1). Maintaining independence to do various everyday activities requires multidimensional support, especially from the neighborhood and built environment, as the physical context wherein older adults are aging in place (Mois & Rogers, 2023). Some objectives of this domain include housing and homes, injury prevention, and ensuring that people with disabilities have equal access to spaces (U.S. Department of Health and Human Services, n.d.). In this chapter, we focus on understanding better ways to design safe

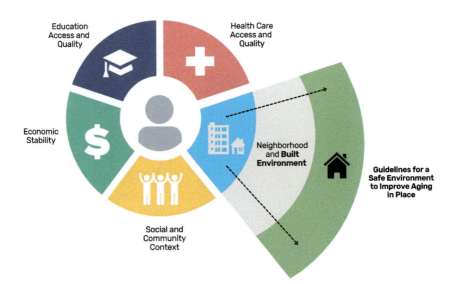

Fig. 9.1 Approach between Guidelines for a Safe Environment to Improve Aging in Place and the SDoH framework. *Source:* Healthy People 2030, as adapted by authors in 2023

home environments for older adults that can reduce the risk of falls and injuries, hence ensuring autonomy and independence in the engagement of various everyday activities.

Designing Safe Housing for Older Adults

The built environment can influence behavior and habits and directly impact physical health (Pinter-Wollman et al., 2018). For older adults, who spend most of their time at home, residential environments are critical to their self-image, identity, independence, and well-being (Tomazzoni, 2011). Unfortunately, accidents at home are also prominent, especially for older adults, whose risks of falls increase with age (Mack et al., 2013). Accidents are the fifth leading cause of death among older adults, with falls accounting for two-thirds of accidental deaths (Lockhart, 2007). When older adults' physical and cognitive abilities are reduced, they need greater support from the built environment to maintain their independence (Fornara et al., 2019). If not supported appropriately, elements of the built environment can become barriers that limit movement, impair perceptions, and cause difficulties in performing daily activities.

The housing design for older adults must adhere to different environmental characteristics, promote independence, and simultaneously transmit a feeling of comfort and safety (Tissot, 2022). The latter—comfort and safety—is grounded in usability and ease of use, also known as ergonomics. Ergonomics addresses postures, body movements, environmental adaptations, perception, and tasks. All these elements, properly combined, create a safe, healthy, and comfortable environment for carrying out everyday activities efficiently and productively (Hazin, 2012). One type of ergonomics is physical ergonomics, which is usually assessed through anthropometric measurements and observations of bodily movements as essential information to determine furniture's depths, heights, and sizes (Daré, 2010). Observing, evaluating, and analyzing the users' difficulties and skills is fundamental to aligning usability requirements with spatial accessibility guidelines, making a harmonious interaction, and ensuring understanding with a perceptive and cognitive approach (Staut, 2014).

It is imperative to design housing with older adults' needs and daily routines in mind. This can be done by understanding activities performed in different spaces of the house. Each activity has tasks or goals to achieve. To carry out a given task, certain means and conditions are needed (Moraes & Mont'alvão 2009). For instance, to dress, there should be enough space for bodily movements, appropriate lighting, and adequate space to hang or put on clothes and other objects. It is also important to note that there is no single way to conduct an activity, as everyone has their habits and behaviors. Therefore, observing the interaction between the person and the environment is important when performing various activities to inform the design of the built environments (Ferreira Filho, 2018).

Making health an explicit component of the planning of physical spaces is fundamental. The built environment must prevent injury and enhance older adults' health and well-being by supporting them in living a healthy, independent, and fulfilling life. We aimed to develop design guidelines for safe housing environments for older adults. The guidelines were developed with the belief that each space must have specific requirements to accommodate diverse everyday activities to promote autonomy and well-being to aging in place.

Risk Factors for Falls in Home Environments

The prevalence of falls has become one of the global health crises. Among other population groups, older adults are at the greatest risk for falls (Nicolussi et al., 2012). Older adults are more exposed to fall risks due to the age-related decline in physical and cognitive abilities (Albuquerque et al., 2018; Staut, 2014). The risk is even higher for inactive older adults, as they are frailer (Sociedade Brasileira De Geriatria E Gerontologia, 2008). The risk of falls also increases for those with a first fall (Antes et al., 2013). Fear of falling is another risk factor that can reduce one's independence and engagement in ADLs to avoid the risk of falls (Lojudice et al., 2010).

There are different fall risk factors, which the World Health Organization classified into four dimensions: biological factors (i.e., age, gender, race, chronic diseases, and physical or cognitive decline), behavioral factors (i.e., use of multiple medications, excessive alcohol intake, or lack of physical activity), environmental factors (i.e., slippery floors and stairs, loose rugs, insufficient lighting, and uneven sidewalks), and socioeconomic factors (i.e., low income, no access to education, access to adequate housing, community services, and social interaction).

Most falls occur in older adults' homes, for instance, bedrooms and bathrooms (Pohl et al., 2015). Factors that contribute to fall incidents are carpets, wet floors, changing furniture arrangements, poor lighting, and stairs (Rodrigues et al., 2014). Additionally, environmental barriers such as lack of grab bars in critical places (i.e., bathrooms), hard-to-reach light switches, floors with height changes, and furniture that obstructs main passages can increase the fall risks of older adults (Neves, 2017). Thus, we must consider which factors in a physical environment will contribute to older adults' safety, as barriers are likely present in the home environment (Hazin, 2012).

Investment in creating and applying design guidelines for safe home environments is crucial, even if they increase the final cost of construction, as they can compensate for future healthcare expenses for medication and care support resulting from fall incidents (Mendes & Côrte, 2009). Currently, there are several instruments, tools, and checklists available to assess and minimize the risks of falls; however, such tools still need to more carefully consider the physical spaces and the activities of users (Nicklett et al., 2017; Fernandes et al., 2007). Therefore, strategies to minimize fall risks should begin by informing the planning and design of the home environment, taking into account the characteristics and needs of older adults. Tools and guidelines are important not only to reduce fall risks and increase safety, but also to support older adults in maintaining their autonomy and independence at home (Tischa et al., 2017).

Design Guideline Development Protocol

We conducted a three-phased approach to develop the design guidelines: (a) literature review, (b) McKechnie Family LIFE Home (in short: LIFE Home) walkthrough and focus groups, and (c) survey with specialists. In the first phase, we conducted a literature review from articles, dissertations, and public policies available on a scientific data basis (SCOPUS, SCIELO, and EBSCO) and official government websites to better understand older adults' needs, aging in place, and the methodologies created to evaluate homes for older adults. The second phase was conducted at the LIFE Home, a research center at the University of Illinois at Urbana, IL, that was designed to mimic the design and layout of a typical home environment.

The research involved four focus groups with a total of 14 participants (i.e., older adults and professionals, including designers and architects) and walkthroughs of LIFE Home environments—a front door, dining and living room, kitchen, laundry, bathroom, bedroom, home office, and garage. Walkthrough helped research participants get to know the place of study while identifying the physical features of the space that might create challenges for older adults to carry out their ADLs.

After the walkthrough, participants were asked to identify activity challenges in various house spaces and then propose possible solutions to be discussed with other participants during the focus group session. Finally, in the third phase, we surveyed 33 built environment specialists (e.g., architects, interior designers, and engineers from Brazil) to identify the importance of physical features of safe home environments compared to the two previous phases to develop the design guidelines for safe housing for older adults.

The three-phased approach contributed to defining all the features related to physical environments that promote active and healthy aging at home and support aging in place. Furthermore, these features were evaluated based on the level of importance from the three-phased approach and contributed to the development of security levels that will be presented later.

Rating Safe Environment Design

Based on the data from the three-phase approach, we rated all the features associated with the physical environment according to the Gravity Urgency Tendency (GUT) Matrix, which is a very useful and important tool developed by Kepner and Tregoe in 1980 to set priorities through gravity, urgency, and tendency (Cesar, 2013). Gravity means the impact of a problem or situation on those involved, including people or processes. Urgency relates to the time or deadline available to solve this problem. If something is very urgent, the resolution deadline should be shorter. Tendency means the problem's likelihood of worsening over time if nothing is done. This research assigns a score from 1 to 5 for each environmental feature or factor, where 1 relates to the lowest severity and 5 to the highest severity (Cesar, 2013).

The priority of each feature was defined according to the three-phase approach which means that more times the features were identified in phase (a) literature review, or mentioned in phase (b) LIFE Home walkthrough and focus groups, and evaluated in phase (c) survey with specialists, the greater its importance for the environment. Furthermore, the scoring scale was defined by authors as 0–24 points, priority level 3 or low; 25–59 points, priority level 2 or medium; and 60–125 points, priority level 1 or high. Thus, if a feature has a high weight, such as 125, it means that its prioritization should be considered number one or high. The value of 125 is obtained by multiplying the severity, urgency, and trend factors. This result indicates that the item has an extremely serious risk, which requires immediate action, and that, if not implemented, the risk can rapidly deteriorate. In this sense, priority 1 is associated with minimum security, as it is understood that if the item is of serious risk, it must be considered basic to guarantee the safety of users.

The GUT Matrix score considered the recurrence of features in the literature review (phase a), LIFE Home walkthrough and focus groups (phase b), and results from the survey with specialists (phase c).

Environmental Design Guidelines for Safe Homes

Environmental design guidelines for safe homes can be used by architects and interior designers planning and developing home projects or anyone who wants to improve their homes. The objective was to provide information about structural changes, regulations, equipment, assistive technologies, constructive materials, furniture, and maintenance that can be implemented or inserted in the housing environment to contribute to older adults' ADLs safely.

Environment Design Guidelines scores from minimum, medium to highest safe homes are distributed in levels of priority for each category's recommendation, according to the GUT Matrix, as shown in examples from Tables 9.1, 9.2, and 9.3, explained below.

Tables 9.1, 9.2 and 9.3 present a summary of the Environmental Design Guidelines. The first column refers to the *Design Categories*. Each design guideline belongs to a type of change or home adaptation. The second column describes all the design guidelines. Columns three to five are points from gravity, urgency, and tendency. The GUT Matrix score is from 1 to 5 and considers the frequency of features in the three-phase approach. The sixth column (score) shows the final score for each design guideline. The score results from the multiplication between gravity, urgency, and tendency. The seventh (front door) to fourteenth (garage) columns represent the housing environment. The marks represent where each feature must be implemented in home spaces. For example, in Table 9.1, Grab bars/Handrails must be in all the home spaces. Table 9.4 presents icons for each category (e.g., structural changes, regulations, etc.).

In this book chapter, only design guidelines for the kitchen and bathroom will be detailed as they were the rooms that had the most design guidelines marked, totaling 11 and 10 guidelines in the three priority levels. The complete design guidelines[1] state that all spaces have 68 features, most of which are related to equipment. Approximately 33.8% referred to equipment, including bars, accessible switches, and lights. Furniture accounted for 19.1%, such as sofas, chairs, and bed types. Constructive materials, such as flooring and colors, accounted for 14.7% of recurrency from design companies. The rest were assistive technologies (11.8%), structural changes (10.3%), maintenance (8.8%), and regulations (1.5%). It is necessary to pay attention to and research the relationship between architecture and aging since our prevalence indicates that most housing configurations for a safe environment for older adults can be implemented during projects because of concerns about architecture and design areas.

Table 9.5 represents a summary of features for the kitchen, based on the level of security.

Table 9.5 presents a summary of features of the kitchen, where basic, instrumental, and EADLs are performed. To achieve minimum safety, emergency lights are needed to help users find their way around and avoid accidents if a power outage occurs. Also, artificial lighting should be positioned and sufficient to perform the tasks. Grab bars should also be available to help with movements. Since food is being handled and cooked in the kitchen, installing smoke and gas sensors would be important. Color contrast is another feature that can be implemented to help identify barriers. Floors, walls, and furniture must have contrasting colors. Designers should avoid using similar colors for floors, walls, and furniture. The floor covering must use non-slip materials.

Moreover, countertops and cupboard heights should consider the user's height and posture. Many professionals or custom furniture stores often work with standard measurements, which may not be suitable for users with physical limitations. User-centered design must be practiced, and measurements and ranges must be considered when specifying the size of kitchen furniture and equipment size. In addition, drawers are more indicated than cabinets, as they require greater movement and effort to access the shelves internally. Rugs should also be avoided because their thickness can become a hazard in addition to accumulating dirt.

To achieve a medium level of safety, a flexible layout is also implemented to ensure adjustments according to the user's needs. Adding windows can allow great daylight penetration and ventilation. If there is no possibility of inserting a window, at least exhaust ventilation should be installed to improve indoor air quality. The position of the kitchen is another important aspect to be considered. The position of

---

[1] The complete table with the design guidelines is described in Tissot (2022).

Table 9.1 Example of *Highest Priority* design features for a safe environment

Categories	Design guidelines	Gravity	Urgency	Tendency	Score	Front door	Living/dining room	Kitchen	Bedroom	Home Office	Bathroom	Laundry	Garage
Equipment	Grab bars/Handrails: help to move around in circulations' areas ensuring support and stability	5	5	5	125	•							
Constructive Materials	Non-slip floor covering	5	4	5	100		•	•	•	•	•	•	•
Maintenance	Rugs (avoid): physical barrier	5	4	5	100		•	•	•	•	•	•	
Structural	Ramps: help during vertical movement and must attend to the regulations	5	4	5	100	•							
Assistive technology	Smoke sensor: avoid major accidents from fire	5	4	5	100			•			•	•	•

Table 9.2 Medium priority design features for a safe environment

Categories	Design guidelines	Gravity	Urgency	Tendency	Score	Front door	Living/dining room	Kitchen	Bedroom	Home office	Bathroom	Laundry	Garage
Equipment	Bench inside the box: to help elderly people who have difficulty staying on their feet for a long time	3	4	3	36						•		
Equipment	Glass box (avoid): preference for half a masonry wall	3	4	3	36						•		
Furniture	Light furniture: helps with movement	2	4	4	32		•	•	•	•			
Furniture	Furniture close to the wall: barrier-free movement	4	4	2	32		•	•	•	•		•	
Equipment	Door opening outwards: in case of a fall inside an environment, if the size of the space is reduced, there is a probability that the person will block the entrance	4	4	2	32						•		

Table 9.3 Minimum priority design features for a safe environment

Categories	Design guidelines	Gravity	Urgency	Tendency	Score	Front Door	Living/ dining room	Kitchen	Bedroom	Home Office	Bathroom	Laundry	Garage
Maintenance	Emergency contact: place in a visible place in case of accidents	2	3	4	24			•	•		•		
Equipment	Emergency alarm: helps in case of accidents and imminent dangers	3	4	2	24			•	•		•		•
Constructive Materials	Easy-care coating: contributes to easy cleaning	3	3	2	18	•		•			•	•	•
Furniture	Variety of furniture allows different users with different abilities, needs, and scopes to use the same environment	3	3	2	18		•		•				

Table 9.4 Icon categories

Icon	Categories
🏠	Structural changes
📄	Regulations
▯	Equipment
📱	Assistive technology
⊟	Constructive material's
🛋	Furniture
✋	Maintenance

Table 9.5 Example of room table for kitchen

Space		Kitchen			
Activities of daily living		Basic ADLs ☑		Instrumental IADLs ☑	Enhanced EADLs ☑
Minimum safe GUT: 60–125 points		Medium safe GUT: 25–59 points		Highest safe GUT: 0–24 points	
▯	– Emergency lights – Artificial lighting (general or task) – Grab bars/handrails	🏠	– Flexibility in layout – Natural lighting – Location of the environment	🏠	– Dimension
📱	– Smoke sensor – Gas sensor	▯	– Exhaust ventilation – Built-in wiring – Built-in lighting in cabinets	▯	– Dishwasher model drawer – Side by side or reverse fridge. – Induction stove – Emergency alarm – Position of appliances

Table 9.5 (continued)

Space		Kitchen			
[icon]	– Color contrast – Non-Slip floor covering	[icon]	– Light furniture – Furniture close to the wall – Handle handles – Tables and chairs with straight legs	[icon]	– Simplified buttons – Water thermostat
[icon]	– Adequacy range – Use of drawers			[icon]	– Easy-care coating – Heat-resistant surfaces – Water-resistant materials
[icon]	– Rugs (avoid)			[icon]	– Overhead cabinets with glass door – Efficient hinges/slides
				[icon]	– Emergency contact – Storage of chemicals (avoid) – Curtains (avoid)

this environment close to the dining room, for example, prevents users from traveling long distances while carrying objects in their hands. Lights in cabinets are another strategy that complements artificial lighting. Often, overhead cabinets or the person themself can shade and make the task difficult when using the workbench. Care must also be taken so that the wiring is not apparent and becomes a physical barrier for users. Installing the furniture close to the walls makes circulation free of barriers, but in the case of the kitchen environment, where a series of equipment is inserted, a thorough study of the layout is necessary. Additionally, if there is loose furniture such as chairs or stools, it is important that they have straight, fixed feet and that they are easy to move. For cabinets and drawers, preference should be given to handle-type handles, as they are easier to use since older adults may have problems with arthritis or arthrosis in their hands. Concerning microwaves and other small appliances, they must be in an easily accessible location, within reach of the user, as well as other equipment in this environment.

When considering a highest safety level, features to implement are dimensions with sufficient measures to accommodate activities, users, and equipment. Essential strategies for the kitchen are, for example, the dishwasher must have a front opening or a drawer-type opening. Even in the case of the standard refrigerator, it can be replaced by a side by side or inverse model. These appliances have a higher investment value than standard models of people's homes. However, when considering a higher level of security, financial investments may also be more significant in some cases.

The induction cooker has more benefits than those with gas. Users with cognitive deficits may forget to bake or fry food, so these sensors help to prevent major accidents. In addition, the position of appliances such as the stove and washing machine requires specific water, sewage, and gas installations. However, they can be installed in other places without significant modifications. In the case of the refrigerator, depending on the model, handling can be facilitated. Thus, the indication is that the position of these items is functional for the performance of the activity.

The button's simplification is another aspect to consider, as the equipment has panels with a series of drives, which can cause confusion and stress for those who use it. Thus, investing in equipment with this simplification contributes to the more efficient use of the device. Additionally, voice assistants can turn on, off, or even program specific equipment. It is emphasized that assistive technology should only be inserted if the user is familiar with its use. Otherwise, it would not be recommended.

For faucets, the models must have handles that facilitate handling, with levers, and that preferably have a hot or cold-water indicator. It is still possible to purchase thermostats on the market that visually indicate the temperature of the water. They are easy to install and do not require structural changes. Heat-resistant and water-resistant surfaces, cabinet coverings, and easy-care floors should be considered.

Regarding kitchen furniture, there are hinges and slides with dampers and models that make opening the cabinets more accessible. Additionally, regarding the kitchen cabinets, in case there are overhead cabinets, which would not indicate, one possibility would be to use glass doors to make it easier for the user to see what is inside. In the case of tall or overhead cabinets, it is worth noting that risky behavior must be avoided to ensure a person's safety in an environment. In this way, the presence of a closet can make older adults want to access it and climb stairs or even a chair. Other strategies that can be adopted and that do not depend on structural modifications, and therefore are easy and costless to implement, would be to avoid storing chemicals or cleaning products in cabinets in this environment and keeping an emergency contact and alarm in a visible place. If the curtains cannot be avoided, at least choose a material that is not flammable.

Table 9.6 presents a summary of the characteristics of the bathroom. To ensure a minimum safety level, it is necessary to install a drain inside and outside of the shower so the water can run off in case of a leak. Avoiding gaps is another relevant point. Differences between wet and dry areas must be overcome with a floor layout that indicates the fall to the drains. Attention should also be paid to regulations to respect the minimum dimensions (consult countries' regulations) for users using a wheelchair (turning area).

Artificial lighting is relevant and essential for carrying out tasks. It is also important to highlight the importance of having enough light in the environment and in the shower area. Emergency lighting is necessary to prevent accidents during a power outage. Another essential item for the bathroom is the installation of grab bars, mainly for access to and within the shower and the toilet. For furniture, it is recommended to use drawers and respect the reach of users. Regarding the shower, another indication would be that the position of the mixers for opening the shower should be close to the shower entrance to prevent the person from getting wet when opening and, in the case of boiling water, to avoid accidents. The height of the toilet must also respect the users' physical abilities and restrictions.

Items such as smoke, gas, and humidity sensors help with security. In addition, as in other environments, color contrast helps to identify barriers. On the coatings, it is ideal that they are always non-slip. Another item that should be avoided is loose

Table 9.6 Example of room table for bathroom

Space		Bathroom			
Activities of Daily Living:		Basic ADLs ☑		Instrumental IADLs ☐	Enhanced EADLs ☐
Minimum safe GUT: 60–125 points		Medium safe GUT: 25–59 points		Highest safe GUT: 0–24 points	
🏠 📊	– Drain – Unevenness (avoid) – Regulations for measurements	🏠	– Location of the environment – Natural lighting	🏠	– Dimension
▯	– Emergency lights – Support bars – Artificial lighting (general or task) – Use of drawers – Adequacy range	▯	– Seats inside the box – Glass box (avoid) – Door opening outwards – Handle (lever type) – Handle model handle -Exhaust ventilation	▯	– Electric towel rails – Anti-fog mirror – Protection of electrical outlets – Lock that opens the door on both sides – Flexible shower
📟	– Smoke sensor – Gas sensor – Humidity sensor	📟	– Intercom/communicator	📟	– Device to open the door automatically – Emergency alarm – Emergency contact – Water thermostat
▭	– Color contrast – Non-slip floor covering	▭	– Smaller coatings	▭	– Easy maintenance coating – Water-resistant materials
✋	– Rugs (avoid)	✋	– Built-in wiring		

mats, as their use is associated with falls. As the bathroom environment is a place where this item is often necessary, the ideal is to try to embed it in the floor or use adhesives to fix it to the floor.

Concerning the criteria for the average safety level, in addition to the items listed above, the use of more minor coatings, mainly in the shower area, so that there is more grout and, therefore, more adherence for the user.

Issues such as air exhaustion and natural lighting from large openings contribute to safety, as natural ventilation is essential due to humidity. Another point to discuss is the location of the bathroom with the other rooms in the house. This environment is close to the bedrooms in most homes, but this is only sometimes the case. Thus, it should be noted that the environments in the house must be closed due to complementary uses. Thus, it is indicated that the bathroom is close to the bedroom.

Over the shower, it is recommended to install a fixed or retractable bench to help older adults who have difficulty standing up for a long time. This way, it avoids placing a bath chair and having to move it. Another strategy is to avoid a glass box or even a curtain if the room size allows it. As a replacement, the ideal would be the construction of a masonry wall with an approximate height of 1.20 m, and above this wall, a fixed glass to avoid splashing water in the environment and consider, according to the user, an access width that is free of barriers to access the shower.

The door is indicated to open outwards, considering there is enough external space for a whole opening. If this is not possible, the solution is sliding doors or the door opening inwards, but with the addition of opening systems. The indicated doorknobs are of the lever type and handle for the handle model furniture. Regarding assistive technologies, an intercom/voice communicator and an emergency alarm were installed inside the bathroom. Bathroom wiring, due to humidity, must be recessed into the wall.

For the highest security level, there are questions related to the dimension. Much has been said about the reduced size of this environment in the case of both environmental adaptations and new projects. In this way, it would be essential to consider the dimension as a criterion to verify if it is adequate or adapt it according to the equipment and activities that will be carried out in this space. The layout and equipment position, such as the tub, toilet, and shower, must be positioned to the detriment of tasks. The installation of electric towel rails is also recommended to facilitate the drying of towels, anti-fog mirrors, and electrical protection sockets to avoid possible accidents in cases of excess humidity.

In addition to the possibility of opening the door outside the environment, the installation of locks that allow opening on both sides and assistive technologies that allow opening the door are indicated. Alarm and emergency contact in visible places so that in the event of an accident, measures can be requested quickly. The thermostat is another device that can be placed on the sink faucet and shower to view the water temperature. The installation of a flexible shower to assist the bath is also indicated inside the shower. Finally, using materials that are resistant to water also helps avoid furniture degradation and possible accidents and is easy to maintain.

Conclusions

With the growing older adult population, numerous efforts are necessary by researchers from different disciplines to explore a better way to support successful aging from biological, behavioral, social, spatial, and more dimensions of aging. The five domains of SDoH—economic stability, education access and quality, healthcare access and quality, social and community context, neighborhood and built environment—are also conditions that impact the health, functioning, and quality of life of older adults who are aging in place. This study aimed to propose a spatial understanding and support of successful aging through design guidelines to create safer environments for older adults, hence achieving successful aging in place. By implementing a three-phase approach to systematically address, categorize, and organize critical design elements of aging in place, we developed a comprehensive and replicable framework for developing environmental design guidelines needed for healthy aging.

There are several determinants of active aging, and the physical environment is one of them. In this chapter, we focus on understanding better ways to design safe home environments for older adults that can reduce the risk of falls and injuries, hence ensuring autonomy and independence in the engagement of various everyday activities. In addition to public policies aimed at social, economic, and health issues, the space in which older adults live is highly relevant in this process. In this way, understanding this population's difficulties is directly related to improving the conditions of use of this space so that it is safe. For safer and more accessible environments to be possible, a greater understanding of the risk factors present in the performance of routine tasks, or ADLs, of older adults is necessary.

Understanding the difficulties, abilities, and deficiencies is the first step in designing suitable environments for this population. Physical space must meet the demands for anyone to have an independent and autonomous life. Thus, basic, instrumental, and engagement activities can develop to guarantee minimum conditions for active and healthy aging. An adequate environment influences the degree of user satisfaction, consequently influencing the level and perception of well-being. Providing conditions for activities to occur safely, with minimized risks, is to provide conditions for active and healthy aging.

It is understood that the design guidelines proposed may not be accessible to everyone, so through this, public policies could be established to improve the safety of older adults in their own homes since now it is known what characteristics and physical features can safely improve aging in place. The results from the protocol indicated seven categories of design: structural changes, assistive technology, regulations, appliances, constructive materials, furniture, and maintenance for the kitchen, bathroom, laundry, garage, dining and living room, bedroom, and home office. The kitchen had the most cited features, and guidelines point to minimum safety needs, for example, having emergency and artificial light, sensors for gas and smoke, contrast of color, non-slip floor, suitability of furniture ranges, drawers, and avoidance of rugs. It must consider a flexible layout and improve natural light to improve home safety. Consider space dimension, appliances with assistive technology, and avoiding blinds for maximum safety. Considering human factors and design characteristics, we provide minimum to maximum safe recommendations. The design guidelines can be used by designers when projecting housing environments or for older adults to improve the environment in their own homes.

Summary Statement of Conclusions

Physical environment is a determinant of successful aging in place.

- Understanding older adults´ difficulties and abilities is the first step in designing sui` environments.
- The kitchen and bathrooms were presented as holding more adaptations to make.
- The proposed design guidelines can create safer environments for older adults to improve aging in place.
- Public policies could be established since there are known characteristics and physical features that can safely improve older adults´ aging in place.

References

Albuquerque, D. S., Amancio, D. A. R., Gunther, I. A., & Higuchi, M. I. G. (2018). Theoretical contributions on aging from the perspective of person-environment studies. *Psychology USP, 29*(3), 442–450.

Antes, D. L., D'orsi, E., & Benedetti, T. R. B. (2013). Circumstances and consequences of falls in elderly people in Florianópolis. Epifloripa elderly 2009. *Brazilian Journal of Epidemiology, 16*(2), 469–481.

Cesar, F. I. G. (2013). *Quality management tools* (1st ed.). Library 24 hours, Seven System International Ltd.

Daré, A. C. L. (2010). Lighting design for the elderly. *Reação Magazine: Technical and Scientific Section, 14*(77), 8–11. https://www.researchgate.net/publication/301199042_O_Design_de_Iluminacao_voltado_aos_idosos

Fernandes, M. G., Santos, M., & S. R. (2007). Public policies and rights of the elderly: Challenges of the social agenda of contemporary Brazil. *Achegas Net, 34*, 49–60.

Ferreira Filho, N. (2018). Topics in ergonomics and workplace safety. *Poisson Belo Horizonte, 2,* 219.

Fornara, F., Lai, A. E., Bonaiuto, M., & Pazzaglia, F. (2019). Residential Place Attachment as an Adaptive Strategy for Coping With the Reduction of Spatial Abilities in Old Age. *Frontiers in psychology, 10*, 856. https://doi.org/10.3389/fpsyg.2019.00856

Hazin, M. M. V. (2012). *Residential spaces in the perception of active elderly people*. Masters dissertation (p. 144). Postgraduate Program in Design, Federal University of Pernambuco.

Katz, S., Chinn, A. B., Cordrey, L. J., Grotz, R. C., Newberry, W. B., Orfirer, A. P., Wischmeyer, E. J., Kelly, A., Maso, R. E., Ryder, M. B., Bittman, M., Conley, C. C., Hayward, M., Hofferberth, A. O., Holman, J., Robins, L. M., Sherback, M. A., Ritchie, S. W., & Takacs, S. J. (1959). Multidisciplinary studies of illness in aged persons: II. A new classification of functional status in activities of daily living. *Journal of Chronic Diseases, 9*(1), 55–62. https://doi.org/10.1016/0021-9681(59)90137-7

Kirkwood, T. B. L. (2017). Why and how are we living longer? *Experimental Physiology, 102*(9), 1067–1074. https://doi.org/10.1113/EP086205

Lawton, M. P., & Brody, E. M. (1969). Assessment of older people: Self-maintaining and instrumental activities of daily living. *The Gerontologist, 9*(3), 179–186.

Lockhart, T. E. (2007). Fall accidents among the elderly. *International Encyclopedia of Ergonomics and Human Factors,* 2626–2630.

Lojudice, D. C., Laprega, M. R., Rodrigues, R. A. P., & Rodrigues Júnior, A. L. (2010). Falls of institutionalized elderly: Occurrence and associated factors. *Brazilian Geriatrics and Gerontology, 13*(3), 403–412.

Mack, K. A., Rudd, R. A., Mickalide, A. D., & Ballesteros, M. F. (2013). Fatal unintentional injuries in the home in the U.S., 2000–2008. *American Journal of Preventive Medicine, 44*(3), 239–246. https://doi.org/10.1016/j.amepre.2012.10.022

Mendes, F. R. C., & Côrte, B. (2009). The old age environment in the country: Why plan? *Kairós Magazine, 12*(1), 197–212.

Mois, G., & Rogers, W. A. (2023). Advancements in technology to promote safety and support aging in place. In R. Wolf, B. S. Eckert, & A. Ehrlich (Eds.), *A comprehensive guide to safety and aging: Minimizing risk, maximizing security.* CRC Press/Taylor & Francis Group, LLC.

Moraes, A. D., & Mont'alvão, C. (2009). *Ergonomics: Concepts and applications* (2nd ed.). A. de Moraes.

Neves, V. L. S. (2017). Fall risk in elderly. *Assessment Instrument, 2015*(30), 23–29.

Nicklett, E. J., Lohman, M. C., & Smith, M. L. (2017). Neighborhood environment and falls among community-dwelling older adults. *International Journal of Environmental Research and Public Health, 14*(2).

Nicolussi, A. C., Fhon, J. R., Santos, C. A., Kusumota, L., Marques, S., & Rodrigues, R. A. (2012). Quality of life in elderly people who have suffered falls: Integrative literature review. *Science & Collective Health, 17*(3), 723–730. https://doi.org/10.1590/s1413-81232012000300019

Penney, L. (2013). The uncertain bodies and spaces of aging in place. *Anthropology and Aging Quarterly, 34*(3), 113–125. https://doi.org/10.5195/aa.2013.12

Pinter-Wollman, N., Jelić, A., & Wells, N. M. (2018). The impact of the built environment on health behaviors and disease transmission in social systems. *Philosophical Transactions of the Royal Society,* B3732017024520170245. https://doi.org/10.1098/rstb.2017.0245

Pohl, P., Sandlund, M., Ahlgren, C., Bergvall-Kåreborn, B., Lundin-Olsson, L., & Melander Wikman, A. (2015). Fall risk awareness and safety precautions taken by older community-dwelling women and men—A qualitative study using focus group discussions. *PLoS One, 10*(3), e0119630. https://doi.org/10.1371/journal.pone.0119630

Ramadhani, W. A. (2023). *Food-related activities among older Indonesian women: Understanding personal and built-environmental adaptation for independence in later life [Dissertation].* University of Illinois Urbana-Champaign.

Ramadhani, W. A., & Rogers, W. A. (2022). Understanding home activity challenges of older adults aging with long-term mobility disabilities: Recommendations for home environment design. *Journal of Aging and Environment,* 1–23. https://doi.org/10.1080/26892618.2022.2092929

Rodrigues, I. G., Fraga, G. P., & De Azevedo Barros, M. B. (2014). Falls in the elderly: Associated factors in a population-based study. *Brazilian Journal of Epidemiology, 17*(3), 705–718.

Rogers, W. A., Blocker, K. A., & Dupuy, L. (2020a). Current and emerging technologies for supporting successful aging. In A. K. Thomas & A. Gutchess (Eds.), *The Cambridge handbook of cognitive aging: A life course perspective* (pp. 717–736). Cambridge University Press.

Rogers, W. A., Mitzner, T. L., & Bixter, M. T. (2020b). Understanding the potential of technology to support enhanced activities of daily living (EADLs). *Gerontechnology, 19*(2), 125–137. https://doi.org/10.4017/gt.2020.19.2.005.00

Rogers, W. A., Ramadhani, W. A., & Harris, M. T. (2020c). Defining aging in place: The intersectionality of space, person, and time. *Innovation in Aging, 4*(4), igaa036. https://doi.org/10.1093/geroni/igaa036

Rowe, J. W., & Kahn, R. L. (1997). Successful aging. *The Gerontologist, 37*(4), 433–440. https://doi.org/10.1093/geront/37.4.433

Rowe, J. W., & Kahn, R. L. (2015). Successful aging 2.0: Conceptual expansions for the 21st century. *The Journals of Gerontology Series B: Psychological Sciences and Social Sciences, 70*(4), 593–596. https://doi.org/10.1093/geronb/gbv025

Sociedade Brasileira De Geriatria E Gerontologia. (2008). *Falls in the elderly: Prevention* (pp. 1–10). Brazilian Medical Association and Federal Council of Falls Medicine.

Staut, A. L. V. (2014). *Universal usability in architecture: Assessment method based on heuristics.* Federal University of Campinas. Faculty of Civil Engineering and Architecture and Urban Planning. (Masters dissertation).

Tischa, J. M., Van Der Cammen, A. A., Voute, E., & Molenbroek, J. F. M. (2017). New horizons in design for autonomous aging. *Age Aging, 46*(1), 11–17. https://doi.org/10.1093/ageing/afw181

Tissot, J. T. (2022). *Aging in place: Protocol with design guidelines for safe living environments for older adults.* Federal University of Santa Catarina. Postgraduate Program in Architecture and Urbanism. Doctoral thesis, Retrieved from https://repositorio.ufsc.br/handle/123456789/241000

Tomazzoni, A. M. R. (2011). The art of living alone and being happy in old age. *Kairós Gerontology, 13*, 109–123.

U.S. Department of Health and Human Services, Office of Disease Prevention and Health Promotion. (n.d.). *Healthy people 2030.* Retrieved April 29, 2023, from https://health.gov/healthypeople/priority-areas/social-determinants-health

Chapter 10
Socially Assistive Technologies for Older Adults in Their Residential Environment

Bruna Luísa Poffo Nobre and Lizandra Garcia Lupi Vergara

Introduction

Older people represent a rapidly growing consumer segment in the coming decades, in the context of globalization and rapid innovation, significantly complicating the product creation process due to their diverse and evolving characteristics. According to Czaja et al. (2019), the increase in the world population of people over 60 years old could make this group reach 16% of the world population in 2050. The emerging demographic trends toward an aging population demand novel solution to improve the quality of life of elderly individuals, including prolonged independent living, improved health care, and reduced social isolation (Koceska et al., 2019). Preserving interpersonal relationships for older adults is one of the principal aspects of quality of life, according to Vecchia et al. (2005). One of the many ways to help older people in these relationships is with emerging technologies, e.g., the Internet of Things (IoT) in the domain of Ambient Assisted Living (AAL), which tends to promote the quality of life and independence of seniors (Gulati & Kaur, 2021).

Technology has become a part of human lives for decades, expanding the need for human-computer interaction (HCI) research, an important area of work involving assistive technology (AT) and medical systems (Ganokratanaa & Pumrin, 2018). Using AT concepts and practices to develop friendly products can help older adults expand their communication, mobility, control of their environment, learning, and work skills (Bersch, 2014). By engaging with specific user cohorts, such as older adults and/or persons with disabilities, AT offers advantages such as greater social inclusion and independence for its users.

This chapter focuses on exploring how emerging technologies are meeting the needs of the older adult, identifying potential applications and barriers of socially

B. L. P. Nobre (✉) · L. G. L. Vergara
Production and Systems Department, Federal University of Santa Catarina,
Florianópolis, Santa Catarina, Brazil

© The Author(s) 2026
M. A. Kolak, I. K. Moise (eds.), *Place and the Social-Spatial Determinants of
Health*, Global Perspectives on Health Geography,
https://doi.org/10.1007/978-3-031-88463-4_10

assistive technologies for healthy aging, and classifying technologies to understand the current scenery better. It focuses on the various needs that an older adult may present and how social assistive technologies can address those needs. The research is exploratory, and we use Social Determinants of Health (SDoH) objectives to determine if these technologies cover one or more of these domains and their objectives.

Characteristics of Aging

The life expectancy of the world's population has increased due to numerous factors, resulting in a significant rise in the number of older adults. According to Czaja et al. (2019), this global phenomenon is driven by increased longevity and declining birth rates. In this context, "older adult" typically refers to individuals aged 65 and above, a commonly accepted definition based on aging research and demographic studies. Furthermore, societal changes have led to new patterns in family structures, including a rise in nontraditional relationships, geographically dispersed families, and a growing number of adults choosing to remain single or childless. Czaja et al. (2019) emphasize these shifts, noting that contemporary family dynamics often involve a complex interplay of geographical and relational factors. Additionally, it's important to consider how disability intersects with aging, as older adults are more likely to experience disabilities, which can affect their quality of life and the types of support they need. Understanding these aspects is crucial for developing effective interventions and support systems for this demographic. Furthermore, the risk of vulnerability is higher in aging individuals due to various factors, including biological decline, social and psychosocial determinants, and interaction with sociocultural processes (Machado & Alvarenga, 2019). For Czaja et al. (2019), the aging process is multidimensional, as it deals with physical, psychological, social, and economic changes as well as social roles (e.g., retired or caregiver of someone older) or living arrangements (e.g., changing for a smaller, more manageable home). Therefore, instead of dividing a group by age, it is categorized by physical, cognitive, or psychographic capacity, as shown in Table 10.1.

The most important factor associated with good quality of life among older people is a stable social relationship (Ferraz & Peixoto, 1997). Research also shows that having a sense of security and familiarity with one's environment is essential (Wiles et al., 2011), especially as we may need to adapt our surroundings to make daily activities easier (Czaja et al., 2019).

Due to demographic changes worldwide, Czaja et al. (2019) project an increase in the number of older adults using technology. Consequently, with AT, there is a need for home automation to promote greater independence in the place of residence and the care of the older person, protection, and education, including for those with intellectual disabilities or dementia (Bersch, 2014). Therefore, there will be an even greater need to create technologies that help with older adults' health

Table 10.1 Summary of the limiting characteristics of aging

	Category	Characteristics
1	Psychography	**(a)** lack of motivation (low expectations); **(b)** social integration (lack of confidence, loneliness, and isolation); **(c)** willingness to adopt innovations (acceptance or openness to experiences – interactions with technology and adherence to guidance treatments)
2	Sensation and perception	**Vision:** **(a)** progressive loss of visual perception; **(b)** decreased peripheral vision; **(c)** decreased visual accommodation; **(d)** decreased depth perception; **(e)** slow processing of visual information; **(f)** difficulties in examining a location **Hearing:** **(a)** difficulty in detecting high-pitched sounds (high frequencies); **(b)** speech discrimination, especially in noisy places; **(c)** impairs the ability to carry out daily activities; **(d)** auditory handicap (inactive social life or social isolation; risk of depression); **(e)** development of chronic diseases **Haptic and Kinesthetic:** **(a)** accidental falls and postural instability; **(b)** integrative capacity of the brain (dizziness or vertigo); **(c)** reduced perception of one's own body movement; **(d)** acuity in locating one's own body parts; (e) identification of temperature and vibrations **Taste and Smell:** Reduced ability to perceive tastes (sweet, sour, bitter, and salty) and odors, having difficulty distinguishing between various foods or odors, can initiate malnutrition
3	Cognition	**(a)** reducing the rate at which information is processed; **(b)** limited attention; **(c)** reduced memory, except procedural memory, not about a cognitive disease (dementia, Alzheimer's); **(d)** ability to mentally manipulate images or patterns; **(e)** ability to interpret verbal information; **(f)** action planning, problem-solving, and response inhibition
4	Mobility	**(a)** smaller in stature; **(b)** reduced movement control (wide and precise); **(c)** decrease in resistance and handgrip strength; **(d)** loss of muscle mass. Discrepancy: since little change in anthropometry to more significant changes

(emotional, mental, cognitive) and independence to increase their quality of life and reduce hospitalizations.

Socially Assistive Technologies

Assistive technology encompasses products, resources, methodologies, strategies, practices, and services that aim to promote functionality related to the activity and participation of people with disabilities or reduced mobility, aiming at their autonomy, independence, quality of life, and social inclusion (Bersch, 2014). When united with the concept of HCI of information technology, socially assistive technologies would incorporate all technology encompassing HCI and social signals. These include socially assistive robots, social robots, combining social robots with

Ambient Assisted Living (AAL), and domotics (i.e., smart objects) linked to the internet of things (IoT). Moreover, virtual reality (VR) and gamification sometimes create a motivating and rewarding experience that includes monitoring, lifestyle, activities, and assistance (Gulati & Kaur, 2021). Additionally, devices connected to technology, environment, or even the user could improve some characteristics of the senior and/or persons with disabilities or reduced mobilities. Artificial intelligence (AI) and machine learning may improve the interaction between the user, the product or service, and its functionality.

When researching the term "Socially Assistive Technology," concepts of "Assistive Technologies" focused on social interaction or connection, "Socially Assistive Robots" or "Socially Assistive Systems" emerge in scholarly searches. It is important to note that the scientific community has yet to reach a consensus on a specific term relating to socially assistive technologies, due to its emerging nature. This chapter reviews assistive technologies when presented with any social interaction tied to HCI or social signals.

Technological Solutions for Older Adults and SDoH

There are various technological solutions available to aid older adults in their homes. These include safety monitoring systems, communication devices, cognitive stimulation tools, telehealth monitoring systems, home assistance robots, and services such as transporting heavy objects, grocery delivery, and garbage collection (Sefcik et al., 2017). However, the cost of these interventions impedes access for older adults with limited financial resources (Aneke et al., 2018). Consequently, organizations and public policies may create guidelines, measures, and actions to improve public health. The social determinants of health (SDoH), non-medical factors that can influence health outcomes, are crucial for contextualizing these potential technological solutions.

The SDoH are "the conditions in the environments where people are born, live, learn, work, play, worship, and age that affect a wide range of health, functioning, and quality-of-life outcomes and risks" (Healthy People 2030, 2023). According to the *Healthy People* framework, SDoH comprises five domains; each has goals and objectives to promote healthy choices and impact people's well-being and quality of life. For this research, we will work with two SDoH domains: "healthcare access and quality" and "social and community context." Based on Healthy People 2030, the first domain of interest aims at "increasing access to comprehensive and high-quality health care services," while the second domain aims to "increase social and community support." We selected both due to their proximity to the primary goal of this scoping review, which focused on the needs and perceptions of emerging technologies and on identifying potential applications and barriers of socially assistive technologies for healthy aging, including classifying technologies to better understand those aged 60 and older. For example, some technological interventions seek to enhance access to healthcare professionals and improve communication, whether in person or remotely.

Review Method

This scoping review included articles published between 2017 and October 2022 captured from two databases: *Web of Science* and *Scopus*. First, we defined approaches for the scoping review according to the central theme and testing keywords. Before searching, we explored the most common keywords related to the topic using titles, abstracts, and keywords. *Scopus* and *Web of Science* databases were used for search because they provide full-text journals and conference proceedings related to AT, social interactions, and their relationships. Furthermore, these databases did not focus on extreme illnesses.

Since there is still no consensus on the definition of SAT, we filtered articles using keywords related to "older person," "assistive technology," and "social interaction." We tested each aspect separately to identify the most effective search terms. The first category included the keywords: "elderly" OR "elder," OR "senior" OR "oldster," OR "old-aged," OR "aging population." The second included: "Assistive technology*" OR "assistive device*" OR "assistive product*" OR "assistive application" OR "technical aid" OR "assisted living" OR "self-help device." Furthermore, for the last aspect, the words "human-computer interaction" OR "social interactions"OR "social activity" OR "social connectedness" OR "social connectivity" OR "social isolation" OR "socially OR "interpersonal relation" were considered.

We identified articles based on the scoping review's question of interest: "Which technologies show opportunities to increase the quality of life of older adults in their homes?" Next, inclusion and exclusion criteria were defined using PICO (Population, Intervention, Comparison, Outcome) characteristics to guide the key elements we were working on in the articles, complemented with source type, language, and other elements, as shown in Table 10.2. The PICO tool characterizes the complete articles, as Galvão and Ricarte (2019) suggested. Therefore, this chapter only worked with articles that exhibit products or systems that help the older person develop social interactions.

Finally, we segmented the resulting articles into two central categories: main and supporting technologies. The first category of more robust technologies includes robotics, virtual reality, gamification, integration, and monitoring. Supportive technologies provide opportunities for future improvement of the functionalities presented in the first (main) category, such as improvements in the communication system and system usability. Furthermore, we found links to the different needs that an older person may present, thus helping to address some of the minor aspects of SDoH related to technology usage.

Results

Of the 414 articles gathered by scoping review, 90 were duplicates, while 4 had publication dates outside the research's limit. The remaining 320 articles underwent a thorough reading process. The selection of emerging technologies relied on

Table 10.2 Search Strategy and Study Selection with PICO Tool plus Inclusion and Exclusion criteria

Key element	Inclusion criterion	Exclusion criteria
Population	Older person	Articles that evaluate rehabilitation or support specific situations/diseases such as (1) Persistent vegetative state; (2) spill; (3) obesity; (4) Alzheimer's or high-grade dementia; (5) Cerebral Vascular Accident (CVA); (6) sequels of Covid-19; (7) Mental disorders; (8) Amyotrophic Lateral Sclerosis (ALS)
Intervention	Assistive Technology Social interaction	Not propose an assistive technology; Technologies without social interactions; Technologies for outdoor/monitor activities in the outdoor environment
Comparison	Any	None
Outcome	Increase in the quality of life of the population	None
Source type	Frameworks, Case of Study, field experience and any study which presents a technological product use	Theoretical or conceptual articles, including literature reviews, scope reviews, systematic reviews, and book reviews
Language	English, Spanish, Portuguese	Others
Others	–	Duplicate studies in different sources

inclusion and exclusion criteria, as shown in Table 10.2. We noted that many emerging technologies focused on specific diseases unrelated to age—furthermore, most research stood in controlled environments, not private residences. Then, we filtered by technologies that focus on older adults' basic needs (Table 10.1) and did not exclude the vast majority carried out in facilities. After eliminating 284 articles, only 36 met the established criteria (Table 10.2). Based on the minor objectives of each SDoH goal, we observed that the articles fit into more than one objective or none, opening space for creating new major and minor objectives. Therefore, we classified the 36 articles into *main* and *supporting* technology categories.

As shown in Table 10.3, each category has a sub-classification, identifying which necessities of older adults the articles address.

Category A: Main Technologies for the Senior Person

Category A encompasses articles focused on creating opportunities for interaction between older adults and technology, or between older adults and other people through technology. This category is divided into four subcategories: "Assistive Robot," "Games and Virtual Reality," "Mobile Applications," and "Technology Integration," which involves combining multiple technologies to interact with and monitor seniors.

Table 10.3 Technologies categories from the reviewed literature

Category	Subcategory	Reference	Older adults needs			
			1	2	3	4
A: Main technologies for the senior person	Assistive Robot	Cruz-Sandoval and Favela (2019)	x			
		Di Napoli et al. (2022)		x		x
		Fan et al. (2021)	x		x	x
		Fattal et al. (2022)	x			
		Fitter et al. (2020)	x	x	x	x
		Khosla et al. (2016)	x			
		Koceska et al. (2019)	x		x	
		Lin et al. (2022)	x		x	
		Obayashi et al. (2020)	x		x	
		Trovato et al. (2019)	x		x	
	Games & Virtual Reality	Pedrozo et al. (2022)	x		x	
		Lin et al. (2018)	x			x
		Ricci et al. (2022)		x		x
		Rings et al. (2020)		x		x
		Unbehaun et al. (2020)	x		x	x
	Mobile Applications (APP)	Goumopoulos et al. (2017)	x		x	
		Ray et al. (2017)	x		x	
		Yurkewich et al. (2018)	x		x	
	Integration of technologies	Aneke et al. (2018)			x	
		Bui and Chong (2018)	x			x
		Davis-Owusu et al. (2019)	x		x	x
		Kearney et al. (2018)	x		x	x
		Kyritsis et al. (2018)	x		x	
		Luperto et al. (2022)	x	x	x	x
		Pinto et al. (2019)	x		x	
B: Supporting technologies	Potential of improvement	Caranica et al. (2017)		x		
		Chang et al. (2018)	x	x		
		Desai and Desai (2017)		x		x
		Esposito et al. (2019)	x			
		Ganokratanaa and Pumrin (2018)		x	x	x
		Gulati & Kaur (2021)		x		
		Li et al. (2017)		x		x
		Mahmud et al. (2020)	x	x		
		Sinha and Caverly (2020)	x	x		
		Valsamakis and Savidis (2017)	x	x	x	x
		Werner et al. (2020)		x	x	x

1 = Psychography, 2 = Sensation and perception, 3 = Cognition, 4 = Mobility

Assistive Robot

Research in this subcategory highlights the development and implementation of robots designed to assist older adults with daily tasks, improve safety, and enhance social interaction. Key themes include the integration of advanced navigation and manipulation capabilities, as well as the adaptation of robots to meet individual needs and preferences. These findings demonstrate the potential of assistive robots to significantly improve the quality of life for seniors by providing personalized support and fostering engagement.

Research in this area underscores the effectiveness of robots in enhancing interactions and therapeutic activities for older adults, particularly those with dementia. Cruz-Sandoval and Favela's (2019) results suggest that individuals with dementia find conversations with the robot Eva as enjoyable as listening to music and singing. Their study analyzed an intervention that combined music and conversation therapy, which are favored recreational activities for people with dementia (PwD). Additionally, Obayashi et al. (2020) investigated the use of two different robots, Cota and Palro, each designed for distinct purposes but sharing features like cloud computing and programmed alerts. These findings illustrate how assistive robots can support both social engagement and practical functions, highlighting their potential to enrich the lives of older adults. Cota could communicate and join a monitoring system, while Palro could play music and lead exercises. Participants received one robot each, placed on their bedside table, where the com-robots could help prevent dementia progression by encouraging activity and sociability. Fattal et al. (2022) assessed how senior individuals and other potential users perceive cohabitation with a social robot (Pepper) for 7 days, nights, and days, considering future home integration. They reasonably and realistically integrated the technological components into a relevant and operational environment. Fitter et al. (2020) developed eight exercise games involving physical contact between humans and a human-sized robot named Baxter. Researchers enlisted the help of specialists during design to propose safe, fun, and beneficial interactions with different music modalities and difficulty levels, aiming to maintain interest in the robot throughout multiple interactions. Di Napoli et al. (2022) designed and implemented robotic applications in unsupervised environments that can be personalized and adapted to each specific user according to a service-oriented approach in terms of the composition of services. These efforts highlight the importance of designing adaptable and engaging robotic systems to enhance user experience and satisfaction in various contexts.

Koceska et al. (2019) conducted on-site experiments in a private elderly care center to evaluate software architecture, a robot's navigation, and manipulator capabilities. Their findings indicated that the shared-control paradigm improved robot navigation and reduced obstacle collisions. Similarly, Lin et al. (2022) developed a robotic architecture system within a commercially available socially assistive robot to engage pairs of older adults in multimodal activities across six sessions over 3 weeks. Using various engagement measures—visual, verbal, and behavioral—they assessed human-human and human-robot interactions in two assisted living facilities. The results showed increased interaction levels, though the engagement measures differed depending on the activity type. Overall, these studies underscore the effectiveness of advanced robotic systems in enhancing interaction and engagement among older adults, demonstrating their potential to improve quality of life and safety.

Trovato et al. (2019) created a prototype called DarumaTO, which stimulates cognition and social connections through games that link aesthetics to Buddhist and Shinto concepts. They demonstrated that interactive elements like touching and caressing can help release stress and alleviate anxiety and pain, thereby reducing the risks of cognitive decline and providing a range of activities. Ro-Tri, developed by Fan et al. (2021), automatically gathered quantitative interaction data, including head pose, vocal sounds, and physiological signals, to assess older adults' activity and social engagement. Khosla et al. (2016) investigated the engagement and acceptability of PwD with RSA Matilda, a system capable of recognizing voices, human faces, gestures, emotions, and acoustic speech, allowing participants to interact through various modalities such as touch control or voice commands. Together, these innovations illustrate the growing potential of interactive and adaptive technologies to enhance cognitive, emotional, and social well-being in older adults.

Games and VR

Research on games and virtual reality (VR) reveals their mixed effects on enhancing the lives of older adults. While incorporating digital games into computer classes did not further reduce loneliness beyond what was achieved by the classes alone, VR technologies showed promising results in other areas. VR platforms designed for visual impairment provide realistic simulations for training and education, while seated VR exergames have proven effective for integrating fall prevention exercises into daily routines. Additionally, video game-based training systems have been beneficial in keeping people with dementia active and engaged, with positive responses from caregivers. Overall, these findings highlight the potential of digital games and VR to support various aspects of well-being and therapeutic needs for older adults, as shown by the studies presented below.

Pedrozo et al. (2022) aimed to assess whether incorporating digital games into computer classes could alleviate loneliness among older adults. Their findings indicated that while computer classes did indeed reduce general loneliness for individuals aged 50 and older, the addition of digital games did not enhance this effect. Similarly, Ricci et al. (2022) developed a platform to evaluate the effectiveness of a custom visual impairment simulation, providing a realistic virtual reality experience crucial for training and educating on the daily challenges faced by individuals with visual impairments in a safer manner. Additionally, Rings et al. (2020) introduced a seated VR exergame, created in collaboration with clinical experts and therapists, designed to offer fall prevention exercises that can be easily integrated into daily routines. VR exergames could provide safe and effective therapies to reduce the risk of falling for older adults. In addition, the study by Unbehaun et al. (2020) found that a video game-based training system can help people with dementia (PwD) stay active and engaged in social activities, with caregivers positively receiving the system and supporting structured training activities like exergames. Lin et al. (2018) developed a VR system featuring 360° images or videos related to travel and relaxation in familiar places, incorporating brief introductions and training to enhance user confidence and safety. Collectively, these studies highlight the potential of VR technologies to not only improve physical safety and activity but also support mental and social well-being among older adults.

Mobile Applications

Mobile applications have increasingly been designed to support various aspects of daily life for older adults, aiming to enhance their social engagement, safety, and overall well-being.

Goumopoulos et al. (2017) developed the "Senior App Suite," which includes features such as social event notifications, emergency buttons, fall detection, and medication reminders to help older individuals stay connected and safe. Similarly, Yurkewich et al. (2018) evaluated tablet-based communication technology tailored for individuals with mild cognitive impairment (MCI), modifying the Google Gmail interface to improve usability and reduce complexity. The study found that female participants were more interested in using the technology for social communication, while males preferred it for information gathering. Ray et al. (2017) introduced a digital photo frame system designed to monitor well-being and foster social connections, utilizing a three-stage design methodology known as "Silvercare" for education and support across different countries. Collectively, these examples illustrate the diverse applications of mobile technology in addressing the needs of older adults, demonstrating its potential to enhance safety, connectivity, and engagement in their daily lives.

Technology Integration

This subcategory encompasses the integration of various technologies, including those mentioned previously, with the Internet of Things (IoT) in Ambient Assisted Living (AAL) environments. The goal is to combine these technologies to enhance the functionality and effectiveness of care and support systems for older adults. These technologies are often integrated with the Internet of Things (IoT) in AAL systems. For example, a study by Davis-Owusu et al. (2019) explored how incorporating everyday objects, such as Philips Hue lamps and LED walking sticks, into

AAL environments can enhance social connectedness, encourage physical activity, and foster better social interactions.

Kyritsis et al. (2018) developed a tablet platform for seniors to assist with physical activity, tracking, games, gamification, and social interaction. Their study highlighted the importance of ensuring that older users are fully aware of technology's communication capabilities. Similarly, Luperto et al. (2022) introduced MoveCare, a platform integrated with AAL and Socially Assistive Robotics (SAR) (Giraff-X) to provide comprehensive monitoring, assistance, and stimulation for seniors living alone and at risk of falling. This system includes a mobile robot for interaction and software functions to manage intelligent objects. Additionally, GameAAL, as detailed by Pinto et al. (2019), employs big data and machine learning to monitor daily activities, promote a healthy lifestyle, and adapt cognitive training based on performance. The system's interface provides a detailed overview of daily tasks, progress, and educational information, achieving 93% accuracy in performance validation. The integration of these technologies within AAL environments demonstrates their potential to create more cohesive and supportive systems for older adults, enhancing their overall quality of life through advanced and interconnected solutions.

The concept of ubiquitous healthcare, or U-Health, introduced by Aneke et al. (2018), aims to provide accessible healthcare services anytime and anywhere by overcoming traditional barriers. U-Health systems utilize sensors to collect patient health data, which are transmitted wirelessly via technologies such as 3G, WLAN, LAN, or GPRS/SMS to various monitoring stations, including hospitals and ambulatory services. These systems encompass vital health sign monitoring, activity planning, and multimedia reminders, allowing for real-time emergency alerts and remote physiological data monitoring by practitioners. In pursuit of making healthcare more affordable for low-income communities, Aneke et al. (2018) propose the iHealthBag system, which leverages cost-effective Arduino-based devices and applications to deliver essential health services efficiently. The system includes remote monitoring, goal setting, planning of activities, personal profile, and status of environmental sensors. Bui and Chong (2018) used a humanoid robot named Pepper and a smart care home interface to improve nursing care in aging societies. Pepper has human-like social skills and can access the iHouse network to provide user-requested data and services. They used an open platform for an Ambient Assisted Living (AAL) solution called Universal, which connects various components, devices, and services in the proposed software framework to enhance the quality of life for smart home residents. Kearney et al. (2018) developed SARA, a Socially Assistive Robotics solution designed for older people with MGI, which aids in daily living tasks such as health monitoring. However, SARA solutions face significant technological challenges, including robust security, privacy, and safety requirements, device and data interoperability, challenges in artificial intelligence, and network management. Innovative technologies addressed by SEMIOTICS could prove invaluable in overcoming these obstacles and advancing the effectiveness of smart home solutions for aging populations. Ultimately, addressing these challenges through innovative approaches will be crucial for developing more reliable and user-friendly technologies for older adults.

Category B: Supporting Technologies

This category focuses on technologies designed to address fundamental needs of older adults, particularly those identified in Table 10.1. It aims to overcome limitations associated with advanced age by improving technologies previously discussed. These supporting technologies target key areas such as psychography, sensation and perception, cognition, and mobility. By analyzing the constraints imposed by aging, this category seeks to enhance existing solutions and offers new opportunities for improving the quality of life for older adults. The results highlight a broad spectrum of technological innovations that address specific challenges related to cognitive decline, sensory impairments, and mobility issues, demonstrating the potential for tailored solutions to meet the diverse needs of the aging population.

Potential for Improvement

Voice-controlled smart homes are now possible thanks to speech recognition technology. Caranica et al. (2017) experimented with building acoustic and grammar models for Romanian using distant speech recognition scenarios. Their results were promising, with the best power-normalized cepstral coefficient model offering up to a 55% gain compared to other acoustic models.

Creating new grammar models is also essential, along with studying vocal cues in different cultures and languages, since the voice of the interface is essential for acceptance and engagement. Chang et al. (2018) show how social robots can have personalities by changing their voice. Their participants, tech-savvy baby boomers in Taiwan, were split into groups based on demographics, lifestyle, and personality. HRI can take many forms, and social robots with AI can adjust their voice cues to cater to tasks or detect users' moods. A limitation of Chang et al.'s (2018) study is that it did not consider hearing loss, given that hearing loss is a common disability in older adults, affecting speech-in-noise perception and contributing to poor cognitive health, social isolation, and loneliness (Mahmud et al., 2020); hearing loss is crucial for effective communication.

Mahmud et al. (2020) developed a framework to distinguish normal hearing from mild hearing impairment using machine learning. Their findings could aid in developing assistive devices to amplify specific speech sounds for hearing-impaired individuals and detect hearing loss early in clinical settings. Another solution is to use other types of signals. Desai and Desai (2017) developed an algorithm that identifies hand gestures and controls home appliances. They used a Kinect sensor and Arduino to assist physically challenged and senior citizens in operating appliances with minimal effort. Ganokratanaa and Pumrin (2018) created a hand gesture recognition technique for older people where their algorithm does not require devices attached to the user, making it comfortable for seniors. The success of HRI using gestures depends on both the robot's ability to recognize gestures and the user's ability to perform them (Werner et al., 2020). Training can be provided in gestural commands to enhance the performance of older adults. A 10–15-min training session on gestural commands measured participants' performance and the robot's ability to recognize commands before and after the activity. This training session can also benefit those with mild-to-moderate cognitive impairment or those who initially struggle with gestural performance and have negative attitudes toward technology.

Li et al. (2017) introduced a new way for people with motion impairment to interact with robots using it. By looking at a specific object in the physical world, the user can communicate their desired task to the robot. Participants in the study used 3D gaze-based interaction to control a robotic arm and grasp objects, showing that this method of HRI is practical and intuitive. Their framework could improve the control of gaze signals in complex robotic tasks.

Valsamakis and Savidis (2017) developed an AAL framework using IoT technologies that allows for customized automation based on changing user requirements. The framework runs on smartphones, and developers have fully developed it in JavaScript." Sinha and Caverly (2020) developed "Eyehear," a device that helps people with hearing impairments locate speech through speech recognition glasses. The glasses have a rectangular microphone and a visual display that shows the real-time direction and transcription of speech. The aim is to increase user safety and communication and improve spatial awareness for hard-of-hearing users. The glasses offer a unique way for users to communicate and understand speech without guessing what was said. Esposito et al. (2019) validated virtual agents to improve well-being and found that healthy seniors prefer female agents, and those with prior tech experience found them less engaging. Gulati & Kaur (2021) created a system prototype called Friend-CareAAL using the "Home Sensor Simulator," which predicts the well-being of elderly individuals. The proposed system effectively handles emergencies and remote health monitoring, but challenges such as scalability and reliability arise in real smart home environments.

Discussion

Technology has played a significant role in the healthcare industry for many years because it can help seniors maintain independence while combating decline and social isolation. Older people generally prefer to live in their homes as long as possible; for instance, a study by Farber et al. (2011) highlights that most seniors choose to age in place due to a desire for familiarity and comfort. Technologies such as remote health monitoring systems, smart home devices, and telehealth services can significantly enhance well-being and delay institutionalization.

For example, remote health monitoring systems allow for continuous tracking of vital signs and health metrics, enabling timely interventions and reducing the need for frequent hospital visits. A study by Hassanalieragh et al. (2015) demonstrated that such systems improve the management of chronic conditions and increase seniors' confidence in managing their own health. Smart home devices, including fall detection sensors and automated lighting, can enhance safety and convenience. These technologies reduce the risk of accidents and help seniors perform daily activities more easily, thereby supporting their preference to stay at home. Telehealth services provide access to medical consultations without the need for travel, which is particularly beneficial for those with mobility issues. Evidence from a study by Ma et al. (2022) shows that telehealth can improve access to care and reduce hospital readmissions by enabling regular check-ins and consultations. Overall, the integration of these technologies not only supports the independence and well-being of older adults but also plays a crucial role in delaying or avoiding the need for institutional care. This approach aligns with the preference of most seniors to remain in their homes while receiving the support they need.

Themes identified in smart home environments, especially for older adults, scalability and reliability are essential. Ensuring that new devices can be seamlessly integrated and that the system can expand to meet evolving needs is vital. Additionally, the system must maintain high reliability to function correctly, particularly for critical safety features. Providing accessible support and maintaining robust security to protect user privacy are also key considerations. Several themes have emerged as key areas of concern for research in technology for aging. First, there is considerable variability in older people's circumstances, with two people of the same age having attributes, living environments, and situations that may be entirely different (Di Napoli et al., 2022). This variability is illustrated and explored further by Fitter et al. (2020), Khosla et al. (2016), and Davis-Owusu et al. (2019). The variability in older people's circumstances presents significant challenges for designing effective technology for aging. Considering the variation in health and abilities, for example, older adults have diverse health conditions and physical abilities. Two individuals of the same age might have vastly different needs due to variations in chronic illnesses, mobility, or sensory impairments. Personal preferences are another aspect to consider. Preferences for technology use, from types of devices to interaction methods, can vary widely. Some may prefer voice commands, while others might find touchscreens more intuitive.

Second, small sample size for between-subject comparisons in many papers identified in this scoping review presents a challenge for making inferences on relationships between socially assistive technologies and seniors. Unbehaun et al. (2020) had the most significant participants from the selected articles (52 PwD and 25 caregivers). Cruz-Sandoval and Favela (2019) could not have total control over their attendance, showing one of the challenges of working with participants with mild-to-moderate stage dementia. Therefore, researchers must be cautious in drawing a defined conclusion (Obayashi et al., 2020; Pedrozo et al., 2022).

Third, designing peripheral technologies that match the user's mental ideal and establish system trust is challenging (Davis-Owusu et al., 2019). Matilda achieved this with anthropomorphism, and Pinto et al. (2019) achieved this with GameAAL weekly updates (ensuring better reliability and security given actual behavior and routines). Third, designing peripheral technologies that match the user's mental ideal and establish system trust is challenging (Davis-Owusu et al., 2019). Achieving this involves creating interfaces and interactions that align closely with users' expectations and cognitive models. Matilda achieved this with anthropomorphism, giving the technology human-like qualities that made interactions more intuitive and relatable for users. This approach can foster emotional connections and trust, making users more comfortable and willing to engage with technology. Similarly, Pinto et al. (2019) achieved success with GameAAL by providing weekly updates that enhanced the system's reliability and security. These updates ensured that the technology remained aligned with the actual behaviors and routines of its users. By continuously adapting and improving based on user feedback and usage patterns, GameAAL maintained relevance and trustworthiness, addressing the evolving needs and expectations of older adults. Both cases highlight the importance of designing technologies that are not only functional but also resonate with users' mental models and experiences. This requires ongoing user research, iterative design processes, and a commitment to understanding and addressing the unique challenges and preferences of older adults.

Fourth, using user data in assistive technologies for older adults is a sensitive topic (Goumopoulos et al., 2017), challenging research in this area. Kyritsis et al. (2018) have shown that many are willing to share their data with family members and professionals, but sometimes, there is uncertainty about what "data sharing" means. For example, one should consider security, privacy, dependability, and safety. Moreover, the environmental condition should encompass the quality of electricity and the Internet (Aneke et al., 2018).

Fifth, even though SAR at home could provide remote monitoring of their users' well-being (physical and psychological support), private home environments are particularly difficult for SAR due to their unstructured and dynamic nature, which often contributes to robot failures. Researchers have developed several SAR prototypes for caring for older adults for these reasons, but they have yet to effectively commercialize and use them in real environments (Luperto et al., 2022). Furthermore, some questionnaires can often exhibit bias because the items typically rely on the researcher's experience and assumptions, causing participants to select a response that does not reflect their opinions.

Dimensions of Social Determinants of Health

When considering the social determinants of health, it is important to note that one domain can complement the other and share similar minor objectives. When considering the subdomain theme of *health care access and quality,* goals are to enhance healthcare accessibility by utilizing telemedicine and gathering, sharing, and analyzing health data. By utilizing monitoring technologies like Aneke et al. (2018),

Pinto et al. (2019), and Luperto et al. (2022) shown, for example, we can collect information on the health of older adults and provide evidence-based preventative healthcare. Home automation and robotics systems can help older people maintain their autonomy and delay the need for social and health service interventions. Pinto et al. (2019) also identified the participants' preference for using intelligent objects (e.g., IoT items) when associated with SAR. Many older people have used tested technology with someone by their side since, in general, they have worked with people with cognitive and physical problems for correct, consistent content operations for all participants (Lin et al., 2018). Moreover, the acceptance of these technologies may depend on previous experiences, ease of use, and accessibility (Chang et al., 2018; Sinha & Caverly, 2020). Training and connections to the IoTs improve acceptability (Rings et al., 2020). Gulati & Kaur (2021) stressed the importance of social connections between smart devices for IoT integration in AAL.

When considering the SDoH subdomain of *social and community context,* we consider goals to enhance the support provided to society and community. Numerous articles support the technologies mentioned in the outcomes and the initial column of "older adults needs" (Table 10.3) promoting social interactions among individuals. It can reduce family concerns and enable seniors to discuss their health with their loved ones, like in the previous domain. It aims to reduce the number of people with intellectual and developmental disabilities residing in institutional settings with more than seven individuals. One effective way to accomplish this is also increasing the proportion of adults who utilize information technology to monitor their healthcare data or communicate with their healthcare providers. The cultural context of a community affects the attribution of personality to robots and the degree of anthropomorphism, expectations, and preferences about their role in society and how they should be (Trovato et al., 2019). Their physical appearance influences the acceptance of robots, so exposed metal parts, cables, and wires could also reduce product reliability (Luperto et al., 2022).

Conclusion

This chapter explored various applications of assistive technologies for older adults in their home environment, identifying barriers and facilitators. Technology can potentially reduce hospitalization rates and emergency cases, resulting in better quality care delivery from primary care providers who can establish more effective communications with their patients (Pugh et al., 2021). Adults can increase their online access to medical records, and doctors can access necessary information electronically. However, challenges arise, such as ensuring technologies comply with data protection laws. Furthermore, the product's usability depends on the user's language proficiency and acceptance. This is crucial when considering the variability among older adults, as identified in our findings, where technological integration must cater to diverse needs. Additionally, the design of these technologies needs to be intuitive to promote widespread adoption and effective use among older adults.

Assistive robots aim to improve cognitive, physical, and interpersonal skills through various activities and exercises. Analyzing social signals involves user engagement assessments tailored to individual user profiles. The studies identified in this review highlighted the importance of innovative approaches to promoting well-being in different population groups through technology, as well as the importance of incorporating social signals into the design of promising technologies. Sensory perception, like hearing, plays a crucial role in communicating and interacting with people. The results of this review reinforce the importance of these technologies for the older person, as they have demonstrated the ability to improve physical and cognitive resources, encourage social interactions, and provide support, also relieving the burden on caregivers. However, challenges and limitations persist, such as the impact of poverty and limited access to the Internet, which can hinder the use and effectiveness of these technologies. Many older adults may not have the financial resources to afford the necessary devices or the stable Internet connection required for optimal use. Additionally, the digital divide often leaves marginalized communities without adequate access to these beneficial technologies, exacerbating existing health disparities. Addressing these barriers is crucial for ensuring that the advantages of assistive technologies are accessible to all older adults, regardless of their socioeconomic status. In this context, they play a fundamental role in improving the quality of life of the older adult, promoting social interactions, supporting daily activities, and contributing to healthy aging—also covering some of the Social Determinants of Health Objectives—SDoH objectives.

Considering the social-spatial determinants of health, this chapter examines how the residential environment impacts the effectiveness and accessibility of these technologies. The conclusion points to the potential of these technologies to improve the quality of life of older adults, promote social interactions, and support healthy aging, aligning with the SDoH objectives. By incorporating place-based perspectives, it becomes clear that the residential context plays a crucial role in the adoption and success of assistive technologies, highlighting the need for tailored solutions that consider the unique spatial and social dynamics of older adults' living environments.

References

Aneke, J., Ardito, C., Caivano, D., Colizzi, L., Costabile, M. F., & Verardi, L. (2018). A low-cost flexible IoT system supporting elderly's healthcare in rural villages. In *A low-cost flexible IoT system supporting elderly's healthcare in rural villages*. https://doi.org/10.1145/3283458.3283470

Bersch, R. (2014). Tecnologia assistiva ou tecnologia de reabilitação. *Simpósio Internacional de Tecnologia Assistiva, 1*, 45–50.

Bui, H.-D., & Chong, N. Y. (2018). An integrated approach to human-robot-smart environment interaction interface for ambient assisted living. In *2018 IEEE workshop on Advanced Robotics and its Social Impacts (ARSO)*. IEEE. https://doi.org/10.1109/arso.2018.8625821

Caranica, A., Cucu, H., Burileanu, C., Portet, F., & Vacher, M. (2017). Speech recognition results for voice-controlled assistive applications. In *2017 International Conference on Speech Technology and Human-Computer Dialogue (SpeD)*. IEEE. https://doi.org/10.1109/sped.2017.7990438

Chang, R. C.-S., Lu, H.-P., & Yang, P. (2018). Stereotypes or golden rules? Exploring likable voice traits of social Robots as active aging companions for tech-savvy baby boomers in Taiwan. *Computers in Human Behavior, 84*, 194–210. https://doi.org/10.1016/j.chb.2018.02.025

Cruz-Sandoval, D., & Favela, J. (2019). Incorporating conversational strategies in a social robot to interact with people with dementia. *Dementia and Geriatric Cognitive Disorders, 47*(3), 140–148. https://doi.org/10.1159/000497801

Czaja, S. J., Boot, W. R., Charness, N., & Rogers, W. A. (2019). *Designing for older adults: Principles and creative human factors approaches* (3rd ed.). Taylor & Francis Group.

Davis-Owusu, K., Owusu, E., Marcenaro, L., Regazzoni, C., Feijs, L., & Hu, J. (2019). Towards a deeper understanding of the behavioural implications of bidirectional activity-based ambient displays in ambient assisted living environments. *Lecture Notes in Computer Science*, 108–151. https://doi.org/10.1007/978-3-030-10752-9_6

Desai, S., & Desai, A. A. (2017). Human computer interaction through hand gestures for home automation using microsoft kinect. *Advances in Intelligent Systems and Computing*, 19–29. https://doi.org/10.1007/978-981-10-2750-5_3

Di Napoli, C., Ercolano, G., & Rossi, S. (2022). Personalized home-care support for the elderly: A field experience with a social robot at home. *User Modeling and User-Adapted Interaction, 33*(2), 405–440. https://doi.org/10.1007/s11257-022-09333-y

Esposito, A., Amorese, T., Cuciniello, M., Esposito, A. M., Troncone, A., Torres, M. I., Schlögl, S., & Cordasco, G. (2019). Seniors' acceptance of virtual humanoid agents. *Lecture Notes in Electrical Engineering*, 429–443. https://doi.org/10.1007/978-3-030-05921-7_35

Fan, J., Ullal, A., Beuscher, L., Mion, L. C., Newhouse, P., & Sarkar, N. (2021). Field testing of Ro-Tri, a robot-mediated triadic interaction for older adults. *International Journal of Social Robotics, 13*(7), 1711–1727. https://doi.org/10.1007/s12369-021-00760-2

Farber, N. J. D., Shinkle, D., Lynott, J., Fox-Grage, W., & Harrell, R. (2011). Aging in place: A state survey of livability policies and practices. In *National conference of state legislatures*. AARP Public Policy Institute. ISBN 978-1-58024-645-3.

Fattal, C., Cossin, I., Pain, F., Haize, E., Marissael, C., Schmutz, S., & Ocnarescu, I. (2022). Perspectives on usability and accessibility of an autonomous humanoid robot living with elderly people. *Disability and Rehabilitation: Assistive Technology, 17*(4), 418–430. https://doi.org/10.1080/17483107.2020.1786732

Ferraz, A. F., & Peixoto, M. R. B. (1997). Qualidade de vida na velhice: Estudo em uma instituição pública de recreação para idosos. *Revista da Escola de Enfermagem da USP, 31*, 316–338.

Fitter, N. T., Mohan, M., Kuchenbecker, K. J., & Johnson, M. J. (2020). Exercising with Baxter: Preliminary support for assistive social-physical human-robot interaction. *Journal of Neuroengineering and Rehabilitation, 17*(1). https://doi.org/10.1186/s12984-020-0642-5

Galvão, M. C. B., & Ricarte, I. L. M. (2019). Revisão sistemática da literatura: Conceituação, produção e publicação. *Logeion: Filosofia da informação, 6*(1), 57–73. https://doi.org/10.21728/logeion.2019v6n1.p57-73

Ganokratanaa, T., & Pumrin, S. (2018). The algorithm of static hand gesture recognition using rule-based classification. *Advances in Intelligent Systems and Computing*, 173–184. https://doi.org/10.1007/978-3-319-70016-8_15

Goumopoulos, C., Papa, I., & Stavrianos, A. (2017). Development and evaluation of a mobile application suite for enhancing the social inclusion and well-being of seniors. *Informatics, 4*(3), 15. https://doi.org/10.3390/informatics4030015

Gulati, N., & Kaur, P. D. (2021). FriendCare-AAL: A robust social IoT based alert generation system for ambient assisted living. *Journal of Ambient Intelligence and Humanized Computing*. https://doi.org/10.1007/s12652-021-03236-3

Hassanalieragh, M., Page, A., Soyata, T., Sharma, G., Aktas, M., Mateos, G., Kantarci, B., & Andreescu, S. (2015). Health monitoring and management using Internet-of-Things (IoT) sensing with cloud-based processing: Opportunities and challenges. In *2015 IEEE International Conference on Services Computing*. https://doi.org/10.1109/scc.2015.47

Healthy People 2030 (2023). *U.S. Department of Health and Human Services, Office of Disease Prevention and Health Promotion*. Retrieved [03 April 2023], from https://www.cdc.gov/about/sdoh/index.html.

Kearney, K. T., Presenza, D., Sacca, F., & Wright, P. (2018). Key challenges for developing a Socially Assistive Robotic (SAR) solution for the health sector. In *2018 IEEE 23rd international workshop on computer aided modeling and design of communication links and networks (CAMAD)*. IEEE. https://doi.org/10.1109/camad.2018.8515005

Khosla, R., Nguyen, K., & Chu, M.-T. (2016). Human robot engagement and acceptability in residential aged care. *International Journal of Human–Computer Interaction, 33*(6), 510–522. https://doi.org/10.1080/10447318.2016.1275435

Koceska, N., Koceski, S., Beomonte Zobel, P., Trajkovik, V., & Garcia, N. (2019). A telemedicine robot system for assisted and independent living. *Sensors, 19*(4), 834. https://doi.org/10.3390/s19040834

Kyritsis, A. I., Nuss, J., Holding, L., Rogers, P., O'Connor, M., Kostopoulos, P., Suffield, M., Deriaz, M., & Konstantas, D. (2018). User requirement analysis for the design of a gamified ambient assisted living application. *Lecture Notes in Computer Science*, 426–433. https://doi.org/10.1007/978-3-319-94274-2_61

Li, S., Zhang, X., & Webb, J. (2017). 3-D-Gaze-based robotic grasping through mimicking human visuomotor function for people with motion impairments. *IEEE Transactions on Biomedical Engineering, 64*(12), 2824–2835. https://doi.org/10.1109/tbme.2017.2677902

Lin, C. X., Lee, C., Lally, D., & Coughlin, J. F. (2018). Impact of virtual reality (VR) experience on older adults' well-being. In *Human aspects of IT for the aged population. Applications in health, assistance, and entertainment: 4th international conference, ITAP 2018, Held as Part of HCI International 2018, Las Vegas, NV, USA, July 15–20, 2018, Proceedings, Part II 4* (pp. 89–100). Springer International Publishing. https://doi.org/10.1145/3167132.3167456

Lin, Y.-C., Fan, J., Tate, J. A., Sarkar, N., & Mion, L. C. (2022). Use of robots to encourage social engagement between older adults. *Geriatric Nursing, 43*, 97–103. https://doi.org/10.1016/j.gerinurse.2021.11.008

Luperto, M., Monroy, J., Renoux, J., Lunardini, F., Basilico, N., Bulgheroni, M., Cangelosi, A., Cesari, M., Cid, M., Ianes, A., Gonzalez-Jimenez, J., Kounoudes, A., Mari, D., Prisacariu, V., Savanovic, A., Ferrante, S., & Borghese, N. A. (2022). Integrating social assistive robots, IoT, virtual communities and smart objects to assist at-home independently living elders: The movecare project. *International Journal of Social Robotics*. https://doi.org/10.1007/s12369-021-00843-0

Ma, Y., Zhao, C., Zhao, Y., Lu, J., Jiang, H., Cao, Y., & Xu, Y. (2022). Telemedicine application in patients with chronic disease: A systematic review and meta-analysis. *BMC Medical Informatics and Decision Making, 22*(1). https://doi.org/10.1186/s12911-022-01845-2

Machado, J. D., & Alvarenga, M. R. M. (2019). Acuidade visual diminuída decorrente do processo de envelhecimento. *Barbaquá, 3*(6), 57–64. https://periodicosonline.uems.br/index.php/barbaqua/article/view/4870

Mahmud, M. S., Ahmed, F., Yeasin, M., Alain, C., & Bidelman, G. M. (2020). Multivariate models for decoding hearing impairment using EEG Gamma-band power spectral density. In *2020 International Joint Conference on Neural Networks (IJCNN)*. https://doi.org/10.1109/ijcnn48605.2020.9206731

Obayashi, K., Kodate, N., & Masuyama, S. (2020). Measuring the impact of age, gender and dementia on communication-robot interventions in residential care homes. *Geriatrics & Gerontology International, 20*(4), 373–378. https://doi.org/10.1111/ggi.13890

Pedrozo, T., Bandeira, C., Crocetta, T. B., Yohanna, J., Cardoso, N., Batista, J., Thais, R., Guarnieri, R., Patricio, A., Raimundo, R. D., & de Abreu, L. C. (2022). Digital games in the computer classes to reduce loneliness of individuals during aging. *Current Psychology, 42*(15), 12857–12865. https://doi.org/10.1007/s12144-021-02521-w

Pinto, M., Pereira, M., Raposo, D., Simões, M., & Castelo-Branco, M. (2019). Artificial intelligence gamified AAL solution. *IFMBE Proceedings*, 977–982. https://doi.org/10.1007/978-3-030-31635-8_119

Pugh, J., Penney, L. S., Noël, P. H., Neller, S., Mader, M., Finley, E. P., Lanham, H. J., & Leykum, L. (2021). Evidence based processes to prevent readmissions: More is better, a ten-site observational study. *BMC Health Services Research, 21*(1), 189. https://doi.org/10.1186/s12913-021-06193-x

Ray, P., Li, J., Ariani, A., & Kapadia, V. (2017). Tablet-based well-being check for the elderly: Development and evaluation of usability and acceptability. *JMIR Human Factors, 4*(2), e12. https://doi.org/10.2196/humanfactors.7240

Ricci, F. S., Boldini, A., Beheshti, M., Rizzo, J.-R., & Porfiri, M. (2022). A virtual reality platform to simulate orientation and mobility training for the visually impaired. *Virtual Reality*. https://doi.org/10.1007/s10055-022-00691-x

Rings, S., Steinicke, F., Picker, T., & Prasuhn, C. (2020). Seated immersive exergaming for fall prevention of older adults. In *2020 IEEE conference on virtual reality and 3D user interfaces abstracts and workshops (VRW)*. https://doi.org/10.1109/vrw50115.2020.00063

Sefcik, J. S., Johnson, M. J., Yim, M., Lau, T., Vivio, N., Mucchiani, C., & Cacchione, P. Z. (2017). Stakeholders' perceptions sought to inform the development of a low-cost mobile robot for older adults: A qualitative descriptive study. *Clinical Nursing Research, 27*(1), 61–80. https://doi.org/10.1177/1054773817730517

Sinha, I., & Caverly, O. (2020). EyeHear: Smart glasses for the hearing impaired. In *HCI International 2020—late breaking papers: Universal access and inclusive design* (pp. 358–370). https://doi.org/10.1007/978-3-030-60149-2_28

Trovato, G., Kishi, T., Kawai, M., Zhong, T., Lin, J.-Y., Gu, Z., Oshiyama, C., & Takanishi, A. (2019). The creation of DarumaTO: A social companion robot for Buddhist/Shinto elderlies. In *2019 IEEE/ASME International Conference on Advanced Intelligent Mechatronics (AIM)*. IEEE. https://doi.org/10.1109/aim.2019.8868736

Unbehaun, D., Aal, K., Vaziri, D. D., Tolmie, P. D., Wieching, R., Randall, D., & Wulf, V. (2020). Social technology appropriation in dementia: Investigating the role of caregivers in engaging people with dementia with a videogame-based training system. In *Proceedings of the 2020 CHI conference on human factors in computing systems*. https://doi.org/10.1145/3313831.3376648

Valsamakis, Y., & Savidis, A. (2017). Visual end-user programming of personalized AAL in the Internet of Things. In *Ambient intelligence: 13th European Conference, AmI 2017, Malaga, Spain, April 26–28, 2017, proceedings* (pp. 159–174). https://doi.org/10.1007/978-3-319-56997-0_13

Vecchia, R. D., Ruiz, T., Bocchi, S. C. M., & Corrente, J. E. (2005). Qualidade de vida na terceira idade: Um conceito subjetivo. *Revista Brasileira de Epidemiologia, 8*, 246–252. https://doi.org/10.1590/S1415-790X2005000300006

Werner, C., Kardaris, N., Koutras, P., Zlatintsi, A., Maragos, P., Bauer, J. M., & Hauer, K. (2020). Improving gesture-based interaction between an assistive bathing robot and older adults via user training on the gestural commands. *Archives of Gerontology and Geriatrics, 87*, 103996. https://doi.org/10.1016/j.archger.2019.103996

Wiles, J. L., Leibing, A., Guberman, N., Reeve, J., & Allen, R. E. S. (2011). The meaning of "Aging in Place" to older people. *The Gerontologist, 52*(3), 357–366. https://doi.org/10.1093/geront/gnr098

Yurkewich, A., Stern, A., Alam, R., & Baecker, R. (2018). A field study of older adults with cognitive impairment using tablets for communication at home. *International Journal of Mobile Human Computer Interaction, 10*(2), 1–30. https://doi.org/10.4018/ijmhci.2018040101

Chapter 11
Teaching and Learning About Geo-Social Determinants of Health

Sigrid Van Den Abbeele, Daniel Grafton, Madison Avila, Esaú Casimiro Vieyra, Brianna Chan, Gabrielle Husted, Sofia Kaloper, Ben Moscona, and Susan Cassels

Introduction

Given the significance of a spatial perspective for the social determinants of health frameworks on health disparities, we must ensure that these concepts are taught and included in graduate coursework in social science and health-related disciplines. A graduate seminar entitled "Geo-Social Determinants of Health & Health Disparities" presented a good opportunity to reflect on the importance of a social-spatial determinants of health framework in the teaching and learning of population health and health disparities. In this chapter, authored by the students and professor of the course, we first explore themes common to the course content and our discussions about our own teaching and learning. We then include three short essays in which we rely on a foundational concept from the class to discuss our learning in the course and apply our knowledge of the concept to a case study of a health disparity. The last section of the chapter reflects on equity in teaching and learning, concluding with suggestions to optimize approaches to teaching and learning about geo-social determinants of health.

The learning objectives of the course included an understanding of the (a) importance of geo-social factors in health exposures and outcomes, (b) purpose and place of social determinants of health within the broader disciplines of geography, demography, and public health, and (c) central questions and theory of geo-social determinants of health. This graduate course is offered in a geography department but draws an interdisciplinary group of students with various academic experiences. After introducing definitions of health disparities, subsequent weeks of the course covered topics related to socio-spatial determinants of health, including race, gender,

S. Van Den Abbeele (✉) · D. Grafton · M. Avila · E. C. Vieyra · B. Chan · G. Husted ·
S. Kaloper · B. Moscona · S. Cassels
University of California, Santa Barbara, Santa Barbara, CA, USA
e-mail: sigrid.van.den.abbeele@geog.ucsb.edu

© The Author(s) 2026
M. A. Kolak, I. K. Moise (eds.), *Place and the Social-Spatial Determinants of Health*, Global Perspectives on Health Geography,
https://doi.org/10.1007/978-3-031-88463-4_11

socioeconomic status, the built environment, and neighborhood context. Weekly readings were crafted to explore each topic while also exposing students to key concepts in the study of place and health, such as fundamental and proximate causes of health. Given the time constraints of the course, the concepts covered were confined to health disparities in a US context. Each week in this seminar course, a student summarizes the assigned readings and facilitates discussion on that week's topic. In the course, students were evaluated on participation in course discussions, facilitation of a week's discussion, the creation of a reading list for a hypothetical week of the course, and a final project. For their final projects, students were given the option of writing a grant proposal or contributing to this book chapter by writing short essays in small groups. No pre/post survey was conducted for this course.

One of the articles covered in this course cites Robert Sampson's Presidential Address to the American Society of Criminology (Sharkey & Faber, 2014). In this address, Sampson highlights the notion of place as "a fundamental context" and calls on the audience to "relentlessly focus on context" (Sampson, 2013). Students in this course have embraced this emphasis on context, frequently setting topics on metaphorical axes of social, spatial, and temporal context during class discussions. By explicitly defining the spatial scale, social identities, and life stage to which findings in an article may be applied, students can better put course concepts into conversation with their work and identify limitations in the literature.

This chapter will demonstrate students' ability to understand a fundamental concept related to health disparities by using course readings to explore the social and spatial context of the topic. These short essays by small groups of students apply their knowledge about a fundamental course concept to a case study that either expands on a topic covered briefly in the course or explores a subject that was not addressed in the course. Students will also comment on how their personal context impacted their learning. One essay investigates food insecurity to better understand the definition of neighborhoods and their role in health disparities. A second essay focuses on life course effects through a case study of heat exposure across the lifespan. The final essay examines fundamental and proximate causes of health disparities through an environmental justice-oriented case study of the impacts of uranium mining on the health of indigenous people. A place-based and social-spatial approach to health disparities is woven throughout the essays.

Food Insecurity

To understand the determinants of nutrition-related health conditions, we must consider the context surrounding an individual's lived experience. The simple question of "why not eat healthy food" leads to far more complex answers relating to food availability (Diez Roux, 2001), food access, and constrained choice (Rieker et al., 2010). These answers suggest that the places in which we live, work, and socialize affect the food we consume because of the spatial variability of access and availability. Without accounting for geographic variability, we cannot provide a nuanced explanation for health disparities. We can better define food accessibility by incorporating a spatial perspective into discussions of nutritional health. Where we live impacts our health, but the impact is not evenly distributed. The determinants that either help or hinder health vary spatially, making a geographic perspective essential for understanding the context of health determinants. In this section, we reflect on the concept of "neighborhoods" and the importance of learning about sociospatial determinants of health using food insecurity as a case study. We weave aspects of our personal and learning contexts to identify the most effective ways to teach and learn social determinants of health.

Neighborhoods are an important concept for contextualizing health disparities in geographically informed research. Broadly, a neighborhood can be defined as the physical and sociocultural features shared by all residents in a locality (Macintyre et al., 2002). Neighborhoods are not independent concrete places but dependent clusters of people. There are more similarities among people within neighborhoods than between neighborhoods (Oakes, 2004). Neighborhoods reflect the socioeconomic status of residents, thereby influencing exposure to both risk factors and mitigating factors that impact the health of residents (Williams & Collins, 2001). Healthy, fresh food is less available in lower-income neighborhoods, while fast-food restaurants and liquor stores are more prevalent than in higher-income neighborhoods (Pampel et al., 2010). Not only do neighborhoods explain factors that directly affect their residents' health, but they also explain the mechanisms through which factors further upstream perpetuate residential health disparities.

The concept of a "neighborhood" is ambiguous, with numerous formal and informal definitions. A litany of definitions of neighborhoods used in the literature were covered in the course. Neighborhoods, understood as "a person's immediate residential environment," can be defined historically, administratively, and socially (Diez Roux, 2001). Even formal definitions of neighborhoods, like the administrative boundaries of a census tract, however, are imprecise and may not be suitable for analyzing health. Students discussed the merits of various definitions of neighborhoods as they related to different topics during class. Students also explored their own informal definitions of neighborhoods and how those definitions influenced their understanding of the material. For instance, one author grew up in a small town in West Virginia where the winding streets and haphazard urban planning handed down from colonial times resulted in ambiguous neighborhood boundaries. For this author, their neighborhood was where they went Halloween trick-or-treating as a child. Another author grew up in a small university town where neighborhood boundaries were marked by physical features such as agricultural fields and roads, creating an immediate distinction between neighborhoods. For this author, their sense of a neighborhood did not include any social context or interaction. They saw neighborhoods as intentional constructs defined by physical objects and markers. Because individuals experience their neighborhoods differently, our lived experiences lead us to create a unique definition of a neighborhood. To better understand what a neighborhood is and the different ways a place can impact health, it is important to acknowledge these different perceptions as a part of learning. In our class, we facilitated student learning by discussing how we individually perceive

neighborhoods allowing our class to gain insight into the ways in which neighborhoods can exist and interact with residents.

The neighborhoods in which people live shape their access to nutritious food
and, therefore, can contribute to food insecurity. Teaching food insecurity as a component of the social determinants should begin by defining food insecurity. Though
food insecurity was never formally defined in our course (as it was not one of the
key topics on the syllabus), there were overviews of food insecurity included in
several key articles touching on food as a human right (Braveman, 2006) and food's
central role as an "absolute" requirement for health (Kawachi et al., 2002). Of note
for the latter was the debate over "absolute poverty" and "relative poverty" and how
food security may be defined differently depending on the society. We found general agreement in the class that understanding both perspectives is critical to
addressing food insecurity in any given locale. Lack of access to grocery stores
selling nutritious foods is linked to numerous health consequences (Andretti
et al., 2023).

A geo-social perspective based on neighborhoods enhances our understanding of
the contexts that cause health disparities and is important in guiding effective health
interventions. A frequent solution to mitigate food insecurity is to increase the availability of healthy food in an area. On the surface, this type of intervention appears
to be an easily applicable solution to the problem. However, an understanding of
neighborhood contexts and the upstream determinants of health reveals this to be an
ineffective intervention. Increasing the availability of healthy food within a neighborhood does not necessarily make healthy food accessible because residents still
need to pay for and prepare it, rather than purchasing something cheaper or easier
to prepare. Similarly, attempting to build an exercise facility to combat obesity is an
ineffective intervention if the shared norms of the neighborhood do not support
exercising (Macintyre et al., 2002). Since the determinants of health vary spatially,
large-scale interventions are unlikely to adequately impact the various mechanisms
affecting health across areas with differing contexts. However, interventions tailored to neighborhoods can target the mechanisms affecting residential health.

Another key theme in our course was the impact of socioeconomic status (SES)
and social hierarchy on food insecurity. This began with the description of social
hierarchies by Michael Marmot, explored in the documentary *Unnatural Causes*
(2006), his classic work *The Status Syndrome* (2004) on the Whitehall Studies in the
United Kingdom, and book *The Health Gap* (2017) (Marmot, 2004, 2006, 2017).
Questions of food insecurity related to the social hierarchy and SES continuously
arose as we developed a richer understanding of the literature on social determinants of health. Using examples of food contamination and subsequent government
warnings for the public to take precautions, Ross and Mirowsky (2010) claim that
by "contextualizing" individually based risk factors, we better understand the social
conditions behind risk factors. Individuals with higher socioeconomic status are
better positioned to take such precautions. Williams and Collins (2001) note the
increased cost of food for racial minorities in the United States and how those living
in low-income neighborhoods have less access to healthy foods. The classic definition of "upstream factors" by McKinlay John (1975), the idea that certain

"manufacturers of illness" are pushing people into unhealthy activities/products, also addresses food insecurity and the growing crisis of declining food quality in America. The widespread adoption of refined sugars and the corresponding increase in obesity, especially childhood obesity, is a sobering example (McKinlay John, 1975). One of the chapter authors, a student parent, noted the continuing advertisement of sugary drinks and foods to children and the difficulty in finding time to provide healthy foods. McKinlay's call for food to "become part of politics" generated an engaging discussion among our class about the role of politics in food and how the lobbying issue might be addressed.

Last, neighborhoods reveal the importance of context in another key manner: the recruitment and distribution of foods at a local level in developing countries such as Brazil (Jacobs & Richtel, 2017). Multinational corporations have increasingly utilized direct sales teams in small communities to expand brand recognition and loyalty in new markets, resulting in massive changes to traditional diets and the widespread adoption of processed foods, often high in sugar. While increased employment opportunities may benefit these communities, there are also far greater issues, such as increasing obesity rates and children developing related illnesses such as hypertension (Jacobs & Richtel, 2017). Global climate change confounds food insecurity, especially in developing parts of the world, contributing to a loss of resilience and a decline in food security (McMichael, 2013). This issue resonated strongly with our class as one in which the USA has exported its issues abroad in the name of increasing profits for multinational corporations even as food swamps, areas with ample food but largely innutritious, replace traditional diets with negative health consequences.

Defining neighborhoods was challenging for our class. We each came into the course with our own definitions of neighborhoods and routinely put them in contrast with more formal definitions from the literature. However, this exercise was beneficial in understanding how neighborhoods add a critical geographical perspective when considering the relationship between socioeconomic status, health disparities, and the unequal burden of food insecurity. Importantly, neighborhoods have both social and material relevance and reflect people's lived experiences. Interventions targeting food insecurity within neighborhoods can harness these neighborhood characteristics to create localized and culturally competent responses to food insecurity.

Life Course and Heat Exposure

Throughout the lifecycle, environmental exposures impact health differently. "Life course effects" capture the phenomenon of how both current and previous living circumstances (such as in utero effects) influence health status at every age for each given birth cohort (Kawachi et al., 2002). Three leading hypotheses connect life course effects to the origins of health inequalities (Kawachi et al., 2002). One, "latent effects" identify how early life environments have far-reaching effects later in life. Two, "pathway effects" explain how early life environments can influence a person's trajectory and health status over time. Three, "cumulative effects" result from the dose-response relationship between the extent of exposure to detrimental environments across many years. These hypotheses frame our understanding of how the timing of exposures over the life course is a fundamental concept in geo-social determinants of health disparities. In this section, we leverage the foundational concept of "life course effects" to explore the implications of heat exposure throughout the life course. This issue resonated with the authors as climate change increasingly alters the planet. We also reflect on how personal contexts and the teaching of this foundational concept impacted our ability to apply our knowledge to this case study.

In the USA, urban areas have experienced increased heat waves over the past few years, with the potential to observe an increased number of days with extreme heat-related events by as much as seventy percent (Hill et al., 2021). These events can be dangerous and potentially life-threatening, especially for vulnerable populations such as young children, elderly individuals, and individuals with preexisting medical conditions (Ruddell et al., 2010). For example, the number of heat-related deaths observed will increase worldwide, but countries with higher rates of low-income and disadvantaged populations will suffer the most (McMichael, 2013). In 2006, California experienced one of the worst heat-related events in its history; several counties reported nearly 600 heat-related deaths during the heatwave that occurred between July 14 and July 30 (Margolis et al., 2008).

We also learned about structural levels of influence (Homan, 2019). We applied this conceptual model to understand the relationship between place and life course effects through our case study of heat exposure across the life course. Homan's conceptual model of structural levels of influence includes the macroscale (institutional systems, policy, norms), the mesoscale (interactions and behaviors or practices within institutions), and the microscale (individual-level identities and ideologies). At the microscale, behaviors to mitigate heat could take the form of household-level adaptations, such as wearing sun-protective clothing, utilizing air conditioning, spending time in cool indoor settings, geographic mobility, and house design (Deschenes, 2014). Mesoscale interventions could include community-level adaptations to extreme heat, communication of upcoming heat waves, outreach systems that comprehensively reach the population, distribution of fans, and establishment, awareness, and access to local cooling centers (Deschenes, 2014). Various macro-level measures can be taken to mitigate the risks of heat exposure, such as implementing heat wave warning systems, establishing community caregiver programs for vulnerable individuals, ensuring policies that establish well-insulated housing, and providing educational advice through primary healthcare providers (McMichael, 2013). Effective strategies to mitigate heat exposure could focus on large-scale interventions informed by research identifying communities or regions most vulnerable to extreme heat, coupled with public health advocacy to enhance awareness and resource availability for preventing heat-related illnesses in underserved communities.

The implications of heat exposure can be traced throughout the life course. Maternal exposure to extreme heat impacts the risk of preterm birth and low birth weight, neurodevelopmental disorders, heart conditions, asthma, and metabolic conditions (Pacheco, 2020). Children are more vulnerable to health effects from extreme heat than adults because of their reduced ability to regulate their body temperature (American Academy of Pediatrics, 2021). Children are at an increased risk of dehydration during extreme heat events (Seattle Children's Hospital, 2023). Several biological factors increase children's risk of dehydration (Antoniadis et al., 2020). Additionally, children are not in control of their own water intake and may be unable to communicate their needs (Zivin & Shrader, 2016). There is far more research on dehydration and cognitive performance within adult populations. However, research among children is increasing, and there appears to be an association between greater hydration and greater cognitive performance among school children (Edmonds & Burford, 2009). Micro- and mesoscale interventions could increase awareness of how heat impacts children and what precautions should be taken in the event of extreme heat to prevent dehydration and other health impacts.

On the other end of the age spectrum, advanced age poses a significant risk for heat-related death. Older adults have diminished thermoregulatory and physiologic heat-adaptation ability, frequently live alone, have fewer social contacts, and have more comorbidities (Luber & McGeehin, 2008). Extreme heat exacerbates comorbidities such as respiratory and cardiovascular diseases, diabetes, and renal disease (World Health Organization, 2018).

The negative impacts of heat exposure across the life course are not evenly distributed across the population. Recent research (Manware et al., 2022) suggests that Black Americans are most susceptible to extreme heat events since they are more likely to live within heat-vulnerable census tracts. An individual's neighborhood may lack the infrastructure to cope with extreme heat events that can impact health inequitably. O'Neill et al. (2005) found that Black and Latino households, compared to White households, were less likely to have central air conditioning within their homes in four cities: Chicago, Detroit, Minneapolis, and Pittsburgh. Williams and Collins (1995, 2001) argue that neighborhood segregation, rooted in the United States's legacy of institutionalized racism, largely explains socioeconomic disparities for Black Americans and subsequent disparities in health. Black Americans have overall higher rates of morbidity and mortality than White Americans and are more likely to have comorbidities that elevate their risk of extreme heat events— high blood pressure, stroke, diabetes, and renal disease (CDC, 2017, 2022; Race, Ethnicity, & Kidney Disease—NIDDK, 2014; World Health Organization, 2018). These differences also apply to socioeconomic differences. Individuals with limited financial resources may be more hesitant to seek medical attention or run central air conditioning in their homes (Gronlund, 2014). Additionally, occupations requiring manual or outdoor labor may create vulnerable conditions for individuals during extreme heat (Gronlund, 2014). However, socioeconomic status and race/ethnicity characteristics are interconnected: Williams and Collins (1995) describe how often individuals with low SES and Black Americans work jobs with greater health risks.

Our varied professional experiences and personal backgrounds motivated our interest in applying some of the foundational concepts we learned in this course to a case study of exposure to heat across the life course. Two chapter authors

(E.C.V. and M.A.) are interested in work-related heat exposure because they have family members who work in occupations that make them vulnerable to heat exposure and illness in their daily work. Another author (G.H.) is a former oncology nurse passionate about learning how people's social and physical environments influence their behavior, exposure, and, consequently, their health. She is fascinated by the health implications of environmental factors and health literacy as they relate to heat and other consequences of climate change.

In-depth case studies comparing the extent to which place affects social determinants of health may help students follow their own interests when learning about the geo-social determinants of health. In addition to studying the relationship between place and health, courses covering the geo-social determinants of health could also investigate the intersection of mobility and health. One's ability to migrate locally and, more broadly, can influence the intensity of impacts from environmental hazards. In our case study of exposure to extreme heat, numerous privileges are required in many strategies for mitigating the impacts of extreme heat, such as the flexibility to move to a cooler place or install and run air conditioning.

The Role of the Built Environment in the Fundamental Causes of Health

In teaching and learning socio-spatial determinants of health, it is crucial to consider the difference between proximate and fundamental causes. Fundamental causes of health inequalities influence multiple disease outcomes through multiple risk factors and have a persistent association with health outcomes. On the other hand, proximate causes influence health outcomes but happen later in the causal pathway (Link & Phelan, 1995). When approaching solutions to health disparities, understanding fundamental causes is essential. Focusing on proximate factors, although they are often easier to analyze, may result in the adoption of less effective interventions compared to addressing the root causes (Link & Phelan, 1995).

Throughout this course, we have come to appreciate that the socio-spatial perspective urges us to look at how fundamental causes mold our normalized perceptions and physical surroundings, which more directly influence how we behave and socialize. Systemic beliefs, notably racism, are commonly considered fundamental causes of health disparities due to their deep-seated roots in our cultural history. The impacts of racism on our built environment, which encompasses the landscapes that humans have significantly transformed, are pervasive—the widely internalized belief that a racial hierarchy serves as a foundation for the segregation of people in space. This explains why underserved groups are more likely to live in areas with less access to opportunities, resources, and communities that support good health (Gee & Payne-Sturges, 2004). Likewise, it rationalizes why minority groups are more exposed to poor environmental conditions, such as noise and air pollution, hazardous waste, and poorly maintained infrastructure (Gee & Payne-Sturges,

2004). By framing racism as a fundamental cause, we can see more clearly how racist ideologies have created inequities in space, further perpetuating internalized stigmas and biases that place an undue burden on underserved groups. This section will analyze how systemic racism is reflected in our natural and built environments and discuss why these matters in the geo-social determinants of the health framework. By sharing a case study about the health legacy of uranium mining on Native Americans in the American Southwest, we will contextualize how racism ripples into several pathways for health disparities. We will conclude our discussion with commentary on how the socio-spatial lens has both enriched our learning from a classroom context and allowed us to engage more deeply with the real applications of course material.

Our backgrounds in public health and environmental science, respectively, motivate our interest in investigating the manifestations of systemic racism in both the built and natural environments. The following case study, in which the Navajo Nation is assigned to be a sacrifice zone, contextualizes one pathway in which racism acts as a fundamental cause of environmental and health disparities. A sacrifice zone is one example of how humans can alter the environment in a way that does not favor all people equitably. It is an area that is "sacrificed" by a small group of people to help a larger group of people achieve some economic or social benefit (Fox, 1999). However, the smaller group of people may be marginalized in decision-making processes that dictate the siting of the sacrifice zone in the first place. The decisions about which lands are deemed acceptable as sacrifice zones are often motivated by societal conceptions of what constitutes a built environment with inherent value or a high-quality built environment; what seems a barren desert to the outward observer may be full of built cultural sites and transportation networks. Thus, we can begin to reform decision-making processes and offer solutions that do not further marginalize or suppress the voices of minority groups. From this example, we may speculate that a key principle of environmental justice and health equity is local engagement in the decision-making process and transparency around the health impacts of a proposed development (Endres, 2012; Lerner, 2012). Here, we argue that racism and persecution of the Navajo Nation are fundamental causes of the high cancer rates seen to this day due to uranium mining. Some proximate causes are geographic isolation, lack of water purification and conveyance infrastructure, and low political prioritization. Furthermore, these proximate causes are features or effects of the built environment that were caused to some degree by racism and persecution (see Fig. 11.1).

Uranium mining on indigenous lands in the American Southwest began in 1943 (Johnston et al., 2010): today, there are approximately 1200 uranium mines on the 27,000 square miles of the Navajo Nation (Dawson & Madsen, 2011). Navajo uranium miners and mining communities experienced exceptionally high levels of lung and other cancers with effects that persist today (Dawson & Madsen, 2011). Furthermore, many people in Navajo Nation do not have access to clean water, so many use unregulated sources of water that are more likely to be contaminated with uranium or other heavy metals (Corlin et al., 2016; Credo et al., 2019). Lack of access to clean water increases the health impacts of uranium mining, and the built

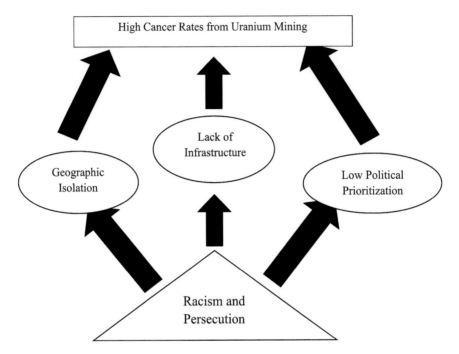

Fig. 11.1 Fundamental and proximate causes of high cancer rates from uranium mining. This figure demonstrates how the proximate causes of geographic isolation, lack of water purification and conveyance infrastructure, and low political prioritization of the Navajo Nation, all follow from the racism and persecution of the Navajo Nation, both in the past and the present

environment mediates this through a lack of water conveyance and purification infrastructure. While uranium mining continues to this day, hundreds of abandoned mines that have not been cleaned up continue contributing to disease in Navajo Nation and poor water quality (Rock & Ingram, 2020).

The geographic isolation of the Navajo Nation is a result of systemic racism because of a lack of quality transportation infrastructure within and near the reservation: indigenous lands in the USA have less transportation infrastructure than in other parts of the country (Anding & Fulton, 1993; Khan & Levy, 2003). The geographic isolation of the Navajo Nation has constrained labor markets. Finding non-uranium mining jobs would have been challenging without moving away, increasing the negative health impacts of mining. Furthermore, moving away is significantly more challenging when travel is more onerous. Isolation and opportunity acting as socio-spatial determinants of health is a common thread we noticed throughout this course.

Low political prioritization of cleaning up abandoned uranium mines constitutes another proximate cause of the health impacts of uranium mining in Navajo Nation. The Western cultural conception of the Navajo Nation (or any environment that has not been heavily urbanized) as a barren desert rather than a built environment has been used to justify the ongoing neglect of mine cleanup efforts. This makes efforts to allocate public resources to more densely populated areas a potential equity concern. Differing cultural understandings and definitions of terms such as the built environment affect our narratives when learning about geo-social health. Our lived histories and perspectives shape how we learn. Therefore, one of the key points that we have become more sensitive to through our learning is trying to incorporate diverse perspectives when teaching and learning about socio-spatial determinants of health.

Often when teaching population health, the nuanced narratives of those who have been historically marginalized are omitted. The socio-spatial approach can focus on missing voices by holistically analyzing social and spatial contexts, including the varying perceptions of what defines and constitutes the built environment. As we saw in the Navajo case study, the holistic character of the socio-spatial perspective allowed us to identify the role that systemic racism has in molding our environment and determining health outcomes while highlighting the narratives of those most impacted. Therefore, a complete understanding of social context allows us to respond with more effective solutions targeted at the root origins of health outcomes rather than the surface-level or intermediary ones.

Understanding how the built environment serves as the context of health disparities contributes to a much better understanding of fundamental causes. This holistic perspective allows us to realize that many health issues are rooted in the same problem—systemic racism and biases ingrained in our society. When approaching health solutions, we must acknowledge that reversing fundamental causes, rather than the short-term treatment of proximate causes, will lead to the most significant reform in our health systems. The socio-spatial approach allows us to truly understand health contexts and rebuild the foundations of our health systems to minimize health disparities and uphold equity values. Learning about fundamental and proximate causes of health and the socio-spatial determinants of health framework has equipped us with analytical tools that we can apply to our respective work on the health implications of the built and natural environments.

Conclusion

This chapter has surveyed teaching pedagogies for engaging student learners on the importance of a social-spatial determinants of health framework of population health and health disparities. Of critical importance was an embrace of context, both within the literature and by establishing a welcoming space for students to express their own individual perspectives. Three case studies illustrate the benefits of this approach in which students brought their own contextual experiences together to achieve a broader understanding of key health issues from the course.

A second common thread throughout the course and this chapter was a running emphasis on fundamental causes or upstream factors. Whether investigating neighborhood food choices and the impact of an upstream food multinational, the negative effects of heat waves on Black Americans and the fundamental issue of income inequality, or the Navajo Nation sacrifice zone and causal force of racism, it is of critical importance for geographers and other students of social determinants of health research to approach these issues with an eye toward the "manufacturers of illness." An advantage in adopting this approach was the diverse backgrounds of the class, allowing for a wide range of opinions and ideas when discussing the source of health disparities.

One of the key lessons learned from teaching the course was the importance of finding ways for student learners to connect individually with topics. Not only does this connection amplify interest and passion in a topic, but it also improves understanding and communication. When a student can take a learned concept and explain to others, in an applied way, why they should care, learning is deeper. Therefore, it seems that the best way to teach a course like this is to start with a number of key learning objectives and topics but also hold space to explore related topics and perspectives that are unique to the learners.

Issues relating to diversity, equity, and inclusion (DEI) were a central consideration throughout the course. The disciplinary perspective of the authors of the work discussed in the course and the time the articles were published were often offered as caveats for perceived limitations of the work and its conclusions. However, as highlighted in the book *Inclusive Teaching*, simply including diverse perspectives in course materials is not enough to create an inclusive classroom (Sathy & Hogan, 2022). This course challenged students to ask who was missing in specific articles (i.e., the limitations of a particular sample or study design) and our discussions overall (i.e., a lack of coverage of sexual and gender minorities and people with disabilities). Reflecting on the omission of some groups from the course allowed students to bring their own personal contexts into discussions and served as motivation for some of the case studies above. Finally, an asset-based framework was emphasized in this course, both in the course content and in teaching and learning. In pedagogy, an asset-based framework emphasizes the variety of strengths that a diverse group of students brings to a classroom and considers how to incorporate them into teaching and learning in a culturally sustainable way (California Department of Education, 2022). We were also careful to take an asset-based approach when discussing health disparities, particularly at the neighborhood level. Emphasizing the role of social networks and community organizations in building community agency during our discussions of how to overcome local inequities often pushed conversations beyond existing health disparities to the merits of various policy approaches. Considering an asset-based approach in the teaching and learning of health disparities can help create an inclusive environment for students from an array of backgrounds.

Reflecting on our own teaching and learning helped us identify the strengths of the social-spatial determinants of health framework and apply it to topics that were relevant to our interests. The changing demographics of higher education mean that students will bring increasingly diverse perspectives and strengths into their classrooms and campus communities. Teaching useful content will not be enough to make higher education accessible and meaningful for all students. Careful course design and intentional consideration of the scholarship on teaching and learning will help students develop knowledge and skills they can employ in their future endeavors. We hope our chapter emphasizes the importance of reflecting on the teaching and learning of the social-spatial determinants of health.

References

American Academy of Pediatrics. (2021). *An * indicates a reading included in the course syllabus. Extreme temperatures: Disaster management resources.* Homeopathy. Retrieved April 21, 2023, from https://www.aap.org/en/patient-care/disasters-and-children/disaster-management-resources-by-topic/extreme-temperatures/.

Anand, P., Kunnumakkara, A. B., Sundaram, C., Harikumar, K. B., Tharakan, S. T., Lai, O. S., Sung, B., & Aggarwal, B. B. (2008). Cancer is a preventable disease that requires major lifestyle changes. *Pharmaceutical Research, 25*(9), 2097–2116. https://doi.org/10.1007/s11095-008-9661-9

Anding, T. L., & Fulton, R. E. (1993). *Assessing transportation needs on Indian Reservations.* (MPC Rept No. 93-21). Article MPC Rept No. 93-21. Retrieved from https://trid.trb.org/View/379881.

Andretti, B., Cardoso, L. O., Honório, O. S., de Castro Junior, P. C. P., Tavares, L. F., da Costa Gaspar da Silva, I., & Mendes, L. L. (2023). Ecological study of the association between socioeconomic inequality and food deserts and swamps around schools in Rio de Janeiro, Brazil. *BMC Public Health, 23*(1), 120. https://doi.org/10.1186/s12889-023-14990-8

Antoniadis, D., Katsoulas, N., & Papanastasiou, D. K. (2020). Thermal environment of Urban schoolyards: Current and future design with respect to children's thermal comfort. *Atmosphere, 11*(11), Article 11. https://doi.org/10.3390/atmos11111144

Boardman, J. D., Daw, J., & Freese, J. (2013). Defining the environment in gene–environment research: Lessons from social epidemiology. *American Journal of Public Health, 103*(S1), S64–S72. https://doi.org/10.2105/AJPH.2013.301355

Braveman, P. (2006). Health disparities and health equity: Concepts and measurement. *Annual Review of Public Health, 27*(1), 167–194. https://doi.org/10.1146/annurev.publhealth.27.021405.102103

California Department of Education (2022). *Asset-based pedagogies—educator excellence.* Retrieved April 18, 2023, from https://www.cde.ca.gov/pd/ee/assetbasedpedagogies.asp

CDC. (2017, July 3). *African American Health.* Centers for Disease Control and Prevention. Retrieved from https://www.cdc.gov/vitalsigns/aahealth/index.html.

CDC. (2022, October 25). *By the numbers: Diabetes in America.* Centers for Disease Control and Prevention. Retrieved from https://www.cdc.gov/diabetes/health-equity/diabetes-by-the-numbers.html.

Cohen, D. A., Inagami, S., & Finch, B. (2008). The built environment and collective efficacy. *Health & Place, 14*(2), 198–208. https://doi.org/10.1016/j.healthplace.2007.06.001

Corlin, L., Rock, T., Cordova, J., Woodin, M., Durant, J. L., Gute, D. M., Ingram, J., & Brugge, D. (2016). Health effects and environmental justice concerns of exposure to uranium in drinking water. *Current Environmental Health Reports, 3*(4), 434–442. https://doi.org/10.1007/s40572-016-0114-z

Credo, J., Torkelson, J., Rock, T., & Ingram, J. C. (2019). Quantification of elemental contaminants in unregulated water across Western Navajo Nation. *International Journal of Environmental Research and Public Health, 16*(15), Article 15. https://doi.org/10.3390/ijerph16152727

Dawson, S. E., & Madsen, G. E. (2011). Psychosocial and health impacts of uranium mining and milling on Navajo Lands. *Health Physics, 101*(5), 618. https://doi.org/10.1097/HP.0b013e3182243a7a

Deschenes, O. (2014). Temperature, human health, and adaptation: A review of the empirical literature. *Energy Economics, 46*, 606–619.

Diez Roux, A. V. (2001). Investigating neighborhood and area effects on health. *American Journal of Public Health, 91*(11), 1783–1789.

Edmonds, C. J., & Burford, D. (2009). Should children drink more water? The effects of drinking water on cognition in children. *Appetite, 52*(3), 776–779. https://doi.org/10.1016/j.appet.2009.02.010

Endres, D. (2009). The rhetoric of nuclear colonialism: Rhetorical exclusion of American Indian arguments in the Yucca Mountain nuclear waste siting decision. *Communication and Critical/ Cultural Studies, 6*(1), 39–60. https://doi.org/10.1080/14791420802632103

Endres, D. (2012). Sacred land or national sacrifice zone: The role of values in the Yucca Mountain participation process. *Environmental Communication, 6*(3), 328–345. https://doi.org/10.108 0/17524032.2012.688060

Fox, J. (1999). Mountaintop removal in West Virginia: An environmental sacrifice zone. *Organization & Environment, 12*(2), 163–183. https://doi.org/10.1177/1086026699122002

Gee, G. C., & Payne-Sturges, D. C. (2004). Environmental health disparities: A framework integrating psychosocial and environmental concepts. *Environmental Health Perspectives, 112*(17), 1645–1653. https://doi.org/10.1289/ehp.7074

Göcke, K. (2014). Indigenous peoples in the nuclear age: Uranium mining on indigenous' lands. In J. L. Black-Branch & D. Fleck (Eds.), *Nuclear non-proliferation in international law—volume I* (pp. 199–223). TMC Asser Press. https://doi.org/10.1007/978-94-6265-020-6_8

Gordon-Larsen, P., Nelson, M. C., Page, P., & Popkin, B. M. (2006). Inequality in the built environment underlies key health disparities in physical activity and obesity. *Pediatrics, 117*(2), 417–424. https://doi.org/10.1542/peds.2005-0058

Gronlund, C. J. (2014). Racial and socioeconomic disparities in heat-related health effects and their mechanisms: A review. *Current Epidemiology Reports, 1*(3), 165–173. https://doi.org/10.1007/s40471-014-0014-4

Hill, A. C., Babin, M., & Baumgartner, S. (2021). *A world overheating.* Council on Foreign Relations. Retrieved from https://www.cfr.org/article/climate-change-world-overheating-how-countries-adApt-extreme-temperature.

Homan, P. (2019). Structural sexism and health in the United States: A new perspective on health inequality and the gender system. *American Sociological Review, 84*(3), 486–516. https://doi.org/10.1177/0003122419848723

Jacobs, A., & Richtel, M. (2017, September 16). How big business got Brazil hooked on junk food. *The New York Times.* Retrieved from https://www.nytimes.com/interactive/2017/09/16/health/brazil-obesity-nestle.html.

Johnston, B. R., Dawson, S. E., & Madsen, G. E. (2010). Uranium mining and milling: Navajo experiences in the American Southwest. In *Uranium mining and milling.* John Wiley & Sons.

Kawachi, I., Subramanian, S. V., & Almeida-Filho, N. (2002). A glossary for health inequalities. *Journal of Epidemiology & Community Health, 56*(9), 647–652. https://doi.org/10.1136/jech.56.9.647

Khan, S., & Levy, D. (2003). Linking economic development to highway improvements: Pine Ridge Reservation, South Dakota. *Transportation Research Record, 1848*(1), 106–113. https://doi.org/10.3141/1848-15

Lantz, P. M., House, J. S., Lepkowski, J. M., Williams, D. R., Mero, R. P., & Chen, J. (1998). Socioeconomic factors, health behaviors, and mortality: Results from a nationally representative prospective study of US adults. *JAMA, 279*(21), 1703. https://doi.org/10.1001/jama.279.21.1703

Lerner, S. (2012). *Sacrifice zones: The front lines of toxic chemical exposure in the United States.* MIT Press.

Link, B. G., & Phelan, J. (1995). Social conditions as fundamental causes of disease. *Journal of Health and Social Behavior.* Spec No, 80–94.

Luber, G., & McGeehin, M. (2008). Climate change and extreme heat events. *American Journal of Preventive Medicine, 35*(5), 429–435.

Macintyre, S., Ellaway, A., & Cummins, S. (2002). Place effects on health: How can we conceptualize, operationalize and measure them? *Social Science & Medicine, 55*(1), 125–139. https://doi.org/10.1016/S0277-9536(01)00214-3

Manware, M., Dubrow, R., Carrión, D., Ma, Y., & Chen, K. (2022). Residential and race/ethnicity disparities in heat vulnerability in the United States. *GeoHealth, 6*(12), e2022GH000695.

Margolis, H. G., Gershunov, A., Kim, T., English, P., & Trent, R. (2008). 2006 California heat wave high death toll: Insights gained from coroner's reports and meteorological characteristics of event. *Epidemiology, 19*(6), S363–S364.

Marmot, M. (2004). Status syndrome. *Significance, 1*(4), 150–154.

Marmot, M. (2006). *Unnatural causes... is inequality making us sick?* PBS.

Marmot, M. (2017). The health gap: The challenge of an unequal world: The argument. *International Journal of Epidemiology, 46*(4), 1312–1318. https://doi.org/10.1093/ije/dyx163

McKinlay John, B. (1975). A case for refocusing upstream: The political economy of illness. In *Applying behavioral science to cardiovascular risk: Proceedings of a conference* (pp. 7–17). Seattle, WA.

McMichael, A. J. (2013). Globalization, climate change, and human health. *New England Journal of Medicine, 368*(14), 1335–1343. https://doi.org/10.1056/NEJMra1109341

Montez, J. K., Mehri, N., Monnat, S. M., Beckfield, J., Chapman, D., Grumbach, J. M., Hayward, M. D., Woolf, S. H., & Zajacova, A. (2022). U.S. state policy contexts and mortality of working-age adults. *PLoS One, 17*(10), e0275466. https://doi.org/10.1371/journal.pone.0275466

O'Brien, D. T., Farrell, C., & Welsh, B. C. (2019). Broken (windows) theory: A meta-analysis of the evidence for the pathways from neighborhood disorder to resident health outcomes and behaviors. *Social Science & Medicine, 228*, 272–292.

O'Dwyer, L. A., Baum, F., Kavanagh, A., & Macdougall, C. (2007). Do area-based interventions to reduce health inequalities work? A systematic review of evidence. *Critical Public Health, 17*(4), 317–335. https://doi.org/10.1080/09581590701729921

O'Neill, M. S., Zanobetti, A., & Schwartz, J. (2005). Disparities by race in heat-related mortality in four US cities: The role of air conditioning prevalence. *Journal of Urban Health, 82*, 191–197.

Oakes, J. M. (2004). The (mis)estimation of neighborhood effects: Causal inference for a practicable social epidemiology. *Social Science & Medicine, 58*(10), 1929–1952. https://doi.org/10.1016/j.socscimed.2003.08.004

Pacheco, S. E. (2020). Catastrophic effects of climate change on children's health start before birth. *The Journal of Clinical Investigation, 130*(2), 562–564.

Pampel, F. C., Krueger, P. M., & Denney, J. T. (2010). Socioeconomic disparities in health behaviors. *Annual Review of Sociology, 36*(1), 349–370. https://doi.org/10.1146/annurev.soc.012809.102529

Race, Ethnicity, & Kidney Disease—NIDDK. (2014, March). *National Institute of Diabetes and Digestive and Kidney Diseases.* Retrieved from https://www.niddk.nih.gov/health-information/kidney-disease/race-ethnicity.

Rieker, P., Bird, C., & Lang, M. (2010). Understanding gender and health: Old patterns, new trends, and future directions. In C. Bird, P. Conrad, A. Fremont, & S. Timmermans (Eds.), *Handbook of medical sociology.* Vanderbilt University Press.

Rock, T., & Ingram, J. C. (2020). Traditional ecological knowledge policy considerations for abandoned uranium mines on Navajo Nation. *Human Biology, 92*(1), 19–26. https://doi.org/10.13110/humanbiology.92.1.01

Roscoe, R. J., Deddens, J. A., Salvan, A., & Schnorr, T. M. (1995). Mortality among Navajo uranium miners. *American Journal of Public Health, 85*(4), 535–540. https://doi.org/10.2105/AJPH.85.4.535

Ross, C., & Mirowsky, J. (2010). Why education is the key to socioeconomic differentials in health. In C. Bird, P. Conrad, A. Fremont, & S. Timmermans (Eds.), *Handbook of medical sociology.* Vanderbilt University Press.

Ruddell, D. M., Harlan, S. L., Grossman-Clarke, S., & Buyantuyev, A. (2010). Risk and exposure to extreme heat in microclimates of Phoenix, AZ. *Geospatial Techniques in Urban Hazard and Disaster Analysis*, 179–202.

Sampson, R. J. (2013). The place of context: A theory and strategy for criminology's hard problems. *Criminology, 51*(1), 1–31. https://doi.org/10.1111/1745-9125.12002

Sampson, R. J., & Raudenbush, S. W. (2004). Seeing disorder: Neighborhood stigma and the social construction of "broken windows". *Social Psychology Quarterly, 67*(4), 319–342. https://doi.org/10.1177/019027250406700401

Sathy, V., & Hogan, K. A. (2022). Inclusive teaching: Strategies for promoting equity in the college classroom. *West Virginia University Press*. https://muse.jhu.edu/pub/20/monograph/book/100502

Seattle Children's Hospital (2023). *Heat exposure and reactions*. Retrieved April 18, 2023, from https://www.seattlechildrens.org/conditions/a-z/heat-exposure-and-reactions

Sharkey, P., & Faber, J. W. (2014). Where when, why, and for whom do residential contexts matter? Moving away from the dichotomous understanding of neighborhood effects. *Annual Review of Sociology, 40*(1), 559–579. https://doi.org/10.1146/annurev-soc-071913-043350

Snodgrass, J. G., Upadhyay, C., Debnath, D., & Lacy, M. G. (2016). The mental health costs of human displacement: A natural experiment involving indigenous Indian conservation refugees. *World Development Perspectives, 2*, 25–33. https://doi.org/10.1016/j.wdp.2016.09.001

Williams, D. R., & Collins, C. (1995). US socioeconomic and racial differences in health: Patterns and explanations. *Annual Review of Sociology, 21*(1), 349–386. https://doi.org/10.1146/annurev.so.21.080195.002025

Williams, D. R., & Collins, C. (2001). Racial residential segregation: A fundamental cause of racial disparities in health. *Public Health Reports, 116*(5), 404.

World Health Organization (2018). *Heat and health*. Retrieved April 18, 2023, from https://www.who.int/news-room/fact-sheets/detail/climate-change-heat-and-health.

Zivin, J. G., & Shrader, J. (2016). Temperature extremes, health, and human capital. *The Future of Children*, 31–50.

Part IV
Methodological Approaches and Techniques

Chapter 12
When to Use What: Methods for Operationalizing Social Determinants of Health Indicators

Imelda K. Moise and Anjali Choudhury

Introduction

Health outcomes are influenced by a myriad of factors that extend beyond the scope of medical treatments. These factors, known as social determinants of health (SDOH), encompass a wide range of social and environmental conditions that determine the quality of life for individuals and communities (Marmot, 2005; Marmot & Wilkinson, 2005; Moise, 2020; Rojas et al., 2020; United States Department of Health and Human Services, 2020b). SDOH include aspects such as socioeconomic status, education, employment, neighborhood conditions, and access to healthcare. These determinants have a significant impact on health disparities and inequities, as they create different opportunities and barriers to health and well-being across various populations and locations.

Operationalizing SDOH indicators is essential for addressing health inequities and developing effective interventions. However, this task is fraught with challenges, including defining and conceptualizing SDOH within established frameworks, selecting the best methods for measurement at individual and community

I. K. Moise (✉)
Department of Medical Education, Dr Kiran C. Patel College of Allopathic Medicine (NSU MD), Nova Southeastern University, Fort Lauderdale, FL, USA
e-mail: imoise@nova.edu

A. Choudhury
University of Washington, Department of Environmental Sciences, Seattle, WA, USA

© The Author(s) 2026
M. A. Kolak, I. K. Moise (eds.), *Place and the Social-Spatial Determinants of Health*, Global Perspectives on Health Geography,
https://doi.org/10.1007/978-3-031-88463-4_12

levels, accessing and utilizing available data sources, and applying analytical techniques to discern the relationships between SDOH and health outcomes. Ensuring the validity, reliability, and ethical integrity of these methods is also paramount. This chapter serves as a comprehensive guide for those seeking to navigate the complexities of SDOH indicators. It is structured to provide clarity and direction in the following areas:

- **Defining and Conceptualizing SDOH:** We begin by defining SDOH within established frameworks such as the WHO model and the Healthy People 2030 framework. The multifaceted and dynamic nature of SDOH and their interactions with health outcomes are explained.
- **Measurement Approaches:** Various approaches for measuring SDOH indicators are reviewed, including direct and indirect methods like surveys, questionnaires, and geospatial data. The strengths and limitations of each approach are discussed, supplemented by examples.
- **Data Sources:** The chapter discusses the available data sources for operationalizing SDOH indicators, ranging from national surveys to big data. The advantages and challenges of these sources, including data quality and privacy considerations, are examined.

In addition to the areas covered above, this book chapter also ventures into the future directions and emerging trends in the field of SDOH indicators. It highlights the advancements in data collection technologies, such as wearable devices and mobile health applications, which are revolutionizing the way we gather health-related data. The chapter discusses the integration of SDOH into electronic health records, which is enhancing the ability to track and address health disparities. It also examines the importance of longitudinal and temporal analysis in understanding the long-term effects of SDOH on health outcomes. A particular emphasis is placed on health equity and intersectionality, ensuring that SDOH research and interventions are inclusive and address the needs of diverse populations.

By providing this guidance, the chapter aims to empower professionals to make informed decisions that will lead to improved health outcomes and reduced disparities across different communities.

Defining and Conceptualizing SDOH within Established Frameworks

The SDOH are the non-medical factors that influence health outcomes, such as the conditions in which people are born, grow, work, live, and age, and the wider set of forces and systems shaping the conditions of daily life (Marmot et al., 2008; Marmot & Wilkinson, 2005). SDOH have a profound impact on health disparities and inequities, as they create differential opportunities and barriers for health and well-being across populations and places (Evans et al., 2021; Wilkinson & Marmot, 2003). To operationalize SDOH indicators, it is important to first define and conceptualize SDOH within established frameworks. Different frameworks may emphasize different aspects or dimensions of SDOH and may have different implications for measurement and analysis. For example, the WHO model of SDOH identifies three layers of determinants: (1) structural determinants, such as income, education, occupation, and gender; (2) ethnicity and intermediary determinants, such as material circumstances, behavioral and biological factors, psychosocial factors, and health systems; and (3) health outcomes, such as mortality, morbidity, and well-being (World Health Organization, 2010). The WHO model also recognizes the role of socioeconomic and political contexts, such as governance, policy, and culture, in shaping the distribution of SDOH and health outcomes.

Another example of a framework for SDOH is the Healthy People 2030 framework, which is a national initiative that sets data-driven objectives for improving health and well-being in the United States (Gómez et al., 2021; Hubbard et al., 2020; Ochiai et al., 2021). Five domains of SDOH are identified in the Healthy People 2030 framework: economic stability, education access and quality, health care access and quality, neighborhood and built environment, and social and community context. Each domain consists of several indicators that measure the status and trends of SDOH and health outcomes. The Healthy People 2030 framework also emphasizes the importance of health equity and the elimination of health disparities.

Combined, these frameworks illustrate the multifaceted and dynamic nature of SDOH and how they interact with each other and with health outcomes. Pointedly, SDOH are not static or isolated factors, but rather complex and interrelated processes that change over time and space. SDOH may also have varying effects on different groups of people, depending on their social position and vulnerability (Brandt, 2023). Thus, operationalizing SDOH indicators requires a comprehensive and nuanced understanding of the conceptual and theoretical foundations of SDOH.

Measurement Approaches for SDOH Indicators

Measuring SDOH indicators is a crucial step for operationalizing SDOH and assessing their impact on health outcomes and disparities. For example, how can we capture the complexity and diversity of SDOH? How can we ensure the validity and reliability of SDOH indicators? How can we balance the feasibility and ethical issues of data collection? How can we compare SDOH indicators across different populations and settings? To address these challenges and choices, various measurement approaches for SDOH indicators have been developed and applied in different contexts and settings. These approaches can be broadly classified into two categories: direct and indirect methods (Davidson et al., 2020; Shokouh et al., 2017). Direct methods involve asking individuals or communities about their SDOH status or experiences, such as through surveys, questionnaires, interviews, or focus groups. Indirect methods involve using existing data sources or proxies to infer SDOH status or experiences, such as through administrative data, geospatial data, or digital data (Loukaitou-Sideris et al., 2019).

Each method has its strengths and limitations, and the choice of method depends on the research or intervention objectives, the availability and quality of data, and the ethical and practical considerations. For example, direct methods may provide more accurate and detailed information on SDOH, but they may also be more costly, time-consuming, and intrusive. Indirect methods may provide more accessible and comprehensive data on SDOH, but they may also be more prone to errors, biases, and confounding factors. Therefore, choosing the most appropriate method for measuring SDOH indicators requires a careful evaluation of the advantages and disadvantages of each method, as well as the trade-offs involved.

To illustrate the applications and benefits of different methods for measuring SDOH indicators, we provide examples in different contexts and settings. For example, one direct method for measuring SDOH is the **SDOH Questionnaire (SDHQ)**, which is a standardized instrument that assesses psychosocial, current behaviors, behavioral capacity, environmental factors, and a measurement of food insecurity that then affects a population's health status (Gadhoke et al., 2018). The Social SDHQ consists of 17 sections and a total of 153 items (in addition to anthropometry items). It primarily includes closed-ended questions related to the three main domains of Albert Bandura's social cognitive theory (Bandura, 1986; Glanz et al., 2008). The SDHQ has been used in various settings, such as primary care clinics, community health centers, and hospitals, to identify and address the SDOH needs and challenges of patients and populations (Lazarou et al., 2012; Matsumoto & Nakayama, 2017; Takatsuka et al., 1997).

An indirect method for measuring SDOH is the **Area Deprivation Index (ADI)**, which is a composite measure of socioeconomic disadvantage at the neighborhood level, based on 17 indicators from the census data, such as income, education, employment, housing, and poverty (Trinidad et al., 2022). The ADI has been used in various settings, such as public health surveillance, health services research, and health policy, to examine and compare the spatial distribution and variation of SDOH and health outcomes across different regions and communities (Balio et al., n.d.; Johnson et al., 2021; Knighton et al., 2016; Rollings et al., 2023).

Data Sources for SDOH Indicators

Data sources are the foundation for operationalizing SDOH indicators, providing the information and evidence needed to measure and analyze SDOH and their impact on health outcomes and disparities. However, accessing and using data sources for SDOH indicators is not an easy task, as it involves multiple challenges and choices. For example, how can we find and access relevant and reliable data sources for SDOH? How can we ensure the quality, validity, and comparability of data sources for SDOH? How can we protect the privacy and confidentiality of data sources for SDOH? How can we integrate and link different data sources for SDOH? To address these challenges and choices, various data sources for operationalizing SDOH indicators have been developed and utilized in different contexts and settings. These data sources can be broadly classified into four categories: national surveys and databases, local and regional data sources, community-based data collection efforts, and the utilization of big data and data linkages (Craig et al., 2021; Harrison & Dean, 2011).

National surveys and databases are large-scale and standardized data sources that collect and provide information on SDOH and health outcomes at the national level, such as the American Community Survey (ACS) (Herman, 2008), the Behavioral Risk Factor Surveillance System (BRFSS) (Mokdad, 2009), and the National Health Interview Survey (NHIS) (National Center for Health Statistics, Division of Health Interview, 1986). These data sources offer the advantages of high quality, validity, and comparability, as well as wide coverage and representativeness. However, these data sources also have some limitations, such as limited availability, timeliness, and granularity, as well as potential biases and errors.

Local and regional data sources are small-scale and customized data sources that collect and provide information on SDOH and health outcomes at the local or regional level, such as the County Health Rankings (Remington et al., 2015), the City Health Dashboard (Gourevitch et al., 2019), and the 500 Cities Project (Garden Grove, 2016). These data sources offer the advantages of high availability, timeliness, and granularity, as well as context-specific and actionable insights. However, these data sources also have some limitations, such as low quality, validity, and comparability, as well as potential gaps and inconsistencies.

Community-based data collection efforts are participatory and collaborative data sources that involve and engage community members and stakeholders in collecting and providing information on SDOH and health outcomes at the community level, such as the Community Health Needs Assessment (CHNA) (Assessment, n.d.), the Community Health Status Indicators (CHSI) (Metzler et al., 2008), and the Community Health Worker (CHW) model (Naimoli et al., 2014). These data sources offer the advantages of high relevance, responsiveness, and empowerment, as well as community ownership and trust. However, these data sources also have some limitations, such as high cost, time, and effort, as well as potential conflicts and challenges.

Recently, the utilization of big data and data linkages represents innovative and emerging data sources (Doğan, 2019; Sagiroglu & Sinanc, 2013). These sources leverage and integrate various types of data from different platforms, including electronic health records (EHRs), social media, mobile devices, and sensors. They provide information on SDOH and health outcomes at both individual and population levels. Examples include the Social Vulnerability Index (SVI) (Mah et al., 2023), the Social Needs Screening Tool (SNST) (Karran et al., 2023), and the Social Determinants of Health Dashboard (SDD) (Petrovskis et al., 2023). These data sources offer the advantages of high volume, variety, and velocity, as well as novel and predictive insights. However, these data sources also have some limitations, such as low veracity, validity, and reliability, as well as potential ethical and legal issues. An example that illustrates the use and benefits of different data sources for operationalizing SDOH indicators is the American Community Survey (ACS), an annual survey conducted by the US Census Bureau that collects and provides information on various aspects of SDOH, such as income, education, employment, housing, and transportation, for every county, city, and neighborhood in the USA, with data presented annually or as 5-year estimates (Bazuin & Fraser, 2013; McNeil, 1994). The ACS data are widely used by researchers, policymakers, and practitioners to measure and compare the SDOH status and trends of different populations and places, and to inform and evaluate policies and programs that address SDOH and health disparities.

One local and regional data source for SDOH indicators is the County Health Rankings, which is an annual report produced by the Robert Wood Johnson Foundation and the University of Wisconsin Population Health Institute that ranks and compares the health status and SDOH of every county in the USA, based on various indicators, such as health behaviors, clinical care, social and economic factors, and physical environment (Hood et al., 2016). The County Health Rankings data are widely used by researchers, policymakers, and practitioners to identify and address the SDOH needs and challenges of different communities, and to inform and evaluate policies and programs that improve health and well-being (Remington et al., 2015).

At the community level, an example of a community-based data collection effort for SDOH indicators is the Community Health Needs Assessment (CHNA) (Assessment, n.d.), which is a systematic process that involves and engages community members and stakeholders in identifying and prioritizing the health and SDOH issues and assets of their community and developing and implementing action plans to address them (Ravaghi et al., 2023). The CHNA is a requirement for nonprofit hospitals under the Affordable Care Act, and a recommended practice for other healthcare organizations and public health agencies (Borders, 2016). The CHNA data are valuable sources of information on the SDOH status and experiences of different communities and can help guide and monitor the impact of community-based interventions and collaborations.

A good example of how big data and data linkages have been utilized for SDOH indicators is the Social Vulnerability Index (SVI). The SVI is a composite measure of the social vulnerability of different populations and places to environmental hazards, disasters, and outbreaks, based on 16 indicators from the census data, such as poverty, disability, minority status, and housing type (Spielman et al., 2020). The SVI is linked with various data sources, such as the National Weather Service, the Centers for Disease Control and Prevention, and the Federal Emergency Management Agency, to provide real-time and dynamic information on SDOH and health risks and impacts of different events and scenarios (Ramesh et al., 2022).

Analytical Techniques for SDOH Indicators

Analyzing SDOH indicators is a key step for operationalizing SDOH and assessing their impact on health outcomes and disparities. However, it involves multiple challenges and choices. For example, how can we account for the complexity and diversity of SDOH? How can we test and interpret the causal relationships between SDOH and health outcomes? How can we account for the spatial and temporal variations of SDOH and health outcomes? How can we handle the large and heterogeneous data sets of SDOH and health outcomes? To address these challenges and choices, various analytical techniques for SDOH indicators have been developed and applied in different contexts and settings. These techniques can be broadly classified into five categories: descriptive analysis, correlation and regression analysis, multilevel modeling, geographic information systems (GIS), and machine learning.

Descriptive analysis is the simplest and most common technique for SDOH indicators, which involves summarizing and presenting the basic characteristics and distributions of SDOH and health outcomes, such as mean, median, standard deviation, frequency, percentage, and range (Lee et al., 2018). Descriptive analysis can help provide an overview and comparison of the SDOH status and trends of different populations and places and identify the gaps and disparities in SDOH and health outcomes. However, descriptive analysis cannot explain the underlying causes and mechanisms of SDOH and health outcomes, nor can it account for the confounding and moderating factors that may influence the relationships between SDOH and health outcomes.

Correlation and regression analysis are the most widely used techniques for SDOH indicators, which involve testing and estimating the strength and direction of the associations between SDOH and health outcomes, such as Pearson correlation, Spearman correlation, linear regression, logistic regression, and Poisson regression (Varbanova & Beutels, 2020). Correlation and regression analysis can help measure and compare the effects of SDOH on health outcomes and identify the significant and influential SDOH factors that predict health outcomes. However, correlation and regression analysis cannot establish the causal relationships between SDOH and health outcomes, nor can they account for the hierarchical and nested structure of SDOH and health data, such as individuals within communities or communities within regions.

Multilevel modeling is an advanced technique for SDOH indicators, which involves modeling and analyzing the relationships between SDOH and health outcomes at different levels of aggregation, such as individual, household, neighborhood, or county level, and accounting for the variability and dependence of SDOH and health outcomes within and between these levels, such as random effects, fixed effects, and cross-level interactions (Evans et al., 2018). Multilevel modeling can help account for the complexity and diversity of SDOH and health outcomes and examine the contextual and compositional effects of SDOH on health outcomes. However, multilevel modeling requires a large and balanced data set of SDOH and health outcomes at different levels, as well as a sophisticated and rigorous statistical approach to ensure the validity and reliability of the results (Evans et al., 2018).

Geographic information systems (GIS) are a powerful technique for SDOH indicators, which involve collecting, storing, manipulating, analyzing, and displaying spatial data of SDOH and health outcomes, such as coordinates, addresses, zip codes, or polygons, using maps, charts, or graphs. GIS can help visualize and explore the spatial patterns and processes of SDOH and health outcomes and identify the hotspots and clusters of SDOH and health disparities (Boggs et al., 2023; Moise & Ruiz, 2016; Ozdenerol, 2016). GIS can also help perform spatial statistics and space-time analysis, such as spatial autocorrelation, spatial regression, spatial interpolation, cluster analysis, or space-time modeling, to account for the spatial dependence and heterogeneity of SDOH and health data (space-time cluster analysis) (Moise et al., 2011). However, GIS requires high quality and accurate spatial data of SDOH and health outcomes, and careful consideration of the spatial scale and resolution of the analysis, such as the modifiable areal unit problem (MAUP) or the ecological fallacy (Sedgwick, 2011; Sui, 2004).

More recently, machine learning has emerged as an innovative technique for SDOH indicators. It involves applying and developing algorithms and models that can learn from and make predictions on large and heterogeneous data sets of SDOH and health outcomes, such as classification, regression, clustering, or deep learning (Segar et al., 2022). Researchers can leverage these tools to handle and analyze the big data and data linkages of SDOH and health outcomes, generating novel and predictive insights on the SDOH and health risks and impacts of different populations and places. Machine learning can also help discover and explain the complex and nonlinear relationships between SDOH and health outcomes to identify the important and relevant SDOH features that influence health outcomes (Kino et al., 2021). However, machine learning requires a high level of technical and computational skills and resources, as well as a careful evaluation and interpretation of the results, such as accuracy, precision, recall, and explainability (Jordan & Mitchell, 2015). Below we provide some to illustrate the use and benefits of different analytical techniques for SDOH indicators.

The first example of descriptive analysis for SDOH indicators is the Health Equity Report, a biennial report produced by the US Department of Health and Human Services that provides a comprehensive overview and comparison of the health status and SDOH of different racial and ethnic groups in the USA, based on various indicators such as mortality, morbidity, health behaviors, health care access and utilization, and social and economic factors (U.S. Department of Health and Human Services, 2020a). The Health Equity Report data are useful sources of information on the SDOH and health disparities and inequities of different populations and can help inform and monitor the progress and challenges of achieving health equity.

Regarding correlation and regression analysis for SDOH indicators, a good example is the study conducted by Braveman and colleagues (Braveman et al., 2021), which examined and estimated the associations between SDOH and birth outcomes, such as preterm birth and low birth weight, among women in California, based on various indicators, such as income, education, race, ethnicity, and neighborhood deprivation. The study found that SDOH was significantly associated with birth outcomes and that the effects of SDOH varied by race and ethnicity. The study also found that neighborhood deprivation had an independent effect on birth outcomes, beyond individual SDOH factors. The study provided evidence and insights on the impact of SDOH on birth outcomes and the need for addressing SDOH and health disparities at multiple levels.

Another study by Diez Roux et al. (2001) modeled and analyzed the relationship between SDOH and cardiovascular disease mortality, such as coronary heart disease and stroke, at the individual and neighborhood levels among adults in four US cities, based on various indicators such as income, education, occupation, smoking, blood pressure, and neighborhood socioeconomic status. The study found that

SDOH at both levels were significantly associated with cardiovascular disease mortality and that the effects of SDOH varied by race. The study also found that neighborhood socioeconomic status had a direct effect on cardiovascular disease mortality, beyond the effects of individual SDOH factors. This study demonstrated the importance and feasibility of multilevel modeling for SDOH indicators and the necessity of considering the contextual and compositional effects of SDOH on health outcomes.

Regarding the use of GIS for SDOH indicators, a good example is the recent study by Moise (2020), who utilized spatial analysis tools to spatially analyze COVID-19 case data from Miami-Dade County at the census block group level to reveal a correlation between SDOH and infection rates. She found that areas with higher social disadvantage showed increased infection rates, while those with better socioeconomic status and lower vulnerability had fewer cases. The infection rates varied significantly across different geographic areas, ranging from 0 to 60.75 per 1000 population. This study underscores the importance of considering social and economic factors in public health responses to the pandemic.

Best Practices and Recommendations for Operationalizing SDOH Indicators

The operationalization of SDOH indicators is vital for shaping health policies and programs that effectively address the social and environmental factors influencing health outcomes. These indicators, which include variables like income, education, and housing, are essential for informing health-related decision-making and identifying community health needs. However, challenges arise in ensuring the relevance, validity, and ethical management of these indicators, including data privacy and the engagement of affected communities. To navigate these challenges, it is recommended to employ a variety of data collection methods, standardize SDOH measurements, and adhere to ethical guidelines. Engaging stakeholders and utilizing research findings are also crucial for implementing SDOH indicators in a way that drives policy and practice changes. By following these best practices, health and social care providers can better understand and address health disparities, ultimately leading to improved health equity and outcomes. Examples of best practices are provided through:

- **The Protocol for Responding to and Assessing Patients' Assets, Risks, and Experiences (PRAPARE)** framework, a comprehensive tool utilized in primary care to assess patients' SDOH through 17 essential and 4 supplementary questions covering aspects like housing and transportation. This tool is designed to seamlessly integrate with EHRs, aiding in the generation of data for quality improvement and research.
- **The Johns Hopkins Center for Population Health Information Technology (CPHIT)** crafted a detailed framework and pathway for incorporating SDOH

data into population health analytics. This includes identifying data sources and domains, outlining the process for data integration, and showcasing projects that merge SDOH data from various systems into healthcare databases.

• The **McKinsey SDOH Report** suggests seven strategic actions for state Medicaid programs to tackle SDOH effectively. These include identifying crucial SDOH, crafting interventions, evaluating their impact, and fostering partnerships. The report highlights the practical application of these strategies in states like Massachusetts, Oregon, and North Carolina.

Future Directions and Emerging Trends in the Field

The landscape of SDOH indicators is dynamic and innovative, continuously shaped by emerging data sources, cutting-edge technologies, and novel methodologies. Staying informed and adaptable in this field is essential for capturing the fluid nature of SDOH and its impact on health outcomes. As we confront the complex health challenges of the twenty-first century, it becomes increasingly important to leverage the power of digital transformation and data-driven innovation within the realms of health and social care. Advancements in health equity and social justice are at the forefront of scientific exploration and healthcare delivery. To foster progress, it is vital to engage in continuous learning and improvement across the various systems and sectors of health and social care. The future of SDOH indicators is marked by several promising directions and trends. The advent of sophisticated data collection technologies, such as mobile devices, wearable sensors, and artificial intelligence, offers unprecedented opportunities to gather real-time, detailed data on SDOH and associated health behaviors and outcomes (Cossio, 2023; Moise et al., 2023). Furthermore, the integration of SDOH data into EHRs is being facilitated by the development of standardized screening tools, coding systems, and interoperability standards (Cantor & Thorpe, 2018). These advancements enable the seamless documentation, exchange, and analysis of SDOH data within and across healthcare settings. Longitudinal and temporal analyses of SDOH indicators are also gaining traction, providing valuable insights into the patterns and trajectories of SDOH over time and their intergenerational influence on health (Ali et al., 2023; Debopadhaya et al., 2021; Feller et al., 2019; Huang et al., 2021).

Additionally, a heightened focus on health equity and intersectionality is driving the field toward more nuanced analyses that consider a range of social identities, including race, ethnicity, gender, sexual orientation, and disability (López & Gadsden, 2017). This approach is crucial for highlighting disparities in health outcomes and experiences among diverse and marginalized populations. Lastly, there is a shift from merely identifying indicators to actively implementing interventions. This involves developing and evaluating evidence-based, scalable interventions that demonstrate the causal impact of addressing SDOH on health outcomes and costs, thereby underscoring the tangible benefits of such initiatives (Kolak et al., 2020;

Rojas et al., 2020). Examples that demonstrate the application of current trends in SDOH indicators across various settings include:

- **The All of Us Research Program** is a significant national effort that seeks to recruit over a million participants from varied backgrounds to collect a wide array of data, including SDOH, genetic, environmental, and lifestyle information (All of Us Research Program, 2019). This initiative aims to propel precision medicine forward and enhance overall health. It employs a range of data collection methods, from web surveys to mobile applications, to gather a detailed and ongoing record of the participants' health and SDOH factors. Additionally, the program incorporates this data into a cloud-based platform, enabling researchers to conduct diverse analyses.
- **The Social Interventions Research and Evaluation Network (SIREN)** that operates nationwide promotes the incorporation of SDOH into healthcare delivery and research (Torres et al., 2017). SIREN offers a suite of resources, including a comprehensive evidence library and a measurement database, to assist healthcare professionals in identifying, implementing, and assessing SDOH interventions. The network actively engages in research and shares findings on the effectiveness of various SDOH interventions, such as programs addressing food insecurity, housing services, and the deployment of community health workers.
- **The Population Health Research Network (PHRN)** in Australia that is dedicated to connecting and evaluating data from multiple sectors, including health, education, social services, and justice, to tackle questions related to population health (Flack & Smith, 2019). Utilizing a distributed data approach, the network enables researchers to work with anonymized and linked data across different regions and fields, ensuring data security and privacy. PHRN also facilitates longitudinal and temporal studies of SDOH indicators to assess their impact on health outcomes, examining factors such as the influence of early childhood education on long-term health, the effects of income support on mental health, and the correlation between incarceration rates and mortality.

Conclusion

In this book chapter, we have explored the critical role and inherent challenges of operationalizing SDOH indicators while also providing best practices and ethical guidelines for their application. We have ventured into the future directions and current innovations in the SDOH indicators field, highlighting their potential to foster health equity and social justice. This chapter aims to offer valuable insights and guidance for researchers, practitioners, policymakers, and stakeholders engaged in the development, implementation, and evaluation of SDOH indicators and interventions. Readers are encouraged to refer to other chapters in this book for more comprehensive information on various methods and tools tailored to operationalizing

SDOH indicators across different domains and settings. By employing the appropriate methods and tools, we can deepen our understanding of the social and environmental factors influencing health, improve health outcomes, and reduce health disparities.

References

Ali, S. M. A., Sherman-Morris, K., Meng, Q., & Ambinakudige, S. (2023). Longitudinal disparities in social determinants of health and COVID-19 incidence and mortality in the United States from the three largest waves of the pandemic. *Spatial and Spatio-temporal Epidemiology, 46*, 100604.

All of Us Research Program, I. (2019). The "all of us" research program. *New England Journal of Medicine, 381*(7), 668–676.

Assessment, A. J. (n.d.). Community Health Needs Assessment.

Balio, C. P., Galler, N., Mathis, S. M., Francisco, M. M., Meit, M. B., & Kate, E. (n.d.). Use of the Area Deprivation Index and Rural Applications in the Peer-Reviewed Literature.

Bandura, A. (1986). Social foundations of thought and action. *Englewood Cliffs, NJ, 1986*(23–28), 2.

Bazuin, J. T., & Fraser, J. C. (2013). How the ACS gets it wrong: The story of the American community survey and a small, inner city neighborhood. *Applied Geography, 45*, 292–302.

Boggs, Z., Beck Dallaghan, G. L., Smithson, S., & Lam, Y. (2023). Teaching social determinants through geographic information system mapping. *The Clinical Teacher, 20*(1), e13553. https://doi.org/10.1111/tct.13553

Borders, S. (2016). Community health needs assessments and the affordable care act: Making the most of the American community survey in understanding population-level data. *Journal of Health Administration Education, 33*(3), 497–510.

Brandt, E. J. (2023). Social determinants of racial health inequities. *The Lancet Public Health, 8*(6), e396–e397.

Braveman, P., Dominguez, T. P., Burke, W., Dolan, S. M., Stevenson, D. K., Jackson, F. M., et al. (2021). Explaining the black-white disparity in preterm birth: A consensus statement from a multi-disciplinary scientific work group convened by the March of Dimes. *Frontiers in Reproductive Health, 3*, 684207.

Cantor, M. N., & Thorpe, L. (2018). Integrating data on social determinants of health into electronic health records. *Health Affairs, 37*(4), 585–590.

Cossio, M. (2023). Digital social determinants of health.

Craig, K. J. T., Fusco, N., Gunnarsdottir, T., Chamberland, L., Snowdon, J. L., & Kassler, W. J. (2021). Leveraging data and digital health technologies to assess and impact social determinants of health (SDoH): A state-of-the-art literature review. *Online Journal of Public Health Informatics, 13*(3).

Davidson, K. W., Kemper, A. R., Doubeni, C. A., Tseng, C.-W., Simon, M. A., Kubik, M., et al. (2020). Developing primary care–based recommendations for social determinants of health: Methods of the US preventive services task force. *Annals of Internal Medicine, 173*(6), 461–467.

Debopadhaya, S., Erickson, J. S., & Bennett, K. P. (2021). Temporal analysis of social determinants associated with COVID-19 mortality.

Doğan, O. (2019). Data linkage methods for big data management in industry 4.0. In *Optimizing big data management and industrial systems with intelligent techniques* (pp. 108–127). IGI Global.

Evans, R. G., Barer, M. L., & Marmor, T. R. (2021). *Why are some people healthy and others not?: The determinants of health of populations*. Walter de Gruyter GmbH & Co. KG.

Evans, C. R., Williams, D. R., Onnela, J.-P., & Subramanian, S. V. (2018). A multilevel approach to modeling health inequalities at the intersection of multiple social identities. *Social Science & Medicine, 203*, 64–73.

Feller, D. J., Zucker, J., Don't Walk, O. B., Yin, M. T., Gordon, P., & Elhadad, N. (2019). Longitudinal analysis of social and behavioral determinants of health in the EHR: Exploring the impact of patient trajectories and documentation practices.

Flack, F., & Smith, M. (2019). The population health research network-population data centre profile. *International Journal of Population Data Science, 4*(2).

Gadhoke, P., Pemberton, S., Foudeh, A., & Brenton, B. P. (2018). Development and validation of the social determinants of health questionnaire and implications for "promoting food security and healthy lifestyles" in a complex urban food ecosystem. *Ecology of Food and Nutrition, 57*(4), 261–281. https://doi.org/10.1080/03670244.2018.1481835

Garden Grove, C. A. (2016). 500 cities project.

Glanz, K., Rimer, B. K., & Viswanath, K. (2008). *Health behavior and health education: Theory, research, and practice*. John Wiley & Sons.

Gómez, C. A., Kleinman, D. V., Pronk, N., Gordon, G. L. W., Ochiai, E., Blakey, C., et al. (2021). Addressing health equity and social determinants of health through healthy people 2030. *Journal of Public Health Management and Practice, 27*(Supplement 6), S249–S257.

Gourevitch, M. N., Athens, J. K., Levine, S. E., Kleiman, N., & Thorpe, L. E. (2019). City-level measures of health, health determinants, and equity to foster population health improvement: The city health dashboard. *American Journal of Public Health, 109*(4), 585–592.

Harrison, K. M., & Dean, H. D. (2011). Use of data systems to address social determinants of health: A need to do more. *Public Health Reports, 126 Suppl 3*(Suppl 3), 1–5. https://doi.org/1 0.1177/00333549111260s301

Herman, E. (2008). The American community survey: An introduction to the basics. *Government Information Quarterly, 25*(3), 504–519.

Hood, C. M., Gennuso, K. P., Swain, G. R., & Catlin, B. B. (2016). County health rankings: Relationships between determinant factors and health outcomes. *American Journal of Preventive Medicine, 50*(2), 129–135. https://doi.org/10.1016/j.amepre.2015.08.024

Huang, Q., Jackson, S., Derakhshan, S., Lee, L., Pham, E., Jackson, A., & Cutter, S. L. (2021). Urban-rural differences in COVID-19 exposures and outcomes in the South: A preliminary analysis of South Carolina. *PLoS One, 16*(2), e0246548.

Hubbard, K., Talih, M., Klein, R. J., & Huang, D. T. (2020). Target-setting methods in Healthy People 2030.

Johnson, A. E., Zhu, J., Garrard, W., Thoma, F. W., Mulukutla, S., Kershaw, K. N., & Magnani, J. W. (2021). Area deprivation index and cardiac readmissions: Evaluating risk-prediction in an electronic health record. *Journal of the American Heart Association, 10*(13), e020466.

Jordan, M. I., & Mitchell, T. M. (2015). Machine learning: Trends, perspectives, and prospects. *Science, 349*(6245), 255–260.

Karran, E. L., Cashin, A. G., Barker, T., Boyd, M. A., Chiarotto, A., Dewidar, O., et al. (2023). The 'what'and 'how'of screening for social needs in healthcare settings: A scoping review. *PeerJ, 11*, e15263.

Kino, S., Hsu, Y. T., Shiba, K., Chien, Y. S., Mita, C., Kawachi, I., & Daoud, A. (2021). A scoping review on the use of machine learning in research on social determinants of health: Trends and research prospects. *SSM Popul Health, 15*, 100836. https://doi.org/10.1016/j.ssmph.2021.100836

Knighton, A. J., Savitz, L., Belnap, T., Stephenson, B., & VanDerslice, J. (2016). Introduction of an area deprivation index measuring patient socioeconomic status in an integrated health system: Implications for population health. *EGEMs, 4*(3), 1238.

Kolak, M., Bhatt, J., Park, Y. H., Padrón, N. A., & Molefe, A. (2020). Quantification of neighborhood-level social determinants of health in the continental United States. *JAMA Network Open, 3*(1), e1919928. https://doi.org/10.1001/jamanetworkopen.2019.19928

Lazarou, C., Karaolis, M., Matalas, A.-L., & Panagiotakos, D. B. (2012). Dietary patterns analysis using data mining method. An application to data from the CYKIDS study. *Computer Methods and Programs in Biomedicine, 108*(2), 706–714.

Lee, J., Schram, A., Riley, E., Harris, P., Baum, F., Fisher, M., et al. (2018). Addressing health equity through action on the social determinants of health: A global review of policy outcome evaluation methods. *International Journal of Health Policy and Management, 7*(7), 581.

López, N., & Gadsden, V. L. (2017). Health inequities, social determinants, and intersectionality. In *Perspectives on health equity and social determinants of health*. National Academies Press (US).

Loukaitou-Sideris, A., Gonzalez, S., & Ong, P. (2019). Triangulating neighborhood knowledge to understand neighborhood change: Methods to study gentrification. *Journal of Planning Education and Research, 39*(2), 227–242.

Mah, J. C., Penwarden, J. L., Pott, H., Theou, O., & Andrew, M. K. (2023). Social vulnerability indices: A scoping review. *BMC Public Health, 23*(1), 1253.

Marmot, M. (2005). Social determinants of health inequalities. *The Lancet, 365*(9464), 1099–1104.

Marmot, M., Friel, S., Bell, R., Houweling, T. A. J., & Taylor, S. (2008). Closing the gap in a generation: Health equity through action on the social determinants of health. *The Lancet, 372*(9650), 1661–1669. https://doi.org/10.1016/S0140-6736(08)61690-6

Marmot, M., & Wilkinson, R. (2005). *Social determinants of health*. Oup Oxford.

Matsumoto, M., & Nakayama, K. (2017). Development of the health literacy on social determinants of health questionnaire in Japanese adults. *BMC Public Health, 17*(1), 30. https://doi.org/10.1186/s12889-016-3971-3

McNeil, J. M. (1994). Census bureau. *Americans with Disabilities, 95*, 70–61.

Metzler, M., Kanarek, N., Highsmith, K., Straw, R., Bialek, R., Stanley, J., et al. (2008). Peer reviewed: Community health status indicators project: The development of a national approach to community health. *Preventing Chronic Disease, 5*(3), A94.

Moise, I. (2020). Variation in risk of COVID-19 infection and predictors of social determinants of health in Miami–Dade County, Florida. *Preventin Chronic Diseases, 17*, E124.

Moise, I. K., Ivanova, N., Wilson, C., Wilson, S., Halwindi, H., & Spika, V. M. (2023). Lessons from digital technology-enabled health interventions implemented during the coronavirus pandemic to improve maternal and birth outcomes: A global scoping review. *BMC Pregnancy and Childbirth, 23*(1), 195. https://doi.org/10.1186/s12884-023-05454-3

Moise, I. K., Kalipeni, E., & Zulu, L. C. (2011). Analyzing geographical access to HIV sentinel clinics in relation to other health clinics in Zambia. *Journal of Map & Geography Libraries, 7*(3), 254–281. https://doi.org/10.1080/15420353.2011.599756

Moise, I. K., & Ruiz, M. O. (2016). Hospitalizations for substance abuse disorders before and after Hurricane Katrina: Spatial clustering and area-level predictors, New Orleans, 2004 and 2008. *Preventing Chronic Disease, 13*, E145. https://doi.org/10.5888/pcd13.160107

Mokdad, A. H. (2009). The behavioral risk factors surveillance system: Past, present, and future. *Annual Review of Public Health, 30*(1), 43–54.

Naimoli, J. F., Frymus, D. E., Wuliji, T., Franco, L. M., & Newsome, M. H. (2014). A community health worker "logic model": Towards a theory of enhanced performance in low-and middle-income countries. *Human Resources for Health, 12*, 1–16.

National Center for Health Statistics. Division of Health Interview, S. (1986). *National health interview survey*. US Public Health Service, National Center for Health Statistics.

Ochiai, E., Kigenyi, T., Sondik, E., Pronk, N., Kleinman, D. V., Blakey, C., et al. (2021). Healthy people 2030 leading health indicators and overall health and Well-being measures: Opportunities to assess and improve the health and Well-being of the nation. *Journal of Public Health Management and Practice, 27*(Supplement 6), S235–S241.

Ozdenerol, E. (2016). *Spatial health inequalities: Adapting GIS tools and data analysis*. CRC Press.

Petrovskis, A., Bekemeier, B., Heitkemper, E., & van Draanen, J. (2023). The DASH model: Data for addressing social determinants of health in local health departments. *Nursing Inquiry, 30*(1), e12518.

Ramesh, B., Jagger, M. A., Zaitchik, B., Kolivras, K. N., Swarup, S., et al. (2022). Flooding and emergency department visits: Effect modification by the CDC/ATSDR social vulnerability index. *International Journal of Disaster Risk Reduction, 76*, 102986.

Ravaghi, H., Guisset, A.-L., Elfeky, S., Nasir, N., Khani, S., Ahmadnezhad, E., & Abdi, Z. (2023). A scoping review of community health needs and assets assessment: Concepts, rationale, tools and uses. *BMC Health Services Research, 23*(1), 44.

Remington, P. L., Catlin, B. B., & Gennuso, K. P. (2015). The county health rankings: Rationale and methods. *Population Health Metrics, 13*, 1–12.

Rojas, D., Melo, A., Moise, I. K., Saavedra, J., & Szapocznik, J. (2020). The association between the social determinants of health and HIV control in Miami-Dade County ZIP codes, 2017. *Journal of Racial and Ethnic Health Disparities, 8*, 763–772. https://doi.org/10.1007/s40615-020-00838-z

Rollings, K. A., Noppert, G. A., Griggs, J. J., Melendez, R. A., & Clarke, P. J. (2023). Comparison of two area-level socioeconomic deprivation indices: Implications for public health research, practice, and policy. *PLoS One, 18*(10), e0292281.

Roux, A. V. D., Merkin, S. S., Arnett, D., Chambless, L., Massing, M., Nieto, F. J., et al. (2001). Neighborhood of residence and incidence of coronary heart disease. *New England Journal of Medicine, 345*(2), 99–106.

Sagiroglu, S., & Sinanc, D. (2013). Big data: A review.

Sedgwick, P. (2011). The ecological fallacy. *BMJ, 343*.

Segar, M. W., Hall, J. L., Jhund, P. S., Powell-Wiley, T. M., Morris, A. A., et al. (2022). Machine learning-based models incorporating social determinants of health vs traditional models for predicting in-hospital mortality in patients with heart failure. *JAMA Cardiology, 7*(8), 844–854. https://doi.org/10.1001/jamacardio.2022.1900

Shokouh, S. M. H., Mohammad, A., Emamgholipour, S., Rashidian, A., Montazeri, A., & Zaboli, R. (2017). Conceptual models of social determinants of health: A narrative review. *Iranian Journal of Public Health, 46*(4), 435.

Spielman, S. E., Tuccillo, J., Folch, D. C., Schweikert, A., Davies, R., Wood, N., & Tate, E. (2020). Evaluating social vulnerability indicators: Criteria and their application to the social vulnerability index. *Natural Hazards, 100*, 417–436.

Sui, D. (2004). Chapter five GIS, environmental equity analysis, and the modifiable areal unit problem (MAUP). *Geographic Information Research: Transatlantic Perspectives, 40*.

Takatsuka, N., Kurisu, Y., Nagata, C., Owaki, A., Kawakami, N., & Shimizu, H. (1997). Validation of simplified diet history questionnaire. *Journal of Epidemiology, 7*(1), 33–41.

Torres, J., Emilia De Marchis, M. D., Fichtenberg, C., & Gottlieb, L. (2017). siren.

Trinidad, S., Brokamp, C., Mor Huertas, A., Beck, A. F., Riley, C. L., Rasnick, E., et al. (2022). Use of area-based socioeconomic deprivation indices: A scoping review and qualitative analysis: Study examines socioeconomic deprivation indices. *Health Affairs, 41*(12), 1804–1811.

U. S. Department of Health and Human Services. (2020a). *Health equity report 2019–2020: Special feature on housing and health inequalities*. Health Resources and Services Administration. Office of Health Equity.

U. S. Department of Health and Human Services. (2020b). *Social Determinants of Health*. https://health.gov/healthypeople/objectives-and-data/social-determinants-health

Varbanova, V., & Beutels, P. (2020). Recent quantitative research on determinants of health in high income countries: A scoping review. *PLoS One, 15*(9), e0239031.

Wilkinson, R. G., & Marmot, M. (2003). *Social determinants of health: The solid facts*. World Health Organization.

World Health Organization. (2010). *A conceptual framework for action on the social determinants of health*. World Health Organization.

Chapter 13
Gentrification and Health: Types, Mechanisms, and Operationalizations

Eileen E. Avery and Danielle C. Kuhl

Introduction

As gentrification continues to proliferate in the United States and globally, scholars (e.g., Acolin et al., 2023; Cole et al., 2023) are increasingly interested in its relationship with health. Gentrification is an increase in the socioeconomic status (SES) of a neighborhood or community, and its varying forms—e.g., residential, tourism, environmental—underlie this fundamental characteristic. It arguably has both positive and negative impacts on residents' health, though the salient aspects of gentrification and their consequences for health across different community types (e.g., urban–rural) are not widely understood.

A social determinants of health (SDoH) framework identifies social conditions that shape health inequalities that are grounded in unequal vulnerabilities at the contextual (e.g., economic stability, access to quality housing, education, health care, needed goods and services, cohesion, and safety) and individual (e.g., age, race, ethnicity, nativity, SES, gender identity, sexual orientation) levels. Both are conditioned by power and privilege, perhaps most consistently captured with the positive relationship between SES and health, such that SES is understood to be a fundamental cause of disease (Link & Phelan, 1995). Likewise, social relationships are a causal factor of health (House et al., 1988). These relationships remain central in understanding the links between myriad aspects of gentrification and health across groups.

Health is affected through both psychosocial and neomaterial SDoH mechanisms. Psychosocial pathways (e.g., loss of sense of belonging, experiences of

E. E. Avery (✉)
Department of Sociology, University of Missouri, Columbia, MO, USA
e-mail: averye@missouri.edu

D. C. Kuhl
Department of Sociology, Bowling Green State University, Bowling Green, OH, USA

© The Author(s) 2026
M. A. Kolak, I. K. Moise (eds.), *Place and the Social-Spatial Determinants of Health*, Global Perspectives on Health Geography,
https://doi.org/10.1007/978-3-031-88463-4_13

racism or classism, frustration with lack of access to previously available goods or services) operate chiefly through the stress response, which leads to a primary physiological response and, primarily when chronic, is associated with a host of longer-term physical and mental health consequences, as well as potential for health-compromising behaviors (e.g., substance use). Neomaterial pathways operate either directly (e.g., access to healthy food, health-promoting services, or recreation space) or indirectly through chronic stress that results from these same situations. A neomaterialist hypothesis is that the "health effects of income inequalities result from differential accumulation of exposures that have their sources in the material world and do not result directly from perceptions and feelings of being disadvantaged" (McDowell, 2023, p. 104). Neighborhood changes, including those from gentrification, can work through both pathways to affect community and individual health and well-being. For example, displacement can be direct (i.e., displaced from residence) or affective, in which residents lose their sense of belonging to their community or no longer feel at home there (Butcher & Dickens, 2016).

In this chapter, we first describe extant research on the ways gentrification is connected to health with a focus on psychosocial and material conditions. We then discuss some types of gentrification and their operationalization, which helps to better understand heterogeneous findings. We then provide an applied descriptive example that describes the distribution of residential gentrification in urban and rural communities and its correlation with mental health distress in the United States leading into COVID-19. We conclude with policy implications and suggestions for future research. Our focus is on the United States, but we incorporate some international examples.

Mechanisms of Connection to Health

There is a vast literature on the link between gentrification and health wherein studies use a multitude of measures, samples, and methodologies. Unfortunately, because they are often motivated by varying (or not) theoretical perspectives, summary takeaways are unclear. Several literature reviews, to be discussed following, have concluded that when gentrification affects health—often there is a weak or null association—its effects are highly heterogeneous across groups. Qualitative studies demonstrate that gentrification exacerbates health inequities, particularly for non-White populations or marginalized groups (Thurber & Krings, 2021; Versey, 2023). Due to mixed results in sociology, economics, geography, and urban planning, scholars advocate for an interdisciplinary approach to studying gentrification in a public health framework and moving away from simply asking whether gentrification affects health and asking instead: "Who does gentrification benefit, or harm, and how?" (Cole et al., 2023, p. 202). To that, we would add: "when?"

The SDoH framework includes a wide range of conditions that influence health and well-being, including but not limited to economic inequality, access to education and health care, the built environment, and social and community contexts, such as neighborhood cohesion or perceptions of safety and individual identities. The neomaterial perspective posits that health inequalities are the result of structural factors, or the "differential accumulation of exposures and experiences that have their sources in the material world" (Lynch et al., 2000, p. 1202). Examples of these sources include the unequal distribution of health services, housing instability, neighborhood infrastructure, such as building designs or access to green space, or residential inequalities in access to affordable nutritious foods. In contrast, the psychosocial perspective argues that health inequalities are produced via affective responses to disadvantage or to relative deprivation. Negative perceptions of relative position within the social structure can "produce negative emotions such as shame and distrust that are translated 'inside' the body into poorer health…via stress induced behaviors such as smoking…. [as well as] 'outside' the individual into antisocial behavior, reduced civic participation, and less social capital and cohesion within the community" (Lynch et al., 2000, p. 1201). The stress response is connected to both perspectives, as is neighborhood change. Thus, scholarship on health outcomes and inequalities has offered support for both neomaterial and psychosocial interpretations of gentrification's effects.

Anguelovski et al. (2021) took the many variables associated with gentrification and created an efficient proposal of four distinct pathways through which gentrification affects health: (1) threats to housing and financial security; (2) sociocultural displacement; (3) loss of public amenities, facilities, and services; and (4) crime and safety. Data analysis from a set of 14 cities/neighborhoods in the United States, Canada, and Western Europe showed gentrification produces "community and individual trauma" [that has] …. clear physical and mental health impacts" (Anguelovski et al., 2021, p. 13). Versey's (2023) study explored these same four pathways using a sample of women of color who were mostly single heads of households who rent. Her findings parallel those of Anguelovski et al. (2021). Examples of their findings include but are not limited to: an association between gentrification and affordable housing, which affects residents' health by creating or increasing chronic stress and by reducing income for healthy food or children's after-school sports activities via increased rent; inadequate sleep due to increases in nightlife noise from tourists; anxiety, suicidal ideation, and depression due to increased social segregation and physically displaced family or friends; PTSD and polysubstance use due to displacement and evictions; and increased crime related to tourism gentrification.

Much extant work on health outcomes focuses on displacement—wherein there is a combination of structural change (i.e., residential moves) and psychosocial consequences that come with it, but clear links to health are debated. Freeman and Braconi (2004) and Freeman, 2005 finds that gentrification can happen without leading to widespread displacement. Thus, we return to the earlier question of "Who does gentrification benefit, or harm, and how?" (Cole et al., 2023, p. 202).

Acolin et al. (2023) analyzed data on gentrifying and nongentrifying lower income neighborhoods in a sample of large metro areas in the United States. They found that between 2006 and 2019, residents of gentrifying neighborhoods experienced worse changes in healthcare access, area deprivation, and walkability but also experienced reduced exposure to air pollution. When comparing movers and stayers, they find that those who stay in gentrifying places experience improvements in deprivation and healthcare access.

A major finding is that gentrification seems to have worse outcomes for non-White communities and individuals. Thus, scholars call for more complex theorizing about the study of gentrification, as a narrow focus on political economy "pays insufficient attention to racialized policies and their resulting inequities" (Thurber et al., 2021, p. 30), which is ironic because "people of color are more likely to live in neighborhoods vulnerable to gentrification and thus are disproportionately harmed" (Thurber et al., 2021, p. 30). Studies report worse health outcomes in gentrified neighborhoods for non-White compared to White residents, including increased hypertension for Hispanic/Latino₍a₎ residents and worse preterm birth and self-rated health for African Americans (Huynh & Maroko, 2014; Gibbons & Barton, 2016; Izenberg et al., 2018).

Upstream structural conditions set the stage for individuals' downstream health consequences. Qualitative scholarship on gentrification in non-White communities points to a history of "housing segregation, discrimination, redlining, property speculation, and government neglect" (Gibson, 2007). Quantitative studies also demonstrate less reinvestment in non-White neighborhoods compared to White neighborhoods. In a study comparing predominantly White, Black, and Hispanic/Latino₍a₎ neighborhoods in Chicago, Hwang and Sampson (2014, p. 746) found that "minority gentrification does not result in substantial neighborhood reinvestment overall" and that non-White neighborhoods experienced a slowed rate of (upward) change or stagnation compared to White neighborhoods, likely due to racial inequalities in wealth or investment agents' biases. Despite these inequalities, some argue that non-White residents welcome gentrification, as it improves their home values and brings in new amenities; thus, among those individuals, gentrification is not "wreaking havoc" and the goal should be "not to stop [it] but to harness it to benefit the existing community" (Gilderbloom, 2019:635). Thus, gentrification is linked to wealth accumulation in non-White communities, with long-term consequences for upward social mobility and thus better health.

Other research points to how identity is tied to building designs, and when gentrification changes the appearance of buildings, this has negative consequences for community connection and health consequences that operate through both psychosocial and material pathways. For example, Hom (2022, p. 222) documents how changes in architectural design in L.A.'s Chinatown led to "disruption of [the] physical expression of community," which produced cracks in residents' sense of place and ethnic identity. Thus, material changes have psychosocial consequences. This affective displacement, or no longer feeling at home (Butcher & Dickens, 2016), is a common well-being outcome of gentrification. Another emotional well-being outcome is distrust. In a study of community change in Camden, NJ, Danley and Weaver (2018) also highlight the skepticism and distrust from Black and Hispanic/Latino₍a₎ residents who fight gentrification largely because it will create White spaces that do not benefit long-term residents. These studies point to the need for more inclusive development so that gentrified areas become "community spaces, not White spaces" (Danley & Weaver, 2018, p. 14).

The heterogeneous effects of gentrification also play out in the context of immigrant status. High-amenity rural areas face a "growing presence and dependence on immigrant labor for the production and maintenance of rural gentrified landscapes" (Nelson et al., 2015, p. 855) such as in Georgia and Colorado, where the economy now depends more on "hierarchies of class, race, and illegality to discipline a contingent, flexible workforce" (2015, p. 843). Unfortunately, there is less scholarship linking these new labor regimes to health outcomes, but some research demonstrates how gentrification in urban areas worsens health in ethnic enclaves by opening expensive food stores. Anguelovski (2015) found that ethnic residents in the Jamaica Plain neighborhood of Boston opposed a Whole Foods opening because White residents were manipulating the idea of a "food dessert" against residents' interests, and after Whole Foods opened, this led to "sources of affordable food vanishing." Future research needs to further investigate how both new labor regimes and environmental gentrification in the form of amenities like grocery stores produce other types of health inequalities for immigrant communities in both rural and urban places.

Types of Gentrification

Urban and Rural Residential Gentrification

Scholars commonly speak of gentrification within a residential geographic context, and most research focuses on neighborhoods within metropolitan statistical areas (MSAs) or cities. Some scholarships expand gentrification to non-urban areas while other work conceptualizes gentrification in terms of the mechanisms that stimulate it, such as tourism and environmental improvements, or its argued consequences (e.g., displacement). While the specific operationalization of gentrification is debated, the fundamental process is characterized by increasing desirability of certain places, which leads to increasing property values and subsequent changes in the socio-demographics of residents—especially changes in SES. Class inequalities that resulted from capitalist urban land policies were at the heart of the original conceptualization of gentrification (Glass, 1964).

Until the last few decades, residential gentrification has referred almost exclusively to urban gentrification. Indeed, Smith (1996, p. 39) describes it as "the class remake of the central urban landscape" and recent reviews of gentrification and health focus exclusively on urban places (Smith et al., 2020). Visible changes in the built environment include, but are not limited to, "a collection of trendy bars and cafes, refurbished luxury lofts and condominiums, upscale boutiques, plentiful entertainment venues, and other trappings of middle-class consumption" (Bryson, 2013, p. 578).

Urban gentrification is partly the result of the devaluing and denigration of suburbanization. Ley (1996) identified urban places as the "landscape of desire," which he contrasted with suburban places as "landscapes of despair." Ley's (1996) research in Canada underscored that residents devalued suburbs and saw them as "too standardized, too homogenous, too bland, too conformist, too hierarchical, too conservative, too patriarchal, too straight" (p. 205). Suburban areas experienced economic decline in the 1990s and 2000s, much like the pattern that affected inner-city neighborhoods in the 1970s, which spurred outward mobility of wealthy White residents.

Racial dynamics play out in these parallel processes of decline in urban and suburban locations. The out-migration of wealthier White residents from cities occurred largely in response to non-White populations moving in when property values became achievable. Similarly, once housing prices in older suburbs started to decline, larger proportions of Black and Hispanic/Latino/a residents moved into these neighborhoods (Markley, 2018). Disinvestment makes a place "gentrifiable" because it has potential for upward mobility via reinvestment. If non-White populations paying lower rents are blamed for area decline, then revitalization efforts may be met with support from economically advantaged White residents (Niedt, 2006).

These racially based processes grounded in settler colonialism and couched in "the colorblind language of entrepreneurial growth strategies… allow more advantaged groups to 'take back' prime suburban spaces from marginalized groups" (Markley & Sharma, 2016, p. 74). For example, in Sydney, long-term older residents were pushed out of their public-housing apartments to make way for wealthy residents and a casino (Morris, 2019). Despite warnings that older residents are at risk of negative health and well-being outcomes, and that displacing older people in the name of "urban renewal" can lead to premature death (Danermark et al., 1996), they were evicted anyway. We note that some countries, such as the United States, Australia, and New Zealand, may have more in common due to similar settler-colonial histories and greater emphasis on privatization of housing markets.

Scholars now advocate for studying rural places. Smith (2002, pp. 390–92) argued that there was a need to "widen the spatial lens" of gentrification studies, because "dramatic population transformations… multiple waves of in-, out-, and intra-migration of relatively affluent and lower income households" (Smith, 2002, p. 386) are occurring in non-urban places as well. Rural racial composition in the United States is still majority White, but there is heterogeneity such that Black residents are the largest non-White group in the lowland South, and Hispanics/Latino/as are the largest group in many Southwestern and Western parts of the United States. Rural gentrification is often characterized by both residential and retail, leisure, and industrial revitalization (Phillips, 1993).

Rural gentrification is often tied to rural rebound (Johnson & Beale, 1995), where, during the 1990s, more than 86% of nonmetro counties adjacent to metropolitan counties gained population (Johnson & Cromartie, 2006). As Nelson et al. (2010, p. 344) noted, just "as urban gentrifiers are attracted to inner-city neighborhoods because of the cultural values associated with urban living and the growing cultural dissatisfaction with suburbia, rural gentrifiers are attracted to rural destinations by what they believe rural living will provide." The economic sectors stimulated by an influx of older, empty nesters are the same ones that draw Hispanic/Latino/a workers in rural areas—construction, food services, and private household services (Nelson et al., 2010, p. 350). Thus, traditional definitions of gentrification in cities have parallels and contrasts in rural places: economic changes, age, ethnicity, and racial re-composition.

Environmental and Tourism Gentrification

Closely tied to gentrification are mechanisms that motivate it—in particular, environmental amenities (we use *environmental* and *green* interchangeably) and tourism. Regarding environmental revitalization, urban gentrification is viewed to maintain amenity inequalities. Dooling (2009) discusses green space provisions as a tool for redevelopment. The environment is recognized as a key component of generating economic growth. For example, water views command significantly more money; referring to lakefront properties in Coeur d'Alene, Idaho, development officials explain that "the view is spectacular, and people are willing to pay for it" (Bryson, 2013, p. 581).

Environmental revitalization also serves gentrification through the "brownfields to green space redevelopment trend" (Bryson, 2013, p. 583). Industrial and contaminated sites have become a high priority for urban renewal. Once these become green spaces, they attract investment from developers and subsequently increase property values. If long-term residents, who have suffered the consequences of health effects of environmental pollution for decades, can no longer afford higher rents, then these green provisions benefit only incoming wealthy gentrifiers. Thus, green gentrification has become a problem of environmental justice (Fox, 2019).

Smith (2002) argues that rural gentrification is fueled by the desire to consume green space. This can lead to similar tensions between long-term and newcomer residents that occur in urban areas; as Leebrick (2015, p. 26) states, "community tensions are often class based and represent a major tension between the consumption of rural landscapes as idyllic vistas and bucolic playgrounds for the enjoyment of the affluent and the use-based needs of the lower classes." However, the lure of health-promoting environmental amenities that spur gentrification varies by the type of environment.

As Darling (2005, p. 1022) points out, "what gets produced in the process of urban gentrification is residential space. What gets produced in the process of wilderness [rural] gentrification is recreational nature." Urban renters are usually planning to stay long-term, and the amenities that attract them are usually linked with entertainment, shopping, food, and drink culture, whereas the amenities drawing rural wilderness renters are part of nature itself— "…lakes, hiking trails, snowmobile trails, charismatic megafauna, open space, and the like" (Darling, 2005, p.1022). These amenities thus offer residents access to recreation areas for physical activity and socialization that benefit health.

The key characteristic that distinguishes tourism from residential gentrification (Gotham, 2005) is that its practitioners are visitors, not permanent residents. Recent research on tourism focuses on the phenomenon of Airbnb rentals as catalysts for this type of gentrification, particularly in several countries outside North America (Cocola-Gant & Gago, 2021; Mermet, 2017). Residential and tourism gentrification are both global and local processes: "tourism is a 'global' industry dominated by international chains.… On the other hand, tourism is a 'local' industry characterized by grassroots cultural production…and localized consumption of place" (Gotham, 2005:1102). Tourism offers a unique lens "on the causes and consequences of gentrification better than existing accounts that focus on identifying the population and demographic variables responsible for residential and commercial change" (Gotham, 2005, p. 1115). Tourism gentrification can affect residents' health by altering the local economy, increasing fear of crime, and producing affective displacement, among other things.

Operationalization of Gentrification

We turn now to the operationalization of gentrification. We discuss residential gentrification and tourism gentrification with a focus on urban and rural distinctions and discuss a contested outcome: displacement. We make notes on green gentrification throughout.

Residential Gentrification

The statistical operationalization of gentrification typically measures socioeconomic ascent, usually with income, education, and/or home values (Firth et al., 2020; Lee & Perkins, 2023). Rural gentrification frequently involves older gentrifiers who may have more wealth than income (Nelson et al., 2010). Population shifts in age, race, and ethnicity is common. Qualitatively, scholars also emphasize changes in the built environment including housing stock and quality, the commercial and service environment, and parks and other built spaces that may promote health or represent a loss of a health-promoting resource. This overlaps with environmental gentrification as it focuses on human enhancements of natural spaces resulting in new or revitalized parks or green spaces.

Statistically operationalizing gentrification begins with identifying the focal community and the reference area. In urban research, focal communities are neighborhoods, typically census tracts (Firth et al., 2020). Both cities and metropolitan areas can be referencing units, as the city ensures urban neighborhoods remain the focus, while MSAs allow inclusion of suburban gentrification. In case studies, qualitative factors can determine the neighborhood and reference area, as well as definitions of gentrification. However, rural focal and reference communities are less clear: census tracts are generally not used, and (small) cities may not be appropriate as reference units. Using states presents a statistical dilemma, as many states have few nonmetropolitan counties for reference, but census divisions are an option

(Nelson et al., 2010). Rural residents' spatial understanding of their community may vary from that of urban residents. Counties are commonly used as focal units in rural research, but findings are limited, and best practices remain unclear.

Generally, scholars use one of three broad ways to capture gentrification: the "two-step" method, data reduction techniques and indices, and qualitative approaches. The two-step method divides focal neighborhoods into two categories: gentrifiable and not gentrifiable, where those that are gentrifiable meet some threshold for identification as lower income (e.g., under the 50th percentile) within the reference area. Those that gentrify meet another threshold for increase in neighborhood SES over some period. There are several different operationalizations of "gentrifiable" and "gentrifying" with scholars (Firth et al., 2020; Lee & Perkins, 2023) varying the threshold and sometimes the metric variable—for example, using family rather than household income.

The second quantitative approach is to use a data reduction technique such as cluster or factor analysis to capture gentrifying communities. As one example, Nelson et al. (2010) included rates of change over a 10-year period for several variables that capture population change, housing changes, household size, home value, home ownership, resident income, and income inequality. They then compared the gentrifying cluster versus all other nonmetropolitan counties on hypothesized components of rural gentrification—seasonal housing, recreation, retirement dependency, natural amenities score, and change in percent of baby boomers and Hispanic /Latino/a population. This approach allows for more complexity via use of a broader array of characteristics than the two-step method. Similarly, indices use several standardized variables and can be interpreted on a continuum. We note that many variables may be biased as Immergluck and Hollis (2023) recently demonstrated with American Community Survey (ACS) data on median home value, which is based on respondents' assessment of their home's value rather than market value assessments.

As a third approach, qualitative methods provide meaningful definitions of gentrification. Zukin et al. (2009) described gentrification change involving the "boutiqueing" of commercial areas as a sign of gentrification, for example. Here, upscale retail, restaurants, food stores, and local chains or independently owned businesses are put in, often replacing national chains. Furthermore, improvements in landscaping and revitalization of parks and greenways are also features of many gentrifying communities. Qualitative approaches can capture gentrification concepts that are not easy to measure quantitatively. For example, in Barcelona, Sánchez-Ledesma et al. (2020) used photovoice to determine themes related to tourism gentrification that residents identified as meaningful for health and well-being. They found that several factors matter, including decline in social networks, loss of identity, built environmental changes, increased pollution (including environmental, visual, and acoustic), changes in services and stores, rising property values and increased evictions, and activism efforts. Experiences of gentrification can also be captured with survey items on social cohesion and attachment. Long-term residents may lose neighborhood identity and social relationships. Interviews and focus groups can provide more in-depth data on residential experiences during or following gentrification.

Tourism Gentrification

Tourism gentrification is a process that continues to play out in both urban and rural areas. Much work examines the European urban context, where increases in tourism, particularly in historic centers ("heritage-led" gentrification (Escobedo, 2020)), come with changes in housing and retail environments that cater to the desires of short-term visitors. Over time, this affects residents' long-term housing and ability to connect and identify with the city and optimally use it for their own health and well-being. To operationalize tourism gentrification, we focus on rising property values and rents, increases in hotels and short-term rentals (STRs) and the consequent reduction in long-term rentals, and demographic shifts in tourists themselves. Many components of tourism gentrification best fit case studies of individual cities due to lack of data for larger samples.

The centrality of lodging shifts, be it STRs or new hotels, cannot be overstated when operationalizing tourism gentrification. González-Pérez (2020) emphasizes the occurrence of STRs, luxury hotels, and historic preservation in Palma, Spain. In response to local activism, Palma took the then rare step to ban STRs in residential areas due to their link to rising property values, eviction, and undesirable changes in the commercial built environment. Although STRs receive much more attention, growth in hotels also leads to increases in the number of tourists using city space, which affects residents' lived experiences in their community. We therefore suggest that scholars consider total accommodations.

In the European context, transnational tourism is considered as the primary cause of gentrification. It occurs through at least two processes. First, Cocola-Gant and Lopez-Gay (2020) describe a process in which increasing numbers of middle-class transnational migrants coalesce in neighborhoods in desirable cities, and international tourists follow. Second, visitors can increase due to efforts to promote economic development through tourism. Scholars describe this happening in numerous cities, including Barcelona, Dubrovnik, and Croatia (Bobic & Akhavan, 2022; Cocola-Gant & Lopez-Gay, 2020).

Tourism gentrification is widespread as tourists' consumption preferences increasingly focus on "unique" or "unusual" tourism experiences, which contribute to this phenomenon in "undiscovered" cities including in middle-sized cities and rural United States locales. Many tourists still seek luxury, uniqueness, particular amenities (including natural resources), and/or the ability to experience a place "like a local" through STR use. As such, increases in international tourists, or in distance traveled by native tourists, are indicators of tourist gentrification.

Displacement

Displacement is a contested component of gentrification, particularly in quantitative work (see Brown-Saracino, 2017, for a nuanced discussion). Extant research largely retains Marcuse's (1986) original definition of residents being forced out of their neighborhoods through pricing, eviction, or similar mechanisms with a consequence of residing in an economically worse neighborhood than where they originated. Rates of displacement are often established by contrasting mobility in gentrifiable neighborhoods that did and did not gentrify. Frequently, the relationship is null. For example, McKinnish et al. (2010) found that Black and Hispanic/Latino/a residents relocated out of gentrifiable neighborhoods at similar rates whether they gentrified or not. Moreover, in cases where residential mobility patterns differ, it may be the more advantaged individuals in gentrifying neighborhoods who move rather than the most vulnerable residents.

Lee and Perkins (2023) conclude that the relationship between gentrification and displacement is unclear; in a national sample, they found a modest positive association between moderate and intense gentrification and displacement, but it was variable across different types of metropolitan areas. Establishing a causal effect of gentrification on displacement requires detailed individual- and neighborhood-level data, including motivation for moves, but such data are difficult to obtain. Furthermore, linking it to health via survey or interview data of the same residents adds another challenge. Many single city/area analyses find that groups were displaced by gentrification with highly detrimental effects (e.g., Weller & Van Hulten, 2012 in Melbourne suburbs). We note that place attachment has been linked to well-being, particularly among older populations (e.g., Wiles et al., 2009 in New Zealand).

Given these challenges, and in line with its connection to health, we advocate for an operationalization of displacement that is consistent with Elliott-Cooper et al.'s (2020) framework of "un-homing." Moving beyond physical displacement to affective displacement, they attend to the psychological aspects of un-homing related to loss of identity and community that are salient to residents, and that have direct and indirect links to health, particularly mental health. Thus, we encourage a nuanced understanding that encompasses psychological aspects and well-being in addition to physical displacement.

Descriptive Application

To briefly illustrate one of the myriad heterogeneous correlations between gentrification and health, we focus on differences in urban and rural locales using one operationalization of gentrification and one of mental health. Specifically, we present descriptive statistics on (1) the extent of gentrification in metropolitan and non-metropolitan counties, and (2) average levels of mental distress among adults in gentrifiable communities that did and did not gentrify across county types. In each case, we use Brummet and Reed's (2019) operationalization of "gentrifiable" and "gentrifying" wherein communities below the reference area's median household income in 2010 are considered "gentrifiable," and among those, those "gentrifying" is defined as being in the top ten percent of change in percent with a bachelor's degree in the reference area from 2010 to 2019. Data are from the ACS 2006–2010 and 2015–2019 5-year estimates (U.S. Census Bureau, 2022). Communities are operationalized as census tracts in metropolitan counties (categories 1–4 in the National Center for Health Statistics [NCHS] urban–rural county classification scheme) and counties in nonmetropolitan context (categories 5 and 6 from NCHS) with complete data using MSAs and census divisions, respectively, as reference areas. Distress, as captured by PLACES data (Centers for Disease Control, 2022), is operationalized as the prevalence of reporting $\geq$14 days of poor mental health in the last 30 days in the community unit.

Figure 13.1 depicts the percent of communities that gentrified in the metropolitan and nonmetropolitan United States from 2010 to 2019. Metropolitan neighborhoods, on average, were more likely to gentrify than rural areas, but there was variation: large central metros gentrified at the highest rate and small metros and micropolitan counties gentrified at the lowest rates. Moreover, there was variation within each broad category (metropolitan versus nonmetropolitan), strongly suggesting that scholars should move beyond an urban–rural dichotomy when examining gentrification and its effects on health. There may well be further heterogeneity within each of the six-county classifications that vary across regions and other attributes.

Figure 13.2 depicts rates of mental health distress across county classifications and gentrification status. Results show mental distress is significantly higher in gentrifying communities in metropolitan counties than it is in those that are gentrifiable but are not gentrifying. This difference was not significant in nonmetropolitan areas.

This brief application of a two-step operationalization of gentrification and mental health distress, as operationalized by national survey data in the United States,

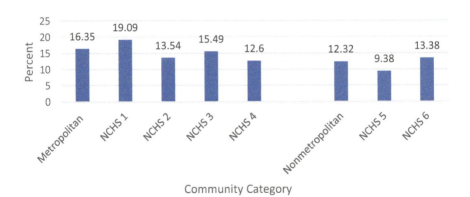

Community Category

Fig. 13.1 Percent of gentrifiable communities undergoing gentrification across metropolitan status, 2010–2019 (We use the 2006–2010 (time 1) and 2015–2019 (time 2) American Community Survey 5-year estimates. Gentrifiable communities were identified at time 1 (yes/no) and gentrification (yes/no) was evaluated at time 2 (Brummet & Reed, 2019). We use census tracts within metropolitan areas and counties within census divisions as the focal community and reference area in metropolitan and nonmetropolitan categories respectively. Counties are classified according to the 2013 National Center for Health Statistics urban–rural county classification scheme where 1 = large central metro; 2 = large fringe metro; 3 = medium metro; 4 = small metro; 5 = micropolitan; and 6 = noncore)

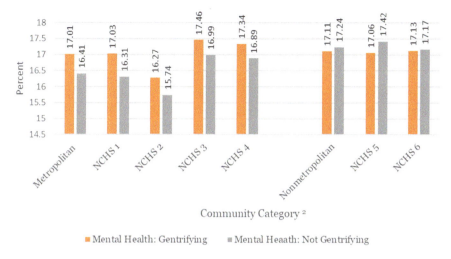

Community Category [2]

■ Mental Health: Gentrifying   ■ Mental Heaath: Not Gentrifying

Fig. 13.2 Percent reporting mental health distress across community gentrification status (We use Centers for Disease Control PLACES data and the 2006–2010 (time 1) and 2015–2019 (time 2) American Community Survey 5-year estimates. Distress is operationalized as the percent of adults reporting poor mental health for at least 14 of the past 30 days. Gentrifiable communities were identified at time 1 (yes/no) and gentrification (yes/no) was evaluated at time 2 (Brummet & Reed, 2019). We use census tracts within metropolitan areas and counties within census divisions as the community and reference area in metropolitan and nonmetropolitan categories respectively. Counties are classified according to the 2013 National Center for Health Statistics urban–rural county classification scheme where 1 = large central metro; 2 = large fringe metro; 3 = medium metro; 4 = small metro; 5 = micropolitan; and 6 = noncore. For the *Community Category,* all differences in distress between gentrifiable and gentrifying communities are significant at the 0.01 level in metropolitan categories. None are significant in nonmetropolitan categories)

illustrates community heterogeneity in both the extent of gentrification and its correlation with mental health distress—both within and between urban and rural counties. Because this is one of many operationalizations of gentrification and one of numerous possible health outcomes, we emphasize that, although some patterns may emerge that suggest there are differences that tend to follow the urban–rural dichotomy, such as those seen in aggregate for mental health distress here, a better way of thinking about patterns, perhaps particularly in the rural context, is as a patchwork that is conditioned on macrostructural and local conditions that also intersect with individual identities. We emphasize that establishing causality requires data over time that is difficult to obtain.

Policy Implications

The policy implications related to gentrification are numerous and varied, and will likely take on greater importance in coming years. Residential gentrification affects affordable housing and zoning, which support stable mixed-income neighborhoods wherein residents can develop cohesion, informal social control, and other health-enhancing characteristics. Another consideration is the need to increase the number of neighborhoods that integrate single-family and multifamily housing. Rent control and tenant protection are crucial as well. In rural places, thoughtful planning is necessary to balance infrastructure—both social (e.g., law enforcement) and material (maintaining and building new roads, sewer lines, etc.)—to support residents' health and well-being as the population grows. Rural communities are particularly sensitive to balancing the necessity of promoting tourism and mitigating its ills.

It is important that policymakers promote social cohesion and access to green space. This could be done through programs for longtime residents that maintain ties to and identity with the community and/or prioritizing inclusive programming that aims to integrate the full community at local events. Zoning provides for areas of affordable housing and mixed uses near parks, paths, water, and other green spaces matter as well. Indeed, environmental/green gentrification policy should focus on ensuring equitable access for all residents to parks and green spaces. The same arguments apply to prioritizing a mix of commercial activity that, overall, meets the needs of and is accessible to all residents.

Policymakers should emphasize sustainable tourism that includes emphasis on historical and cultural preservation, equitable economic development, and balanced hotel and STR stock that fits the needs of their community. The development model should maximize options for non-tourism sectors to the extent possible to minimize overreliance on tourism. Sustainable tourism is particularly important to preserving the natural environment in rural communities undergoing rapid tourism gentrification.

Directions for Future Research

While there are numerous directions for future research related to gentrification and health, we focus on six: attention to temporality, seasonality, climate change, moving beyond urban areas, operationalizing gentrification based on residents' perceptions, and moving to consider its impact on well-being as well as health.

Regarding temporality, a life course approach is desirable (Firth et al., 2020), as it allows researchers to focus on the ages or turning points when health may be most affected. It also allows examination of duration—as how long individuals experience gentrification is likely a factor that affects health. A focus on different stages of gentrification would also be desirable, but there is a need for long-term data over several years in conjunction with residential histories to capture these stages. While a detailed discussion is beyond scope, the stages of gentrification individuals have experienced may also matter for health.

Related to temporality, we suggest that future work should address seasonality. Tourism is often seasonal or episodic, meaning that the short-term visitors who shape the character of the community, often to the detriment of residents, are present during specific time periods. This can cause affective displacement for those residents who lose connection with their community in ways that are heightened by the ebbs and flows of tourist seasons, major events, and the schedules of second homeowners. In rural areas, cyclical visitors displace residents' identities and practices more than their residence (Kocabıyık & Loopmans, 2021). Studentification (e.g., college students) is also applicable.

Third, we join others in advocating for research on the ways that climate change will influence residential and commercial environments, natural spaces, and tourism patterns and their consequences for health. Both urban and rural communities require a better understanding of the unequal impacts of emerging policies that aim to preserve housing and commerce spaces for the affluent in the face of increased flooding, rising sea levels, and other climate-related shifts.

Fourth, despite scholars' calls to attend to heterogeneity in both urban and rural contexts, much of the gentrification literature uses a dichotomy. A better understanding of the ways the process plays out in different suburban, micropolitan, and rural communities will shed light on some of the inconsistent findings in accurately identifying gentrifying communities and inform understanding of health effects across places.

Fifth, researchers should focus more on defining gentrification using measures based on residents' perceptions. It is well known that "perception is reality" in both SDoH and neighborhood effects research. As such, it is individuals' perceptions of negative experiences rather than an outsider's "objective" judgment of the circumstance that is linked to stress. For example, perceptions of cohesion predict self-rated health (Bjornstrom et al., 2013). We suggest that scholars examine perceptions of residential and commercial change due to gentrification and their associations with health. Findings can be contrasted with those that link objective measures of gentrification to health.

Finally, we suggest scholars move beyond standard health outcomes as consequences of gentrification to also examine aspects of well-being. Physical health outcomes often take years to develop, and it can be difficult to make links between outcomes and community characteristics, including gentrification. Mental health is perhaps more straightforward, but well-being, or the extent to which one is happy and/or satisfied with one's life, may flow, in part, more directly from residential experiences, including gentrification.

References

Acolin, A., Crowder, K., Decter-Frain, A., Hajat, A., & Hall, M. (2023). Gentrification, mobility, and exposure to contextual determinants of health. *Housing Policy Debate, 33*(1), 194–223.

Anguelovski, I. (2015). Alternative food provision conflicts in cities: Contesting food privilege, injustice, and whiteness in Jamaica Plain, Boston. *Geoforum, 58*, 184–194.

Anguelovski, I., Cole, H. V., O'Neill, E., Baró, F., Kotsila, P., Sekulova, F., et al. (2021). Gentrification pathways and their health impacts on historically marginalized residents in Europe and North America: Global qualitative evidence from 14 cities. *Health & Place, 72*, 102698.

Bjornstrom, E. E., Ralston, M. L., & Kuhl, D. C. (2013). Social cohesion and self-rated health: The moderating effect of neighborhood physical disorder. *American Journal of Community Psychology, 52*, 302–312.

Bobic, S., & Akhavan, M. (2022). Tourism gentrification in Mediterranean heritage cities. The necessity for multidisciplinary planning. *Cities, 124*, 103616. https://doi.org/10.1016/j. cities.2022.103616

Brown-Saracino, J. (2017). Explicating divided approaches to gentrification and growing income inequality. *Annual Review of Sociology, 43*, 515–539.

Brummet, Q., & Reed, D. (2019, July). The effects of gentrification on the Well-being and opportunity of original resident adults and children. (Federal Reserve Bank of Philadelphia Working Paper No 19-30). Retrieved from https://www.philadelphiafed.org/-/media/frbp/assets/working-papers/2019/wp19-30.pdf?la=en

Bryson, J. (2013). The nature of gentrification. *Geography. Compass, 7*(8), 578–587.

Butcher, M., & Dickens, L. (2016). Spatial dislocation and affective displacement: Youth perspectives on gentrification in London. *International Journal of Urban and Regional Research, 40*(4), 800–816.

Centers for Disease Control. (2022). *PLACES: Local Data for Better Health*. Retrieved May, 2023.

Cocola-Gant, A., & Gago, A. (2021). Airbnb, buy-to-let investment and tourism-driven displacement: A case study in Lisbon. *Environment and Planning A: Economy and Space, 53*(7), 1671–1688.

Cocola-Gant, A., & Lopez-Gay, A. (2020). Transnational gentrification, tourism and the formation of 'foreign only'enclaves in Barcelona. *Urban Studies, 57*(15), 3025–3043.

Cole, H. V., Anguelovski, I., Triguero-Mas, M., Mehdipanah, R., & Arcaya, M. (2023). Promoting health equity through preventing or mitigating the effects of gentrification: A theoretical and methodological guide. *Annual Review of Public Health, 44*, 193–211.

Danermark, B. D., Ekström, M. E., & Bodin, L. L. (1996). Effects of residential relocation on mortality and morbidity among elderly people. *The European Journal of Public Health, 6*(3), 212–217.

Danley, S., & Weaver, R. (2018). "They're not building it for us": Displacement pressure, unwelcomeness, and protesting neighborhood investment. *Societies, 8*(3), 74.

Darling, E. (2005). The city in the country: Wilderness gentrification and the rent gap. *Environment and Planning A, 37*(6), 1015–1032.

Dooling, S. (2009). Ecological gentrification: A research agenda exploring justice in the city. *International Journal of Urban and Regional Research, 33*(3), 621–639.

Elliott-Cooper, A., Hubbard, P., & Lees, L. (2020). Moving beyond Marcuse: Gentrification, displacement and the violence of un-homing. *Progress in Human Geography, 44*(3), 492–509.

Escobedo, D. N. (2020). Foreigners as gentrifiers and tourists in a Mexican historic district. *Urban Studies, 57*(15), 3151–3168.

Firth, C. L., Fuller, D., Wasfi, R., Kestens, Y., & Winters, M. (2020). Causally speaking: Challenges in measuring gentrification for population health research in the United States and Canada. *Health & Place, 63*, 102350.

Fox, S. (2019). Environmental gentrification. *U. Colo. L. Rev., 90*, 803.

Freeman, L. (2005). Displacement or succession? Residential mobility in gentrifying neighborhoods. *Urban Affairs Review, 40*(4), 463–491.

Freeman, L., & Braconi, F. (2004). Gentrification and displacement New York City in the 1990s. *Journal of the American Planning Association, 70*(1), 39–52.

Gibbons, J., & Barton, M. S. (2016). The association of minority self-rated health with black versus white gentrification. *Journal of Urban Health, 93*, 909–922.

Gibson, K. J. (2007). Bleeding Albina: A history of community disinvestment, 1940-2000. *Transforming Anthropology, 15*(1), 3–25.

Gilderbloom, H. J. I. (2019). The future of cities: The end of Marxism and the promise of green urbanism.

Glass, R. (1964). Introduction: Aspects of change, in Centre for Urban Studies. In *London: Aspects of change*. MacGibbon and Kee.

González-Pérez, J. M. (2020). The dispute over tourist cities. Tourism gentrification in the historic Centre of Palma (Majorca, Spain). *Tourism Geographies, 22*(1), 171–191.

Gotham, K. F. (2005). Tourism gentrification: The case of Nnew Orleans' vieux carre (French Quarter). *Urban Studies, 42*(7), 1099–1121.

Hom, L. D. (2022). Symbols of gentrification? Narrating displacement in Los Angeles Chinatown. *Urban Affairs Review, 58*(1), 196–228.

House, J. S., Landis, K. R., & Umberson, D. (1988). Social relationships and health. *Science, 241*(4865), 540–545.

Huynh, M., & Maroko, A. R. (2014). Gentrification and preterm birth in New York City, 2008–2010. *Journal of Urban Health, 91*, 211–220.

Hwang, J., & Sampson, R. J. (2014). Divergent pathways of gentrification: Racial inequality and the social order of renewal in Chicago neighborhoods. *American Sociological Review, 79*(4), 726–751.

Immergluck, D., & Hollis, A. (2023). Different data, different measures: Comparing alternative indicators of changes in neighborhood home values. *Housing Policy Debate*, 1–21.

Izenberg, J. M., Mujahid, M. S., & Yen, I. H. (2018). Health in changing neighborhoods: A study of the relationship between gentrification and self-rated health in the state of California. *Health & Place, 52*, 188–195.

Johnson, K. M., & Beale, C. L. (1995). The rural rebound revisited. Small towns and country homes are growing rapidly in the 1990s. This surprising reversal reveals several basic shifts in US demographic trends. *American Demographics, 17*, 46–46.

Johnson, K. M., & Cromartie, J. B. (2006). The rural rebound and its aftermath: Changing demographic dynamics and regional contrasts. *Population change and rural society*, 25–49.

Kocabıyık, C., & Loopmans, M. (2021). Seasonal gentrification and its (dis) contents: Exploring the temporalities of rural change in a Turkish small town. *Journal of Rural Studies, 87*, 482–493.

Lee, H., & Perkins, K. L. (2023). The geography of gentrification and residential mobility. *Social Forces, 101*(4), 1856–1887.

Leebrick, R. A. (2015). Environmental gentrification and development in a rural Appalachian community: Blending critical theory and ethnography. Dissertation. University of Tennessee.

Ley, D. (1996). *The new middle class and the remaking of the central city*. Oxford University Press.

Link, B. G., & Phelan, J. (1995). Social conditions as fundamental causes of disease. *Journal of Health and Social Behavior, 35*, 80–94.

Lynch, J. W., Smith, G. D., Kaplan, G. A., & House, J. S. (2000). Income inequality and mortality: Importance to health of individual income, psychosocial environment, or material conditions. *BMJ, 320*(7243), 1200–1204.

Marcuse, P. (1986). Abandonment, gentrification, and displacement: The linkages in New York City. In: Smith, N. & Williams, P. (eds) Gentrification of the City. : Allen and Unwin,121–152.

Markley, S. N. (2018). New urbanism and race: An analysis of neighborhood racial change in suburban Atlanta. *Journal of Urban Affairs, 40*(8), 1115–1131.

Markley, S., & Sharma, M. (2016). Gentrification in the revanchist suburb: The politics of removal in Roswell, Georgia. *Southeastern Geographer, 56*(1), 57–80.

McDowell, I. (2023). *Understanding health determinants: Explanatory theories for social epidemiology*. Springer.

McKinnish, T., Walsh, R., & White, T. K. (2010). Who gentrifies low-income neighborhoods? *Journal of Urban Economics, 67*(2), 180–193.

Mermet, A. C. (2017). Airbnb and tourism gentrification: Critical insights from the exploratory analysis of the 'Airbnb syndrome' in Reykjavik. In *Tourism and gentrification in contemporary metropolises* (pp. 52–74). Routledge.

Morris, A. (2019). 'Super-gentrification' triumphs: Gentrification and the displacement of public housing tenants in Sydney's inner-city. *Housing Studies, 34*(7), 1071–1088.

Nelson, P. B., Oberg, A., & Nelson, L. (2010). Rural gentrification and linked migration in the United States. *Journal of Rural Studies, 26*(4), 343–352.

Nelson, L., Trautman, L., & Nelson, P. B. (2015). Latino immigrants and rural gentrification: Race,"illegality," and precarious labor regimes in the United States. *Annals of the Association of American Geographers, 105*(4), 841–858.

Niedt, C. (2006). Gentrification and the grassroots: Popular support in the revanchist suburb. *Journal of Urban Affairs, 28*(2), 99–120.

Phillips, M. (1993). Rural gentrification and the processes of class colonization. *Journal of Rural Studies, 9*(2), 123–140.

Sánchez-Ledesma, E., Vásquez-Vera, H., Sagarra, N., Peralta, A., Porthé, V., & Díez, È. (2020). Perceived pathways between tourism gentrification and health: A participatory Photovoice study in the Gòtic neighborhood in Barcelona. *Social Science & Medicine, 258*, 113095.

Smith, N. (1996). *The new urban frontier: Gentrification and the revanchist city*. Psychology Press.

Smith, D. P. (2002). Extending the temporal and spatial limits of gentrification: A research agenda for population geographers. *International Journal of Population Geography, 8*(6), 385–394.

Smith, G. S., Breakstone, H., Dean, L. T., & Thorpe, R. J. (2020). Impacts of gentrification on health in the US: A systematic review of the literature. *Journal of Urban Health, 97*, 845–856.

Thurber, A., & Krings, A. (2021). Gentrification. In *Oxford Research Encyclopedia of Social Work*. Oxford University Press.

Thurber, A., Krings, A., Martinez, L. S., & Ohmer, M. (2021). Resisting gentrification: The theoretical and practice contributions of social work. *Journal of Social Work, 21*(1), 26–45.

U.S. Census Bureau. (2022). *2006-2010 and 2015-2019 American Community Survey 5-year Public Use Microdata Samples*. Retrieved May, 2023, from https://data.census.gov/.

Versey, H. S. (2023). Gentrification, health, and intermediate pathways: How distinct inequality mechanisms impact health disparities. *Housing Policy Debate, 33*(1), 6–29.

Weller, S., & Van Hulten, A. (2012). Gentrification and displacement: The effects of a housing crisis on Melbourne's low-income residents. *Urban Policy and Research, 30*(1), 25–42.

Wiles, J. L., Allen, R. E., Palmer, A. J., Hayman, K. J., Keeling, S., & Kerse, N. (2009). Older people and their social spaces: A study of well-being and attachment to place in Aotearoa New Zealand. *Social science & medicine, 68*(4), 664–671.

Zukin, S., Trujillo, V., Frase, P., Jackson, D., Recuber, T., & Walker, A. (2009). New retail capital and neighborhood change: Boutiques and gentrification in New York City. *City & Community, 8*(1), 47–64.

Chapter 14
Relative Time and Social-Spatial Determinants of Health

Susan Cassels and Sean C. Reid

Introduction

The places in which we live, grow, play, and work affect our health. The social determinants of health are the non-medical factors that influence health outcomes (World Health Organization, 2023), and these include the places in which people live, grow, play, and work, as well as many other economic, cultural, environmental, and political forces. In fact, the World Health Organization concludes that social determinants influence health more than health care or lifestyle choices (World Health Organization, 2023). The term social-spatial determinants of health acknowledges that place and spatial attributes of social determinants are central to understanding how exposures can lead to inequities in health. This concept is grounded in the social-ecological model, which considers that individuals are nested in relationships, communities, and societies. Thus, in order to improve the health of individuals, we must address factors at multiple levels at the same time (Bronfenbrenner, 1996). The evidence supporting the importance of social-spatial determinants of health is indisputable.

The chapters in this book support the claims that space and place matter as well. More recently, the question has been asked: "Where, when, why or for whom" do these social-spatial exposures matter for health (Sharkey & Faber, 2014)? In this chapter, we consider existing evidence to answer the question of "when do social determinants of health exposures matter," and we propose the idea of relatively vulnerable periods of exposure. Furthermore, we outline the mechanisms in which relative time, in addition to absolute time in one's life course, matters when considering social-spatial exposures. Finally, we offer evidence from our qualitative

S. Cassels (✉) · S. C. Reid
Department of Geography, University of California at Santa Barbara,
Santa Barbara, CA, USA
e-mail: scassels@geog.ucsb.edu

© The Author(s) 2026
M. A. Kolak, I. K. Moise (eds.), *Place and the Social-Spatial Determinants of Health*, Global Perspectives on Health Geography,
https://doi.org/10.1007/978-3-031-88463-4_14

studies of geographic mobility and sexual health among Black and Hispanic sexual minority men in Southern California to support these claims.

Background and Motivation

Many others have considered the importance of timing and social-spatial determinants of health (Berkman et al., 2014; Harris & Schorpp, 2018; Kawachi, 2002). The importance and long-lasting consequences of adverse childhood experiences (ACEs) are grounding examples. Exposure to traumatic events during childhood significantly affects health later in life, ranging from an increased risk of obesity to a higher propensity to suffer from substance use disorder (Burke et al., 2011; Kalmakis & Chandler, 2015).

Social-spatial exposures can have multigenerational impacts as well. The effect of cumulative exposure to racism over time on the incidence of low-birth-weight babies is a key example. The disparity between low-birth-weight babies by the age of mothers demonstrates this association; older Black mothers are more likely to give birth to low-birth-weight babies, whereas the opposite is true for older White mothers (Geronimus, 1996). Low birth weight, in turn, has long-lasting impacts on health into adulthood as well (Forde et al., 2019), leading to the intergenerational health consequences of racism. This is known as the weathering hypothesis (Geronimus, 1996).

The life course perspective is useful to consider the role of time in social-spatial determinants of health (Harris & Schorpp, 2018; Kawachi, 2002) and how social-spatial determinants of health can "get under the skin" and lead to health inequities. Health inequities across the dimension of time can emerge via three distinct pathways (Kawachi, 2002). The latent effects pathways consider how early life social-spatial exposures affect health later in life, independent of intervening experiences. Second, the pathway effects theory posits that social-spatial exposures early in life can set an individual on an alternate life trajectory that may influence health later in life. Third, the cumulative effects pathway suggests that the intensity and duration of social-spatial exposures can adversely affect health according to the dose-response relationship (Harris & Schorpp, 2018; Kawachi, 2002). Indeed, these pathways are not mutually exclusive, and we suggest that more work is needed to disentangle different timings of exposures, as well as the mechanisms by which they lead to adverse health outcomes.

"Sensitive period exposures" is another intuitive way to conceptualize the importance of time in social-spatial determinants of health (Cohen et al., 2010; Gluckman et al., 2008; Harris & Schorpp, 2018; Hayward & Gorman, 2004; Knudsen, 2004; Kuh, 2003; McFarland, 2017; Sampson, 2013; Sharkey & Faber, 2014). The "sensitive period" model examines timing effects in which exposures during sensitive periods of development have stronger effects on health outcomes than they would at other life stages (Cohen et al., 2010; Gluckman et al., 2008; Hayward & Gorman, 2004). Sensitive period effects operate through a "biological embedding" mechanism whereby social exposures during sensitive windows of development have the potential to induce structural and functional changes to the developing individual through biological programming, which cannot be reversed irrespective of intervening experience. This life course model posits that the effect of the sensitive period exposure is typically latent, in that its impact on health outcomes may not appear until later life stages. However, this definition is limited in the sense that it only refers to development periods, while exposures during other relative periods of vulnerability may also have similar long-lasting health consequences.

Additionally, stress is critical to consider when examining the timing of social-spatial exposures, as many have posited that the stress response is a key pathway of biological embedding (Sampson, 2013; Sharkey, 2013; Sharkey & Faber, 2014). The stress paradigm describes how negative social exposures influence physiological processes. Essentially, stress inhibits the body from maintaining physiological stability (i.e., allostasis), and this contributes to the development of chronic diseases (Pearlin, 1989). In this chapter, we rely on theories of the life course, sensitive periods, and the stress paradigm to argue and demonstrate that social-spatial determinants of health research must consider how the relative timing of contextual exposures matters.

Relative Time and Exposures to Social-Spatial Determinants of Health

Health geographers should consider relative time in the same way that we consider the importance of context or quality of space and geospatial exposures. We should identify the context and quality of time lived by individuals throughout their life course. Not all time is experienced equally, and exposures at different periods of vulnerability or in different time contexts can influence health differently. Considering relative time along with individual characteristics and life course stage when examining the importance of geo-social determinants of health may explain variability in health outcomes.

Relative time is defined by its relation to other co-occurring events occurring in an individual's life. Therefore, relative time is different for everyone, as opposed to absolute time, such as age or the period of childhood, which is the same for everyone. Similarly, relative times of vulnerability to exposures occur differentially along the life course, and only a subset of individuals will experience these periods of relative exposure. In this paper, we consider the relative timing of exposures to social-spatial determinants of health. Thus, we define this as a period of relative vulnerability due to stressful physical, social, mental, or emotional distress, in which exposures may have a disproportionate effect on later health.

Individuals are exposed to social-spatial determinants of health in different ways, with differing impacts. The space and context of these exposures matter. Additionally, the relative timing of these exposures, as well as their broader temporal context of the exposures, likely matters as well. One can think of an individual experiencing life, through space and time, on a trajectory. Previous work (cite) has identified "sensitive periods," which one can imagine as a qualitatively different type of trajectory (or one can imagine a dotted or different type of line through space), when in these periods, the individual has different social-spatial exposures, or the consequences of the exposures are different. The relative timing of exposures works in the same way. Instead of absolute time, relative times of exposure could be, for example, during stressful moments, demographic transitions, environmental disasters, disease diagnoses, the birth of a child, the time before, during, and after migration, deportation, marriage/divorce, eviction, or other periods of housing insecurity.

We combine the life course approach with an environmental exposure framework (Belsky, 2013) to consider the quality and magnitude of relative exposures in time. The environmental exposures framework first considers the exposure: the source of the hazard (e.g., pollution), the environmental pathway in which the hazard exists (e.g., air), and the way in which an individual encounters the hazard (e.g., breathing pollution in the air). The second key element is the dose-response relationship, which considers the duration, frequency, and magnitude of the exposure, as well as the body size of the individual. Third, the environmental exposure framework considers individual susceptibility. Usually, individual characteristics that moderate the health impact of exposure are considered, such as pregnancy status, age, weakened immune system, or history of previous exposure.

We propose integrating relative time into the environmental exposure framework or, in other words, incorporating individual temporal context into the framework. We demonstrate this in Fig. 14.1. First, a period of relative risk may change the nature of contact with an exposure or change the quality or magnitude of exposure. This is depicted through the "exposure effect" in Fig. 14.1. For example, the period following an HIV diagnosis is an especially vulnerable time, and individuals may be more likely to engage in substance use or higher-risk sex if they feel fatalistic or uninvested in their future (Pampel et al., 2010). In essence, this relative time frame is associated with different exposure doses. Second, the relative time in a person's life can be thought of as an individual's characteristic, which, according to environmental pathways theory, can alter individual susceptibility (Jirtle & Skinner, 2007). We think of this as the "pathway effect": the same quality or magnitude of exposure during a time of relative susceptibility may have a different effect on health than it would have otherwise. For example, exposure to racism at a time of heightened stress after immigration to the United States may cause more harm than exposure to racism at a less stressful time due to the underlying physiological impacts of stress. Ultimately, relative timing can affect the quality or magnitude of the exposure itself, or it can affect the response or consequence of exposure regardless of whether the dose is different.

The relative time of exposure may influence individuals quite differently depending on their personal characteristics and life course trajectory. Additionally, the social-spatial circumstances of an individual may render that individual more vulnerable during certain temporal periods, or in essence create a period of vulnerable time for additional contextual exposures. This is depicted in Fig. 14.1 as the arrows

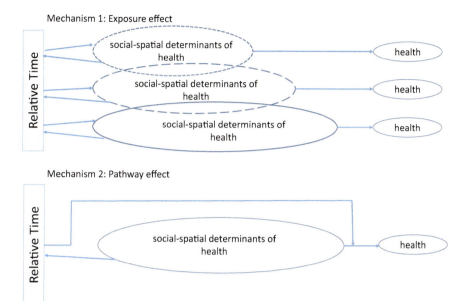

Fig. 14.1 Two pathways in which adverse geo-social exposures during relative times of vulnerability may impact health: the *exposure effect*, in which the quality and/or magnitude of exposures are different during times of relative vulnerability, and the *pathway effect*, in which the effects of social-spatial exposures during a relative time of vulnerability are amplified, regardless of the dose

from "social-spatial determinants of health" back to "relative time." Social and spatial contexts may exacerbate or protect from exposures during relative times, such as social support, social networks, cultural stigma, or isolation. As mentioned before, periods of heightened stress may exacerbate the adverse effects of the exposure. Therefore, accounting for biological mechanisms, such as heightened cortisol levels or allostatic load, is important as well. A better understanding of social-spatial exposures during relative times of vulnerability can provide needed insight into mechanisms driving health disparities.

Examples from Qualitative Research

To demonstrate the importance of the relative timing of social-spatial determinants of health, we present some qualitative findings from our research on geographic mobility and sexual health among Black and Latinx sexual minority men in Southern California. Data for the first study came from 20 in-depth interviews with Black and Latinx SMM living in Los Angeles; data for the second came from 16 semi-structured qualitative interviews with 16 foreign-born Latinx SMM living in San Bernardino, CA (For detailed descriptions of the two studies, the study populations, and recruitment strategies, see Cassels et al., 2023 and Cassels et al., 2020). Racial disparities in HIV, especially among sexual minority men, are large, enduring, and not well explained by individual risk behaviors. Social-spatial and structural determinants of health seem to play an outsized role in perpetuating these disparities. Our work aimed to understand the role of geographic mobility in HIV risk, prevention, engagement, and retention in the HIV care cascade (Cassels et al., 2023, 2020). We highlight three examples of periods of relatively vulnerable time—the time around migration events, housing insecurity, and HIV diagnosis—and suggest how exposures to social-spatial determinants of health during these times may have disproportionate impacts on health.

Migration and the Relative Timing of Social-Spatial Determinants of Health

The time before migration often dictates the reasons for migration. Some individuals in our study samples had agency over the decision to migrate, while others were quite young and moved with their families. Among individuals who had agency over their decision to migrate, their decision was often driven by negative experiences in their home country. The men in our sample reported feeling ostracized by their culture, family, or friends based on their sexual and gender identity. If they acted in ways that were not consistent with stereotypical gender roles in Latin American countries, for example, there were reports of bullying or harassment:

> Ah, I felt that, that I…in my childhood it was a really nice place, but I knew that it was not where I wanted to be my whole life. Because specifically, my persona, with the risk of bullying, a lot of bullying, and my sexual behavior, and with my sexual orientation. A gay person in a town is not very well accepted. There is a lot of bullying in occasions, it was never my case but there were people that received physical abuse even and a lot of emotional and psychological abuse. So, it was not for me, I think that when I had the opportunity at my 17 years of age, when I had the use of reason, I decided to leave my town.

In many cases, this resulted in people being unable to express themselves or communicate aspects of their lives to those around them. The lack of support and openness influenced their mental health and forced them to navigate aspects of sexual health on their own. The lack of acceptance was the primary driver of migration to the United States, where people expected to be more open with their sexual and gender identity. Alternately, there were situations when family and friends were supportive of their sexual and gender identity, and this acted as a protective factor. Having a support system in place allowed for more open expression that improved mental health and opened a channel of communication to discuss questions and concerns related to sexual health and behavior. Either way, the time before immigrating to the United States was a relative time in which individuals in the sample were potentially more vulnerable to contextual health exposures, and family context moderated the effects of exposures.

The time during the act of migration is potentially a vulnerable time in one's life, leading to different exposures or experiencing consequences of exposures differently. Similar to the time before migration, individuals with different social contexts had different experiences with the migration journey. Individuals who had family support often experienced less difficulty in their migration journey, as they could lean on support from family members and leverage networks of people close to them to find stable living situations in the United States. Individuals with less support or a more detrimental home environment experienced greater difficulties in their migration journey, such as beginning their journey at a less ideal time or with minimal resources, which is an example of the pathway effect. A specific example from our study was an individual who went to the US-Mexico border in Juarez and lived in a temporary living shelter before crossing the border. They discussed the traumatic and long-lasting effects of their time in the shelter, where they experienced mental and physical abuse from others based on their sexual and gender identity.

Yes, it was something horrible. I was traumatized from what I saw. I was left with traumas. [From] inside the shelter. I saw lots of bad things. But all thanks to God, I came out physically fine. But there were some really ugly things in there.

Last, the time after migration is also a relatively vulnerable time, when exposures may have disproportionate effects. Individuals may be especially susceptible to negative health outcomes (pathway effect), and the magnitude and/or severity of health insults may be higher than in different time intervals (exposure effect). Integration into the destination location or host community is particularly important at this stage of the migration journey. The ability to integrate is often a function of existing support networks or documentation status. Undocumented individuals face challenges with employment, access to social services, insurance, and healthcare. These experiences during the relative time after migration can compound social-spatial determinants of health.

No, no not in this moment… but there are some barriers on not being a resident or citizen of this country brings barriers to not being able to get the proper medical attention. For example, if you appear from a chronic disease, like HIV, it's more difficult because you have to pay off of your own tab and it is more difficult getting medical insurance for people that are like in my situation.

Family and social networks can act as protective factors during a vulnerable period of relative time after migration. They can share resources, knowledge, and support to integrate into a new community. For example, they can share their healthcare resources or leverage their network to find employment. People who do not have a network in place at the destination are vulnerable to a higher degree.

Right now, I am going through a pretty difficult situation. This is because I am practically living naked where I am currently being lent a room. Right now, I am asking for help from a department to try to obtain my own space. Because right now where I am at, no… But I feel like it is all little by little.

Among individuals in our study, discrimination and stigma often limited their ability to integrate into the destination community after migration. Some individuals in our sample reported turning to drugs and alcohol as a coping mechanism due to the lack of acceptance. Especially when substances were used in combination with sexual activity, health risks were high. This demonstrates the exposure effect mechanism in which vulnerable time periods can interact with social-spatial determinants of health and result in different magnitudes of health exposure as well as different health outcomes.

> Um, well, when I'm behind closed doors I feel like I'm, you know, in the zone, I guess. I mean, um, I've done some really crazy stuff, you know, crazy, crazy stuff with different people. But that's because at one point I felt like, I just didn't care about myself, I just didn't really care, you know?

Housing Insecurity and the Relative Timing of Social-Spatial Determinants of Health

The timing around housing insecurity and homelessness is another example of a potentially vulnerable time when social-spatial exposures may lead to more adverse health outcomes. Housing insecurity is a time of relative stress that can alter the consequences of geo-social exposures. However, the relative timing around housing insecurity also demonstrates how the quality and magnitude of exposures can differ during this vulnerable time, representing the exposure effect. For example, where a person lives and the resources available to them have a direct impact on the extent to which housing insecurity can influence physical and mental health.

The geography of housing insecurity can influence the quality and magnitude of adverse exposures over time. During this vulnerable time, individuals may have differential access to social services, depending on whether they are in urban or rural areas. For people experiencing homelessness, this could determine the amount of time sheltered or unsheltered. For some participants in our sample, relative times of housing insecurity also led to feeling unsafe or being exposed to violence.

> But first, I would not do it with anyone from my neighborhood. I would not do it because to me it is very dangerous. They could say... from there they could get opportunities to hurt someone.

Some participants in our study discussed feeling a sense of detachment from their home locations because of the temporary nature of their living arrangements. Many viewed it just as a place to sleep and would engage in other activities away from their home. This is an example of an exposure effect. Individuals may have qualitatively different geo-social exposures that could lead to adverse health outcomes, such as engaging in substance abuse or sexual acts in unsafe places.

> At this point of my life, it doesn't matter where I live. Yeah, I don't care. Cause right now I am just saving up money cause I want to move to a different neighborhood.
> Well, I do not spend time there. I go from one job to the other, I do not spend time there at home.

Lack of place attachment also resulted in a large distance between daily activities, which could also lead to different adverse exposures during this relative time. Individuals in our study had to travel large distances during the day in a sprawling city such as Los Angeles, where automobile transportation is one of the only reliable options. Therefore, for some, the vulnerable period around housing insecurity was associated with poor health due to less time to engage in medical care or the inability to socialize in accepting LGBTQ environments.

Last, the stress of housing insecurity, and thus the health consequences of adverse exposures during this relatively vulnerable time, is moderated by social networks. If a person has a strong social network through friends or family, they are less burdened to find a safe place to sleep at night as they navigate the difficult situations related to not having a place to live. However, some participants in our study reported living in ethnic enclaves or with families that were not accepting of their sexual and gender identity. Many reported uncomfortable living situations where they could not express their identity for fear of being judged or kicked out of their living arrangements. This demonstrates that the social and cultural context matters for the relative timing of adverse exposures such as housing insecurity (hence the bidirectional arrows in Fig. 14.1): for some, social networks are protective, but for others, they can introduce additional challenges that influence health outcomes.

> ... a bit difficult because again I'll say... I have to be living as though I was a macho, a man, and I don't want to be like how people want me to be. I want to be who I am, and despite being in a country where people have an 'open mind' it is difficult regardless, because it's a thing for Hispanic people, Latinos, to not accept gay people.

Disease Diagnoses and the Relative Timing of Social-Spatial Determinants of Health

Our study sample was composed of people living with and without HIV. For those living with HIV, the time after HIV diagnosis was a key relative time of heightened vulnerability. Again, similar to migration, the level of family support and stability influenced the impact of exposure during this period of relative time.

Highly effective anti-retroviral drugs allow individuals living with HIV to live long and healthy lives and prevent onward transmission to others (Cohen et al., 2016). Unfortunately, there are racial disparities in the uptake of these medications, and the idea of relative time exposures offers an explanation. Three primary factors in the time after diagnosis that influence health are family/social support, documentation status, and discrimination.

After diagnosis, a serious health event, such as acquiring HIV, is a stressful experience and must result in careful health decisions to ensure positive long-term outcomes. A supportive community allows for discussion of options, sharing of resources, and social support and social cohesion. In our studies, individuals who had a supportive family or social network were better able to adjust to the diagnosis and receive the care they needed. Others who were not open about their sexual and gender identity or were not accepted by their family had to overcome additional barriers to adapt to lifestyle changes and engage with needed care. The absence or presence of social networks influences the severity of the diagnosis on long-term health.

> *I have not talked to my family, nor friends, because... there could be some discrimination because of their lack of knowledge on the topic, and... more than anything because of that, to not worry them about my condition, but I think that it's necessary for people to be conscious on that because if something were to happen, if something were to happen to me about my health, they would not know what would be happening.... and that is important, it's a hard decision and in that moment, I would have to talk to them.*

Documentation status also dictated the impact on health in the relative time after diagnosis. Having documentation facilitated access to employment or social services that linked people with healthcare. Living in the United States without documentation can delay or prevent engagement with healthcare services, which can adversely affect health in the short and long term. The lack of easy access to healthcare services also limited knowledge of the healthcare system. This was particularly important in times of health emergencies, such as a recent diagnosis of HIV. Within our qualitative studies, these factors impacted the use of anti-retroviral medication needed to reduce the viral load in an individual's body to prevent autoimmune complications and transmission of HIV.

Last, discrimination was also a particularly important exposure in the relative time after a new health diagnosis. Individuals in our studies reported a lack of understanding and stigma about HIV among family and friends. A specific example reported was that after HIV diagnosis, family members would not share dishes such as plates and cups out of fear of getting infected. This type of prejudice and lack of understanding can ostracize people and even prevent them from seeking treatment based on fear of rejection from family and friends. This example demonstrates both pathways in which relative times of vulnerability may adversely affect health: during this relative time, individuals may experience higher levels of discrimination (exposure effect) and may delay engagement in care (pathway effect). The differential experience of discrimination during the time after diagnosis influenced the severity of the diagnosis of HIV.

> *... because with them... they weren't well informed about HIV, AIDS, and all that, and I am a carrier of HIV, so they weren't well informed on that and... there were some things that I didn't like because they would separate my plates, my cups, and I told her 'it doesn't transmit like that' I told her, no, no, no you don't have to worry, neither about using the restrooms, nor for using my towels are you going to get HIV, I told her it doesn't transmit like that. And I would explain and explain that it can only transmit only by sexual intercourse and blood transfusions.*

Discussion and Conclusion

In this chapter, we have demonstrated the need to consider relative time periods of exposure within the social-spatial determinants of the health framework. Research on life course effects and sensitive periods of absolute exposure has advanced a great deal in the last few decades, as have metrics to quantify their impact on health, such as ACE scores. Health surveys regularly incorporate questions and metrics related to absolute exposures, but few give explicit attention to the relative time of exposures. These exposures are heterogeneous, and researchers must be careful about where and how to incorporate them into new and existing surveys measuring population health. Nonetheless, our research has demonstrated that relative time periods over the life course can significantly affect the quality or quantity of social-spatial exposures or amplify the effects of adverse exposures.

Our findings also suggest that social environments moderate the effects of social-spatial exposures during vulnerable times, likely due to higher social capital and social cohesion, which reduce stress. Although more work is needed to understand the behavioral, social, and biological mechanisms in which social-spatial exposures can "get under the skin" and cause poor health, policies can be implemented. For instance, community-based organizations that support new migrants could lessen the impact of the stressors of acclimation. It is also important to note that relative exposure can be detrimental or protective based on the context and personal characteristics of the individual experiencing the exposure. Thus, health interventions that optimize the relative timing of exposures could achieve greater efficacy as well.

References

Belsky, J. (2013) Differential Susceptibility to Environmental Influences. *ICEP* 7, 15–31.

Berkman, L. F., Kawachi, I., & Glymour, M. M. (Eds.). (2014). *Social epidemiology* (2nd ed.). Oxford University Press.

Bronfenbrenner, U. (1996). *The ecology of human development: Experiments by nature and design.* Harvard University Press.

Burke, N. J., Hellman, J. L., Scott, B. G., Weems, C. F., & Carrion, V. G. (2011). The impact of adverse childhood experiences on an urban pediatric population. *Child Abuse & Neglect, 35*(6), 408–413. https://doi.org/10.1016/j.chiabu.2011.02.006

Cassels, S., Cerezo, A., Reid, S. C., Rivera, D. B., Loustalot, C., & Meltzer, D. (2023). Geographic mobility and its impact on sexual health and ongoing HIV transmission among migrant latinx men who have sex with men. *Social Science & Medicine, 320*, 115635. https://doi.org/10.1016/j.socscimed.2022.115635

Cassels, S., Meltzer, D., Loustalot, C., Ragsdale, A., Shoptaw, S., & Gorbach, P. M. (2020). Geographic mobility, place attachment, and the changing geography of sex among African American and Latinx MSM who use substances in Los Angeles. *Journal of Urban Health, 97*(5), 609–622. https://doi.org/10.1007/s11524-020-00481-3

Cohen, M. S., Chen, Y. Q., McCauley, M., Gamble, T., Hosseinipour, M. C., Kumarasamy, N., Hakim, J. G., Kumwenda, J., Grinsztejn, B., Pilotto, J. H. S., Godbole, S. V., Chariyalertsak, S., Santos, B. R., Mayer, K. H., Hoffman, I. F., Eshleman, S. H., Piwowar-Manning, E., Cottle, L., Zhang, X. C., et al. (2016). Antiretroviral therapy for the prevention of HIV-1 transmission. *New England Journal of Medicine, 375*(9), 830–839. https://doi.org/10.1056/NEJMoa1600693

Cohen, S., Janicki-Deverts, D., Chen, E., & Matthews, K. A. (2010). Childhood socioeconomic status and adult health: Childhood socioeconomic status and adult health. *Annals of the New York Academy of Sciences, 1186*(1), 37–55. https://doi.org/10.1111/j.1749-6632.2009.05334.x

Forde, A. T., Crookes, D. M., Suglia, S. F., & Demmer, R. T. (2019). The weathering hypothesis as an explanation for racial disparities in health: A systematic review. *Annals of Epidemiology, 33*, 1–18.e3. https://doi.org/10.1016/j.annepidem.2019.02.011

Geronimus, A. T. (1996). Black/white differences in the relationship of maternal age to birth-weight: A population-based test of the weathering hypothesis. *Social Science & Medicine, 42*(4), 589–597. https://doi.org/10.1016/0277-9536(95)00159-X

Gluckman, P. D., Hanson, M. A., Cooper, C., & Thornburg, K. L. (2008). Effect of in utero and early-life conditions on adult health and disease. *New England Journal of Medicine, 359*(1), 61–73. https://doi.org/10.1056/NEJMra0708473

Harris, K. M., & Schorpp, K. M. (2018). Integrating biomarkers in social stratification and health research. *Annual Review of Sociology, 44*(1), 361–386. https://doi.org/10.1146/annurev-soc-060116-053339

Hayward, M. D., & Gorman, B. K. (2004). The long arm of childhood: The influence of early-life social conditions on men's mortality. *Demography, 41*(1), 87–107. https://doi.org/10.1353/dem.2004.0005

Jirtle, R., Skinner, M. (2007). Environmental epigenomics and disease susceptibility. *Nat Rev Genet 8*, 253–262

Kalmakis, K. A., & Chandler, G. E. (2015). Health consequences of adverse childhood experiences: A systematic review. *Journal of the American Association of Nurse Practitioners, 27*(8), 457–465. https://doi.org/10.1002/2327-6924.12215

Kawachi, I. (2002). A glossary for health inequalities. *Journal of Epidemiology & Community Health, 56*(9), 647–652. https://doi.org/10.1136/jech.56.9.647

Knudsen, E. I. (2004). Sensitive periods in the development of the brain and behavior. *Journal of Cognitive Neuroscience, 16*(8), 1412–1425. https://doi.org/10.1162/0898929042304796

Kuh, D. (2003). Life course epidemiology. *Journal of Epidemiology & Community Health, 57*(10), 778–783. https://doi.org/10.1136/jech.57.10.778

McFarland, M. J. (2017). Poverty and problem behaviors across the early life course: The role of sensitive period exposure. *Population Research and Policy Review, 36*(5), 739–760. https://doi.org/10.1007/s11113-017-9442-4

Pampel, F. C., Krueger, P. M., & Denney, J. T. (2010). Socioeconomic disparities in health behaviors. *Annual Review of Sociology, 36*(1), 349–370. https://doi.org/10.1146/annurev.soc.012809.102529

Pearlin, L. I. (1989). The sociological study of stress. *Journal of Health and Social Behavior, 30*(3), 241. https://doi.org/10.2307/2136956

Sampson, R. J. (2013). The place of context: A theory and strategy for criminology's hard problems: The place of context. *Criminology, 51*(1), 1–31. https://doi.org/10.1111/1745-9125.12002

Sharkey, P. (2013). *Stuck in place: Urban neighborhoods and the end of progress toward racial equality*. The University of Chicago Press.

Sharkey, P., & Faber, J. W. (2014). Where when, why, and for whom do residential contexts matter? Moving away from the dichotomous understanding of neighborhood effects. *Annual Review of Sociology, 40*(1), 559–579. https://doi.org/10.1146/annurev-soc-071913-043350

World Health Organization. (2023). *Social determinants of health*. Retrieved from https://www.who.int/health-topics/social-determinants-of-health

Chapter 15
A Social-Spatial Network Approach to Characterize Social Determinants of Health in Syndemics Research within an Intersectionality Framework

Ran Xu and Qinyun Lin

Introduction

Syndemics is a widely known concept in social epidemiology to describe clustered, co-occurring epidemics of diseases in concentrated populations and the underlying large-scale social crises (Singer, 2000, 2011). Although the notion of syndemics has been proposed for more than two decades, evidence is still limited on whether and how different epidemics interact with each other, mutually generating more deleterious health consequences (Tsai, 2018; Tsai et al., 2017; Tsai & Venkataramani, 2016). A handful of studies have demonstrated that multiple diseases and social issues tend to occur simultaneously within a particular population or location (e.g., Illangasekare et al., 2013; Monnat et al., 2019; Stall et al., 2003). However, these findings do not necessarily establish evidence of a synergistic or mutually causal relationship between different problems that could exacerbate the health burden for vulnerable populations. Such interplay or interaction is of great significance, as it is closely related to the design of public health interventions. For instance, if two epidemics occur simultaneously but are independent of each other, interventions do not need to consider how the impact of one epidemic depends on the state of the other.

In this chapter, we draw on the intersectionality framework to examine synergy in syndemics or explore such interactions and interdependencies among co-occurring diseases and social problems. By emphasizing the multidimensionality of Black women's experience, especially the unique discrimination they faced, Kimberlé Crenshaw (1989) first proposed intersectionality to develop a Black

R. Xu
Department of Allied Health Sciences, University of Connecticut, Storrs, CT, USA

Q. Lin (✉)
School of Public Health and Community Medicine, University of Gothenburg, Gothenburg, Sweden
e-mail: qinyun.lin@gu.se

© The Author(s) 2026
M. A. Kolak, I. K. Moise (eds.), *Place and the Social-Spatial Determinants of Health*, Global Perspectives on Health Geography,
https://doi.org/10.1007/978-3-031-88463-4_15

feminist criticism in legal studies. The theoretical lens of intersectionality intro-duces a way of examining the interlocking matrices of individual factors across multiple levels of society (Kapilashrami & Hankivsky, 2018), emphasizing social structural forces and patterns (Bowleg, 2012; Phelan & Link, 2015). More impor-tantly, centering on understanding and changing ideologies in relation to power, the intersectionality framework allows us to gain insights into the social dynamic pro-cess within which multifaceted power structures interact to produce and reinforce inequity in health outcomes (Agénor, 2020; Bauer, 2014; Hankivsky et al., 2017; Larson et al., 2016). In other words, intersectionality offers a framework to under-stand the concentration of different diseases within specific populations or regions by illuminating how various factors across multiple levels interact with each other, contributing to the co-occurrence of these diseases and social problems.

An intersectional lens can also guide the development of a social-spatial network approach to examine health inequity in syndemics. In fact, social networks and spatial contexts play critical roles in intersectionality, as they are essential to the construction and formation of ideology, social positions, and other aspects of indi-vidual experiences closely related to discrimination, vulnerability, and inequality in health. At the same time, we underscore the interplay between spatial contexts and social networks, considering the dynamic process in which they interact and con-tribute to health inequity together.

Throughout this chapter, we will focus on the example of opioid-related syndem-ics, which involve opioid misuse, overdose, and other downstreaming health out-comes. For example, people who inject drugs are at a high risk of blood-borne infections, such as human immunodeficiency virus (HIV) and the hepatitis virus (HCV). Using a syndemics framework, we emphasize that each health condition (e.g., overdose, HIV, HCV) involved in the syndemics interacts with one another not only at causes, consequences but also needed responses (Perlman & Jordan, 2018). More importantly, we highlight the importance of examining structural factors in communities and neighborhoods, especially social interactions at the interpersonal level (e.g., injection risk networks) and equitable access to social, economic, health care, and physical or built environmental conditions at the neighborhood level (e.g., access to harm reduction and unstable housing). In what follows, we first review the effects of spatial contexts and social networks in existing research. Then, using the intersectionality framework, we develop a social-spatial network approach to exam-ine health inequity in syndemics research, followed by discussions about method-ological considerations, mainly in terms of quantitative analyses. We conclude with challenges in existing research and avenues for future work.

Spatial Contexts and Social Networks in Existing Syndemics Research

Many studies have been presented to build connections between spatial contexts and health behaviors and outcomes in syndemics research. In particular, the "risk envi-ronment" framework precisely depicts how the syndemics of opioid misuse, overdose, and related health consequences are embedded in the social, physical, economic, and political environment (Ciccarone, 2017; Heimer et al., 2014; Rhodes, 2009). The risk environment framework has been applied in a series of studies to characterize neighborhoods and communities that have been heavily impacted by the syndemic. For example, socially and economically disadvantaged neighborhoods, such as those with lower income, more manual labor industries, and more immigrants, were found to be more seriously impacted in the opioid-related syndemics (Anderson et al., 2019; Cerdá et al., 2017; Cho et al., 2013; Molina et al., 2012). Areas characterized by disadvantageous built environments, such as dirty streets, crumbling buildings, and fragmented communities, are also shown to have a higher risk of harm (Cerdá et al., 2013; Hembree et al., 2005). A recent study developed a socio-built environment framework to identify both urban and nonurban areas with the highest risk for opioid overdose (Tempalski et al., 2022). Residents living in such high-risk communities are more likely to experience exposure to psychological stress, alcoholism, and drug addiction. There is also evidence indicating that communities predominantly affected by opioid-related syndemics have experienced escalated rates of opioid prescriptions and encountered greater challenges in accessing treatment resources, such as naloxone distribution programs, providers for medication for opioid use disorder, and other harm reduction services (Cerdá et al., 2017; García, 2019; Kolak et al., 2020). As Keyes et al. (2014) summarized, a greater number of opioid prescriptions may be associated with increasing availability in illegal markets, leading to nonmedical prescription opioid misuse. In essence, the increased accessibility of opioids coupled with a severe shortage of treatment options exacerbates the exposure and vulnerability of these communities, compounding their preexisting social, economic, and built-environment disadvantages.

Similarly, a few studies have demonstrated relationships between social network characteristics and opioid misuse-related risk behaviors and downstream health outcomes. First, social networks play an important role in offering support and encouraging behaviors that mitigate risks (Bouris et al., 2017; Duncan et al., 2019). For example, discussions with confidants, or someone with a close and trusting relationship, are linked to stronger awareness of prevention intervention (Chen et al., 2019), while poor social support, such as smaller network size and lower satisfaction with social support, is related to a higher possibility of engaging in unprotected sex (Deuba et al., 2016). Social norms are also linked to risk behaviors such as the sharing of syringes and drug preparation equipment (De et al., 2007; Latkin et al., 2010). Second, certain network structures, such as highly cohesive and centralized networks, are related to faster transmission of HIV (Young et al., 2013). Network-based interventions have also been implemented to identify individuals who have recently been infected and allocate resources accordingly (Hunter et al., 2019; Nikolopoulos et al., 2016).

However, there are only a limited number of studies that combine both spatial contexts and social network characteristics to examine syndemics. Chen et al. (2019) and Duncan et al. (2019) examined the effects of both neighborhood and social network characteristics on prevention practices. Some recent studies have explicitly examined the spatial structure of social networks to deepen the understanding of the interconnection between bridging populations and different communities and places they connect (Gesink et al., 2018; Kolak et al., 2021; Youm et al., 2009).

Notwithstanding these endeavors, a significant proportion of current research tends to concentrate either on the neighborhood context in isolation or solely on social network factors. We contend that, apart from limited data availability, another crucial reason for this is the tendency to treat social networks and neighborhood context as independent factors operating at distinct levels within the socioecological model. While the socioecological model provides a powerful framework to consider social determinants of health (SDoH) across multiple levels (Bronfenbrenner, 1994; Roxberg et al., 2020), the intersectionality framework also underscores the fact that each level itself is socially constructed, and other levels could play a role in such a construction process. In our context, this means that social networks and spatial contexts interact with each other to impact health inequity together. Such interaction could even construct the scope or level where they play a role. Incorporating such interactions is especially crucial for syndemics research since they allow us to gain insight into why different diseases and social problems occur simultaneously within concentrated populations.

A Social-Spatial Network Approach Under the Intersectionality Framework

The starting point of the intersectionality framework is how different social categories, such as gender and race/ethnicity, interact and mutually reinforce social inequities (Bowleg, 2013; Collins, 2015; Hill Collins & Bilge, 2020; Sangaramoorthy & Benton, 2022). By focusing on how different social categories converge on the same individual, the intersectionality framework seeks to understand, and more importantly, to dismantle, various systems of oppression, power, and privilege (Chun et al., 2013; Lewis, 2013; MacKinnon, 2013; Mohanty, 2013; Sangaramoorthy & Benton, 2022; Verloo, 2013). To achieve this goal, the key question that researchers must consider in analysis is "how social positions and processes impact health and well-being" (Sangaramoorthy & Benton, 2022).

Spatial contexts and social networks are two crucial components in the formation process of social positions and their influence on health and well-being. Within the framework of intersectionality, we see that all social determinants of health (SDoH), including those that act through spatial contexts and social networks, operate at varying scales. More importantly, by noting that all such scales, for both spatial contexts and social networks, are socially and politically constructed rather than fixed and predetermined (Howitt, 2003; Marston, 2000; Tsing, 2012), intersectionality allows for simultaneous examination of the social reproduction process across multiple scales. With such guidelines, we propose a theoretical framework to examine spatial context and social network effects in the context of syndemics (see Fig. 15.1).

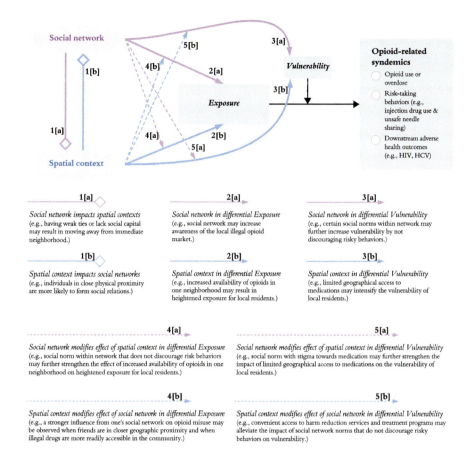

Fig. 15.1 A social-spatial network approach under the intersectionality framework

Interaction Between Spatial Context and Social Networks

First and foremost, spatial contexts and social networks interact with each other, as indicated by arrows 1[a] and 1[b] in Fig. 15.1. On the one hand, social interactions are embedded in specific geographic locations, meaning that geographic locations may also influence social network characteristics and structures, as represented by arrow 1[b] in Fig. 15.1. Most directly, individuals who are in close physical proximity to one another have a greater likelihood of forming social relationships. This is referred to as the "propinquity" effect (Festinger et al., 1950). For instance, public injection spaces may promote the formation of high-risk injection drug use networks (Tempalski & McQuie, 2009). On the other hand, social interactions may also affect individuals' activity spaces and, consequently, their daily spatial contexts (illustrated as arrow 1[a] in Fig. 15.1). Social network members may spatially bridge high- and low-risk areas, such as people traveling through urban and suburban public spaces to inject drugs (Boodram et al., 2018; Wylie et al., 2007). It is also possible that having weak social ties within the community or lacking social capital can result in individuals moving away from their immediate neighborhood or being less likely to be affected by it.

Recognizing the significance of such interactions, a recent study conducted by Kolak et al. (2021) employed spatial analysis to examine different types of social networks and the geographic communities in which they are located. They utilized point pattern analysis to determine the spatial distribution of participants and their network members' community residences. As such, they identified various social-spatial patterns. For example, the authors demonstrated that individuals with partners living in communities that were further away and had more resources experienced larger ego-alter power disparities and faced a higher risk of adverse outcomes. Without a predetermined spatial scale, such a method reflects how social network ties may engage in the construction of the scale at which spatial context exerts its influence. It recognizes that social networks play a role in shaping the spatial boundaries and contexts within which individuals operate, emphasizing the dynamic interplay between social networks and spatial scales.

Considering the mutual reinforcement between spatial contexts and social network effects as a dynamic process, the structural influences stemming from such interaction extend beyond individual behaviors. They can not only concentrate risk behaviors and adverse health outcomes within individuals' networks but also situate these social networks within areas that have limited resources (Brawner et al., 2022). As an example, research has identified spatial clustering of risk behaviors and norms, such as drug use and sexual partnerships, within specific geographic areas (Latkin et al., 2007; Lin et al., 2023). Such clustering may also be associated with perceived "neighborhood disorder," which refers to the concentration of violence, housing problems, economic stress, and drug market activities (Latkin et al., 2013). As such, the interplay between social networks and spatial contexts illustrates the synergistic relationship between different problems that are essential in syndemics research.

Spatial Context and Social Networks in Differential Exposure and Vulnerability

In earlier sections, we briefly summarize how neighborhood/spatial and social network characteristics may impact risk behaviors and health outcomes based on extant literature. Here, we further theorize such an impact by considering their roles in differentiating exposure and differential vulnerability (Diderichsen et al., 2001, 2012; World Health Organization, 2010).

Differential exposure and differential vulnerability are two causal mechanisms through which socioeconomic position influences individuals' health outcomes (Diderichsen et al., 2001, 2012). Differential exposure refers to the fact that the distribution of causes of diseases is uneven; thus, individuals have different likelihoods of being exposed to the cause of disease (e.g., exposed to opioids in our case). Such a "differential exposure" concept could also be hypothesized as a mediation mechanism, in which exposure is regarded as a mediator caused by social positions (Hussein et al., 2018). Differential vulnerability, on the other hand, pertains to varying effects experienced by groups with different social positions, even when exposed to the same risk factor. For example, a more vulnerable population may be more likely to engage in risky behaviors such as syringe sharing and suffer from downstreaming health outcomes like HCV. This differential vulnerability can be modeled using interaction or moderation analysis, as it involves modifying the impact of a specific risk factor. Figure 15.1 illustrates such mechanisms (as black arrows and gray boxes) that consider the causal chain from exposure to health outcomes as moderated by vulnerability. To clarify, we define vulnerability here with a relatively narrow definition that describes the effect of exposure or susceptibility, as recent literature has given vulnerability a broader definition that covers both exposure and susceptibility (Diderichsen et al., 2019; USGCRP, 2016). Another distinct dimension of the broader definition of vulnerability is the capacity of response (Adger, 2006; Diderichsen et al., 2019), which describes the power and resources to modify exposures and effectively adjust to and recover from the consequences. If one considers this capacity of response to modify the effect of exposure, then one could also conceptualize the capacity of response as part of vulnerability in Fig. 15.1.

Both spatial contexts and social networks can change one's exposure to different risk factors associated with opioid-related syndemics. As an example, increased availability of opioids in one neighborhood may result in heightened exposure for residents, as shown by arrow 2[b] in Fig. 15.1. It is also possible that through connections within social networks, individuals can become more aware of the local illegal opioid market, as shown by arrow 2[a].

Moreover, both spatial contexts and social networks can worsen vulnerability and amplify the impacts of these exposures on individuals. For instance, certain social norms within one individual's networks may further increase their vulnerability by not discouraging risky behaviors such as syringe sharing, as represented by arrow 3[a] in Fig. 15.1. Likewise, as mentioned earlier, communities that are exposed to a higher prevalence and accessibility of opioids often face a shortage of available treatment options at the same time. This scarcity further intensifies the vulnerability of residents within these communities, as indicated by arrow 3[b].

Interdependence Between Spatial Context and Social Networks in Differential Exposure and Differential Vulnerability

Importantly, the interplay between spatial contexts and social networks can also contribute to their respective roles in both differential exposure and differential vulnerability. In other words, the effect of social networks on exposure and vulnerability could be further modified or moderated by spatial contexts, and vice versa. For instance, a stronger influence from one's social network on opioid misuse may be observed when friends are in closer geographic proximity and when illegal drugs are more readily accessible in the community. Arrow 4[b] in Fig. 15.1represents such a case that spatial context modifies the effect of social networks in differential exposure. As another example, strong stigma related to medication within social networks may exacerbate the effects of limited geographic access to medications, further increasing the vulnerability, as indicated by arrow 5[a] in Fig. 15.1 to describe social network modifying the effect of spatial context in differential vulnerability. Such interplay between spatial contexts and social interactions further complicates the mechanisms through which spatial contexts and social networks impact health outcomes and reinforce disparities over time in syndemics.

As we noted earlier, being in a certain spatial environment and engaging in social interactions can expose individuals to potential risks. However, when these exposures occur simultaneously, their combined effect may go beyond a simple summation. As a key aspect of intersectionality, various social categories intersect, leading to an integration that surpasses a mere additive outcome but rather generates a distinctive, unique experience (Hancock, 2007; Harari & Lee, 2021). Likewise, when exposures from spatial environments and social networks intersect, such as arrows 4[b] and 4[a] in Fig. 15.1, it may lead to an exceptionally intense exposure scenario. Such an intersection also applies to simultaneous modification effects from spatial contexts and social networks in terms of differential vulnerability, such as arrows 5[b] and 5[a] in Fig. 15.1. By using the term "differential vulnerability," we refer to the possibility that certain spatial contexts and social networks could modify the effects of risk factors. When these modifications occur simultaneously, they can intersect and produce new dynamics.

Although we did not find any existing studies that examine the combined effects of social networks and spatial contexts explicitly with concepts of exposure and vulnerability, there are a few studies that considered and presented evidence for such combined effects from spatial contexts and social networks. For example, in the studies by Mason et al. (2010) and Mennis and Mason (2011), the researchers developed place-based social network measures and examined the relationship between these measures and substance use behaviors among urban adolescents. Specifically, they first identified activity spaces, which are defined by the locations that individuals come into direct contact with because of their daily routines (Ren, 2016). Following this, the researchers established connections between each identified location and the social network members with whom the participants interacted at those specific locations. As such, they integrated both the spatial context and social networks when constructing the place-based social network measure. These measures were then used to investigate the associations with risk behaviors and outcomes.

Methodological Considerations for a Social-Spatial Network Approach

Considering the complexities outlined in the proposed social-spatial network approach, investigating the concurrent and synergistic effects of social networks and spatial contexts can be a challenging task. We assert that it is crucial to leverage methodological advancements from various disciplines.

Exploratory spatial data analysis (ESDA) could play a crucial role in studying the interaction between social networks and spatial contexts. ESDA encompasses techniques for describing and visualizing spatial distributions, discovering patterns of spatial association such as spatial clusters, and suggesting possible spatial heterogeneity. One valuable tool in ESDA is the kernel density estimate (KDE), which enables researchers to estimate the distribution of point data across a continuous surface. Briefly speaking, KDE is a smoothing technique that approximates the distribution of point data on a continuous surface, helping researchers to gain insights into the likelihood of events or the intensity of activities occurring in different areas. Consequently, KDE has been widely utilized to identify "hot regions" with high activity density (Bornmann & Waltman, 2011; Carlos et al., 2010). It has also been employed to determine the spatial distributions of individuals' activity spaces, allowing researchers to examine the spatial context associated with specific social interaction activities (Kolak et al., 2021; Lin et al., 2023). In other words, social networks are then integrated to define the scale at which spatial context impacts individuals. Rather than examining the spatial context solely through predetermined administrative boundaries such as census tracts or zip codes, researchers consider the spatial context in relation to specific social interactions. Such integration then allows for a more nuanced understanding of how social networks and spatial contexts intersect and influence individuals altogether.

Recent advances in social network analysis offer useful tools to effectively explore how spatial contexts influence the formation, structure, and dynamics of social networks. By integrating spatial information into network analysis, researchers can investigate the impact of geographical proximity, spatial constraints, and spatial environment on social connections. For example, latent space models (LSMs) are statistical tools in social network analysis that consider relations between actors in an unobserved social space (Hoff et al., 2002). Such models are useful tools to model tie- or edge-level outcomes in social network analysis, allowing researchers to include both edge- and node-level traits as explanatory variables (Sewell & Chen, 2015). This means that researchers could include the spatial environment at the node/individual level as well as spatial environment proximity at the edge level to examine how spatial contexts may impact social network relations.

Furthermore, tools such as LSM in social network analysis also enable the exploration of potential intersection effects between spatial context and social networks, which together influence individuals' behaviors. LSMs have been employed to better identify social influence or contagion effects in dynamic social networks (Xu, 2018, 2020), based on which researchers could further examine how spatial contexts interact with social networks to shape various aspects related to individuals' health outcomes, such as the spread of health-related knowledge and adoption of risky behaviors.

Agent-based modeling (ABM) is another powerful tool that could be utilized to consider the synergic effects of social networks and spatial contexts on health behaviors and outcomes. As a simulation approach built upon complex systems theory, ABM shows advantages over traditional approaches by allowing feedback loops and multiple layers of complex, interacting components (Silverman et al., 2021). Specifically, individuals are represented by agents associated with places where they interact with other agents. By explicitly modeling individual-level decision-making (e.g., how agents move, act, and interact), ABM enables researchers to gain insights into emergent patterns at the population level that might be unexpected due to the complicated system process (e.g., the social-spatial approach we illustrated in Fig. 15.1). In fact, GIS agent-based models have already been utilized to study contagious disease spread and opioid and other drug use epidemics (Bobashev et al., 2020; Tatara et al., 2019).

Moreover, recent developments in both epidemiologic and econometric decompositions into mediation and interaction (Blinder, 1973; Fortin et al., 2011; Oaxaca, 1973; VanderWeele, 2013, 2014) could also provide versatile tools to help us quantify and decompose the spatial and social network effects into differential exposure and differential vulnerability. As mentioned earlier, differential exposure could be modeled as a mediation analysis (i.e., exposure considered as the mediator of interest), while differential vulnerability is linked to a moderation or interaction effect (i.e., the effect of a given risk factor is modified). This decomposition approach has already gained attention recently in public health (Basu et al., 2015; Hussein et al., 2018; Powell et al., 2012; Sen, 2014). However, more work is needed, both methodologically and empirically, to consider spatial context and social networks in these decomposition analyses. This is essential due to the potential complexities such as (a) interaction between spatial and social network data (e.g., the inclusion of both social network and spatial context may involve three-way interactions), and (b) additional considerations for nonindependence and autocorrelation inherent in social network and spatial data.

Finally, we contend that traditional statistical methods such as multilevel models are still of great importance when examining the combined effects of spatial contexts and social networks. Under the intersectionality framework, syndemics are by essence a phenomenon or consequence of oppression and discrimination across multiple levels. As discussed previously, intersectionality emphasizes how interlocking systems of power and structural inequalities contribute to the co-occurrence of diseases and social problems within specific populations. By examining these interactions, we can better understand how marginalized groups experience compounded health disparities as a result of intersecting social determinants and systemic discrimination. As such, analyses beyond the individual level and considering contextual factors such as spatial and social networks are essential for syndemics research. As a statistical tool, multilevel models allow the estimation of associations across different levels, bridging analyses of ecological and individual-level data (Diez-Roux, 1998; Duncan et al., 1998; Tsai, 2018). In particular, the cross-level interaction in multilevel models facilitates the quantification of the intersection between effects at varying levels (also see Blakely & Woodward, 2000, for three main mechanisms through which contextual factors may impact individual outcomes). By modeling individuals nested within intersectional social strata, recent advances in Multilevel Analysis of Individual Heterogeneity and Discriminatory Accuracy (MAIHDA) (see, e.g., Evans et al., 2018; Green et al., 2017; Merlo, 2018 for details) may also serve as a useful approach to model potential intersectional effects generated by spatial contexts and social networks (i.e., the social strata are then defined based on social network interactions and spatial contexts).

Discussion: Avenues for Future Research

In this chapter, we propose a social-spatial network approach for investigating health inequity in syndemics research. We start with extant literature that substantiates the influence of social networks and spatial contexts on individuals' exposure to risk factors as well as the effect of exposure, which we theorize as their respective roles in differential exposure and differential vulnerability (see Fig. 15.1). Importantly, when employing the intersectional perspective, we underscore the dynamic process in which social interactions and spatial contexts intersect and collectively contribute to health inequity by way of differential exposure and differential vulnerability. This dynamic process can be conceptualized in three steps. To begin with, it is crucial to recognize that social networks and spatial contexts are mutually influential, potentially shaping one another: geographic locations may shape the development and configuration of social interactions, while social networks may likewise affect individuals' activity spaces and, consequently, their daily geographic or spatial contexts (illustrated as arrows 1[b] and 1[a] in Fig. 15.1). Secondly, social networks and spatial contexts can mutually influence their respective roles in differential exposure. The influence of social networks on exposure may be contingent upon spatial contexts, and conversely, the impact of spatial contexts on exposure may be shaped by the characteristics of social networks (illustrated as arrows 4[b] and 4[a] in Fig. 15.1, respectively). Lastly, they may also mutually modify each other's roles in differential vulnerability (as indicated by arrows 5[a] and 5[b] in Fig. 15.1).

We contend that the proposed framework may serve as a general approach to conceptualize SDoH factors that operate through social network and/or spatial context. It provides a means to enhance our understanding of the complex structural factors contributing to health inequity within syndemics. By that we mean researchers do not need to address all of its mechanisms in a single study. If a researcher's interest lies in a specific aspect of the framework, such as the influence of spatial context or neighborhood environment, they may consider using social interaction data when determining the appropriate level and scale at which to assess and analyze the neighborhood effect. Furthermore, researchers may also find it worthwhile to explore the role of social interactions in modifying the impact of spatial contexts on both differential exposure and differential vulnerability. In instances where a researcher seeks to investigate all the mechanisms within the framework and aims to perform a comprehensive decomposition analysis, then additional methodological work is essential to better facilitate the modeling and estimation process.

To address the intricacies inherent in the interactions between social networks and spatial contexts, we additionally delve into methodological considerations that could be helpful for quantitative researchers. These considerations encompass ESDA, tools within the domain of social network analysis such as latent space modeling and ABM, recent advancements in mediation and interaction decomposition techniques, as well as the application of multilevel models.

While our methodological discussions in this chapter primarily center on quantitative analyses, we strongly advocate for the significance of qualitative research in incorporating social network and spatial contexts in syndemics research. As noted earlier, the intersectionality framework starts from a node of convergence, or how different social categories intersect and shape one individual's unique experience. Qualitative research could be extremely helpful when considering the inherent heterogeneity embedded in unique individual experiences, providing guidance and hypotheses for quantitative modeling at a later stage. The proposed framework may also be helpful to conceptualize how social interactions and spatial contexts collectively shape the unique experiences of each individual in qualitative research.

Finally, we contend that it is crucial to consider spatial contexts and social networks within more macro-level structural factors. Recent work highlights the legacy role of laws and practices reinforcing racial health inequalities through processes such as residential segregation, housing policies, and other aspects of institutional racism (see review by Brawner et al., 2022). To attain a deeper understanding of such a complex system and ultimately drive social changes, we advocate for the integration and collaboration of multiple disciplines to comprehend syndemics using an intersectionality framework. We also emphasize the need for engagement from multiple sectors to implement effective actions.

References

Adger, W. N. (2006). Vulnerability. *Global Environmental Change, 16*(3), 268–281. https://doi.org/10.1016/j.gloenvcha.2006.02.006

Agénor, M. (2020). Future directions for incorporating intersectionality into quantitative population health research. *American Journal of Public Health, 110*(6), 803–806. https://doi.org/10.2105/AJPH.2020.305610

Anderson, T. L., Zhang, X., Martin, S. S., Fang, Y., & Li, J. (2019). Understanding differences in types of opioid prescriptions across time and space: A community-level analysis. *Journal of Drug Issues, 49*(2), 405–418. https://doi.org/10.1177/0022042618815687

Basu, S., Hong, A., & Siddiqi, A. (2015). Using decomposition analysis to identify modifiable racial disparities in the distribution of blood pressure in the United States. *American Journal of Epidemiology, 182*(4), 345–353. https://doi.org/10.1093/aje/kwv079

Bauer, G. R. (2014). Incorporating intersectionality theory into population health research methodology: Challenges and the potential to advance health equity. *Social Science & Medicine, 110*, 10–17. https://doi.org/10.1016/j.socscimed.2014.03.022

Blakely, T. A., & Woodward, A. J. (2000). Ecological effects in multilevel studies. *Journal of Epidemiology & Community Health, 54*(5), 367–374. https://doi.org/10.1136/jech.54.5.367

Blinder, A. (1973). Wage discrimination: Reduced form and structural estimates. *Journal of Human Resources, 8*(4), 436–455.

Bobashev, G. V., Hoffer, L. D., & Lamy, F. R. (2020). C14Agent-based modeling to delineate opioid and other drug use epidemics. In Y. Apostolopoulos, M. K. Lemke, & K. Hassmiller Lich (Eds.), *Complex systems and population health*. Oxford University Press. https://doi.org/10.1093/oso/9780190880743.003.0014

Boodram, B., Hotton, A. L., Shekhtman, L., Gutfraind, A., & Dahari, H. (2018). High-risk geographic mobility patterns among young urban and suburban persons who inject drugs and their injection network members. *Journal of Urban Health, 95*(1), 71–82. https://doi.org/10.1007/s11524-017-0185-7

Bornmann, L., & Waltman, L. (2011). The detection of "hot regions" in the geography of science—A visualization approach by using density maps. *Journal of Informetrics, 5*(4), 547–553. https://doi.org/10.1016/j.joi.2011.04.006

Bouris, A., Jaffe, K., Eavou, R., Liao, C., Kuhns, L., Voisin, D., & Schneider, J. A. (2017). Project nGage: Results of a randomized controlled trial of a dyadic network support intervention to retain young black men who have sex with men in HIV care. *AIDS and Behavior, 21*(12), 3618–3629. https://doi.org/10.1007/s10461-017-1954-8

Bowleg, L. (2012). The problem with the phrase women and minorities: Intersectionality—An important theoretical framework for public health. *American Journal of Public Health, 102*(7), 1267–1273. https://doi.org/10.2105/AJPH.2012.300750

Bowleg, L. (2013). "Once you've blended the cake, you can't take the parts back to the main ingredients": Black gay and bisexual men's descriptions and experiences of intersectionality. *Sex Roles: A Journal of Research, 68*, 754–767. https://doi.org/10.1007/s11199-012-0152-4

Brawner, B. M., Kerr, J., Castle, B. F., Bannon, J. A., Bonett, S., Stevens, R., James, R., & Bowleg, L. (2022). A systematic review of neighborhood-level influences on HIV vulnerability. *AIDS and Behavior, 26*(3), 874–934. https://doi.org/10.1007/s10461-021-03448-w

Bronfenbrenner, U. (1994). Ecological models of human development. *International Encyclopedia of Education, 3*(2), 37–43.

Carlos, H. A., Shi, X., Sargent, J., Tanski, S., & Berke, E. M. (2010). Density estimation and adaptive bandwidths: A primer for public health practitioners. *International Journal of Health Geographics, 9*(1), 39. https://doi.org/10.1186/1476-072X-9-39

Cerdá, M., Gaidus, A., Keyes, K. M., Ponicki, W., Martins, S., Galea, S., & Gruenewald, P. (2017). Prescription opioid poisoning across urban and rural areas: Identifying vulnerable groups and geographic areas. *Addiction, 112*(1), 103–112. https://doi.org/10.1111/add.13543

Cerdá, M., Ransome, Y., Keyes, K. M., Koenen, K. C., Tardiff, K., Vlahov, D., & Galea, S. (2013). Revisiting the role of the urban environment in substance use: THE case of analgesic overdose fatalities. *American Journal of Public Health, 103*(12), 2252–2260. https://doi.org/10.2105/AJPH.2013.301347

Chen, Y.-T., Kolak, M., Duncan, D. T., Schumm, P., Michaels, S., Fujimoto, K., & Schneider, J. A. (2019). Neighborhoods, networks and preexposure prophylaxis awareness: A multilevel analysis of a sample of young black men who have sex with men. *Sexually Transmitted Infections, 95*(3), 228–235. https://doi.org/10.1136/sextrans-2018-053639

Cho, Y. I., Johnson, T. P., Fendrich, M., & Pickup, L. (2013). Treatment facility neighborhood environment and outpatient treatment completion. *Journal of Drug Issues, 43*(3), 374–385. https://doi.org/10.1177/0022042612472332

Chun, J. J., Lipsitz, G., & Shin, Y. (2013). Intersectionality as a social movement strategy: Asian immigrant women advocates. *Signs, 38,* 917–940. https://doi.org/10.1086/669575

Ciccarone, D. (2017). Fentanyl in the US heroin supply: A rapidly changing risk environment. *International Journal of Drug Policy, 46,* 107–111. https://doi.org/10.1016/j.drugpo.2017.06.010

Collins, P. H. (2015). Intersectionality's definitional dilemmas. *Annual Review of Sociology, 41*(1), 1–20. https://doi.org/10.1146/annurev-soc-073014-112142

Crenshaw, K. (1989). *Demarginalizing the intersection of race and sex: A black feminist critique of antidiscrimination doctrine, feminist theory and antiracist politics.* 31.

De, P., Cox, J., Boivin, J.-F., Platt, R. W., & Jolly, A. M. (2007). The importance of social networks in their association to drug equipment sharing among injection drug users: A review. *Addiction, 102*(11), 1730–1739. https://doi.org/10.1111/j.1360-0443.2007.01936.x

Deuba, K., Anderson, S., Ekström, A. M., Pandey, S. R., Shrestha, R., Karki, D. K., & Marrone, G. (2016). Microlevel social and structural factors act synergistically to increase HIV risk among Nepalese female sex workers. *International Journal of Infectious Diseases, 49,* 100–106. https://doi.org/10.1016/j.ijid.2016.06.007

Diderichsen, F., Andersen, I., Manuel, C.,. The working group of the danish review on social determinants of health, andersen, A.-M. N., Bach, E., Baadsgaard, M., Brønnum-Hansen, H., Hansen, F. K., Jeune, B., Jørgensen, T., & Søgaard, J. (2012). Health Inequality—Determinants and policies. *Scandinavian Journal of Public Health, 40*(8_suppl), 12–105. doi: https://doi.org/10.1177/1403494812457734

Diderichsen, F., Evans, T., & Whitehead, M. (2001). The social basis of disparities in health. In T. Evans, M. Whitehead, F. Diderichsen, A. Bhuiya, & M. Wirth (Eds.), *Challenging inequities in health: From ethics to action.* Oxford University Press. https://doi.org/10.1093/acprof:oso/9780195137408.003.0002

Diderichsen, F., Hallqvist, J., & Whitehead, M. (2019). Differential vulnerability and susceptibility: how to make use of recent development in our understanding of mediation and interaction to tackle health inequalities. *International Journal of Epidemiology, 48*(1), 268–274. https://doi.org/10.1093/ije/dyy167

Diez-Roux, A. V. (1998). Bringing context back into epidemiology: Variables and fallacies in multilevel analysis. *American Journal of Public Health, 88*(2), 216–222. https://doi.org/10.2105/AJPH.88.2.216

Duncan, D. T., Hickson, D. A., Goedel, W. C., Callander, D., Brooks, B., Chen, Y.-T., Hanson, H., Eavou, R., Khanna, A. S., Chaix, B., Regan, S. D., Wheeler, D. P., Mayer, K. H., Safren, S. A., Carr Melvin, S., Draper, C., Magee-Jackson, V., Brewer, R., & Schneider, J. A. (2019). The social context of HIV prevention and care among black men who have sex with men in three U.S. Cities: The neighborhoods and networks (N2) cohort study. *International Journal of Environmental Research and Public Health, 16*(11), 1922. https://doi.org/10.3390/ijerph16111922

Duncan, C., Jones, K., & Moon, G. (1998). Context, composition and heterogeneity: Using multilevel models in health research. *Social Science & Medicine, 46*(1), 97–117. https://doi.org/10.1016/S0277-9536(97)00148-2

Evans, C. R., Williams, D. R., Onnela, J. P., & Subramanian, S. V. (2018). A multilevel approach to modeling health inequalities at the intersection of multiple social identities. *Social Science & Medicine, 203,* 64–73. https://doi.org/10.1016/j.socscimed.2017.11.011

Festinger, L., Schachter, S., & Back, K. (1950). *Social pressures in informal groups; a study of human factors in housing.* Harper. pp. x, 240.

Fortin, N., Lemieux, T., & Firpo, S. (2011). Chapter 1—Decomposition methods in economics. In O. Ashenfelter & D. Card (Eds.), *Handbook of labor economics* (Vol. 4, pp. 1–102). Elsevier. https://doi.org/10.1016/S0169-7218(11)00407-2

García, M. C. (2019). Opioid prescribing rates in nonmetropolitan and metropolitan counties among primary care providers using an electronic health record system—United States, 2014–2017. *MMWR. Morbidity and Mortality Weekly Report, 68.* https://doi.org/10.15585/mmwr.mm6802a1

Gesink, D., Wang, S., Guimond, T., Kimura, L., Connell, J., Salway, T., Gilbert, M., Mishra, S., Tan, D., Burchell, A. N., Brennan, D. J., Logie, C. H., & Grace, D. (2018). Conceptualizing geosexual archetypes: Mapping the sexual travels and egocentric sexual networks of gay and bisexual men in Toronto, Canada. *Sexually Transmitted Diseases, 45*(6), 368. https://doi.org/10.1097/OLQ.0000000000000752

Green, M. A., Evans, C. R., & Subramanian, S. V. (2017). *Can intersectionality theory enrich population health research?* https://doi.org/10.1016/j.socscimed.2017.02.029

Hancock, A.-M. (2007). When multiplication doesn't equal quick addition: Examining intersectionality as a research paradigm. *Perspectives on Politics, 5*(1), 63–79. https://doi.org/10.1017/S1537592707070065

Hankivsky, O., Doyal, L., Einstein, G., Kelly, U., Shim, J., Weber, L., & Repta, R. (2017). The odd couple: Using biomedical and intersectional approaches to address health inequities. *Global Health Action, 10*(sup2), 1326686. https://doi.org/10.1080/16549716.2017.1326686

Harari, L., & Lee, C. (2021). Intersectionality in quantitative health disparities research: A systematic review of challenges and limitations in empirical studies. *Social Science & Medicine (1982), 277*, 113876. https://doi.org/10.1016/j.socscimed.2021.113876

Heimer, R., Barbour, R., Palacios, W. R., Nichols, L. G., & Grau, L. E. (2014). Associations between injection risk and community disadvantage among suburban injection drug users in southwestern Connecticut, USA. *AIDS and Behavior, 18*(3), 452–463. https://doi.org/10.1007/s10461-013-0572-3

Hembree, C., Galea, S., Ahern, J., Tracy, M., Markham Piper, T., Miller, J., Vlahov, D., & Tardiff, K. J. (2005). The urban built environment and overdose mortality in New York City neighborhoods. *Health & Place, 11*(2), 147–156. https://doi.org/10.1016/j.healthplace.2004.02.005

Hill Collins, P., & Bilge, S. (2020). *Intersectionality* (2nd ed.). Polity Press.

Hoff, P. D., Raftery, A. E., & Handcock, M. S. (2002). Latent space approaches to social network analysis. *Journal of the American Statistical Association, 97*(460), 1090–1098. https://doi.org/10.1198/016214502388618906

Howitt, R. (2003). Scale. In *A companion to political geography* (pp. 132–157). John Wiley & Sons, Ltd. https://doi.org/10.1002/9780470998946.ch10

Hunter, R. F., de la Haye, K., Murray, J. M., Badham, J., Valente, T. W., Clarke, M., & Kee, F. (2019). Social network interventions for health behaviors and outcomes: A systematic review and meta-analysis. *PLOS Medicine, 16*(9), e1002890. https://doi.org/10.1371/journal.pmed.1002890

Hussein, M., Diez Roux, A. V., Mujahid, M. S., Hastert, T. A., Kershaw, K. N., Bertoni, A. G., & Baylin, A. (2018). Unequal exposure or unequal vulnerability? Contributions of neighborhood conditions and cardiovascular risk factors to socioeconomic inequality in incident cardiovascular disease in the multi-ethnic study of atherosclerosis. *American Journal of Epidemiology, 187*(7), 1424–1437. https://doi.org/10.1093/aje/kwx363

Illangasekare, S., Burke, J., Chander, G., & Gielen, A. (2013). The syndemic effects of intimate partner violence, HIV/AIDS, and substance abuse on depression among low-income urban women. *Journal of Urban Health, 90*, 934–947.

Kapilashrami, A., & Hankivsky, O. (2018). Intersectionality and why it matters to global health. *The Lancet, 391*(10140), 2589–2591. https://doi.org/10.1016/S0140-6736(18)31431-4

Keyes, K. M., Cerdá, M., Brady, J. E., Havens, J. R., & Galea, S. (2014). Understanding the rural–urban differences in nonmedical prescription opioid use and abuse in the United States. *American Journal of Public Health, 104*(2), e52–e59. https://doi.org/10.2105/AJPH.2013.301709

Kolak, M. A., Chen, Y.-T., Joyce, S., Ellis, K., Defever, K., McLuckie, C., Friedman, S., & Pho, M. T. (2020). Rural risk environments, opioid-related overdose, and infectious diseases: A multidimensional, spatial perspective. *International Journal of Drug Policy, 85*, 102727. https://doi.org/10.1016/j.drugpo.2020.102727

Kolak, M. A., Chen, Y.-T., Lin, Q., & Schneider, J. (2021). Social-spatial network structures and community ties of egocentric sex and confidant networks: A Chicago case study. *Social Science & Medicine, 291*, 114462. https://doi.org/10.1016/j.socscimed.2021.114462

Larson, E., George, A., Morgan, R., & Poteat, T. (2016). 10 Best resources on… intersectionality with an emphasis on low- and middle-income countries. *Health Policy and Planning, 31*(8), 964–969. https://doi.org/10.1093/heapol/czw020

Latkin, C. A., Curry, A. D., Hua, W., & Davey, M. A. (2007). Direct and indirect associations of neighborhood disorder with drug use and high-risk sexual partners. *American Journal of Preventive Medicine, 32*(6), S234–S241. https://doi.org/10.1016/j.amepre.2007.02.023

Latkin, C. A., German, D., Vlahov, D., & Galea, S. (2013). Neighborhoods and HIV: A social ecological approach to prevention and care. *American Psychologist, 68*(4), 210–224. https://doi.org/10.1037/a0032704

Latkin, C. A., Kuramoto, S. J., Davey-Rothwell, M. A., & Tobin, K. E. (2010). Social norms, social networks, and HIV risk behavior among injection drug users. *AIDS and Behavior, 14*(5), 1159–1168. https://doi.org/10.1007/s10461-009-9576-4

Lewis, G. (2013). Unsafe travel: Experiencing intersectionality and feminist displacements. *Signs, 38*(4), 869–892. https://doi.org/10.1086/669609

Lin, Q., Aguilera, J. A. R., Williams, L. D., Mackesy-Amiti, M. E., Latkin, C., Pineros, J., Kolak, M., & Boodram, B. (2023). Social-spatial network structures among young urban and suburban persons who inject drugs in a large metropolitan area. *medRxiv*, 2023.02.21.23286255. https://doi.org/10.1101/2023.02.21.23286255

MacKinnon, C. A. (2013). Intersectionality as method: A note. *Signs, 38*(4), 1019–1030. https://doi.org/10.1086/669570

Marston, S. A. (2000). The social construction of scale. *Progress in Human Geography, 24*(2), 219–242. https://doi.org/10.1191/030913200674086272

Mason, M. J., Valente, T., Coatsworth, J. D., Mennis, J., Lawrence, F., & Zelenak, P. (2010). Place-based social network quality and correlates of substance use among urban adolescents. *Journal of Adolescence, 33*(3), 419–427. https://doi.org/10.1016/j.adolescence.2009.07.006

Mennis, J., & Mason, M. J. (2011). People, places, and adolescent substance use: Integrating activity space and social network data for analyzing health behavior. *Annals of the Association of American Geographers, 101*(2), 272–291.

Merlo, J. (2018). Multilevel analysis of individual heterogeneity and discriminatory accuracy (MAIHDA) within an intersectional framework. *Social Science & Medicine, 203*, 74–80. https://doi.org/10.1016/j.socscimed.2017.12.026

Mohanty, C. T. (2013). Transnational feminist crossings: On neoliberalism and radical critique. *Signs, 38*(4), 967–991. https://doi.org/10.1086/669576

Molina, K. M., Alegría, M., & Chen, C.-N. (2012). Neighborhood context and substance use disorders: A comparative analysis of racial and ethnic groups in the United States. *Drug and Alcohol Dependence, 125*, S35–S43. https://doi.org/10.1016/j.drugalcdep.2012.05.027

Monnat, S. M., Peters, D. J., Berg, M. T., & Hochstetler, A. (2019). Using census data to understand county-level differences in overall drug mortality and opioid-related mortality by opioid type. *American Journal of Public Health, 109*(8), 1084–1091.

Nikolopoulos, G. K., Pavlitina, E., Muth, S. Q., Schneider, J., Psichogiou, M., Williams, L. D., Paraskevis, D., Sypsa, V., Magiorkinis, G., Smyrnov, P., Korobchuk, A., Vasylyeva, T. I., Skaathun, B., Malliori, M., Kafetzopoulos, E., Hatzakis, A., & Friedman, S. R. (2016). A network intervention that locates and intervenes with recently HIV-infected persons: The transmission reduction intervention project (TRIP). *Scientific Reports, 6*(1), 38100. https://doi.org/10.1038/srep38100

Oaxaca, R. (1973). Male–female wage differentials in urban labor markets. *International Economic Review, 14*(3), 693–709. https://doi.org/10.2307/2525981

Perlman, D. C., & Jordan, A. E. (2018). The syndemic of opioid misuse, overdose, HCV, and HIV: structural-level causes and interventions. *Current HIV/AIDS Reports, 15*, 96–112. https://doi.org/10.1007/s11904-018-0390-3

Phelan, J. C., & Link, B. G. (2015). Is racism a fundamental cause of inequalities in health? *Annual Review of Sociology, 41*(1), 311–330. https://doi.org/10.1146/annurev-soc-073014-112305

Powell, L. M., Wada, R., Krauss, R. C., & Wang, Y. (2012). Ethnic disparities in adolescent body mass index in the United States: The role of parental socioeconomic status and economic contextual factors. *Social Science & Medicine (1982), 75*(3), 469–476. https://doi.org/10.1016/j.socscimed.2012.03.019

Ren, F. (2016). *Activity space.* Oxford University Press. Retrieved from https://oxfordbibliographies.com/view/document/obo-9780199874002/obo-9780199874002-0137.xml

Rhodes, T. (2009). Risk environments and drug harms: A social science for harm reduction approach. *International Journal of Drug Policy, 20*(3), 193–201. https://doi.org/10.1016/j.drugpo.2008.10.003

Roxberg, Å., Tryselius, K., Gren, M., Lindahl, B., Werkander Harstäde, C., Silverglow, A., Nolbeck, K., James, F., Carlsson, I.-M., Olausson, S., Nordin, S., & Wijk, H. (2020). Space and place for health and care. *International Journal of Qualitative Studies on Health and Well-Being, 15*(sup1), 1750263. https://doi.org/10.1080/17482631.2020.1750263

Sangaramoorthy, T., & Benton, A. (2022). Intersectionality and syndemics: A commentary. *Social Science & Medicine, 295*, 113783. https://doi.org/10.1016/j.socscimed.2021.113783

Sen, B. (2014). Using the Oaxaca–Blinder decomposition as an empirical tool to analyze racial disparities in obesity. *Obesity (Silver Spring, Md.), 22*(7), 1750–1755. https://doi.org/10.1002/oby.20755

Sewell, D. K., & Chen, Y. (2015). Latent space models for dynamic networks. *Journal of the American Statistical Association, 110*(512), 1646–1657. https://doi.org/10.1080/01621459.2014.988214

Silverman, E., Gostoli, U., Picascia, S., Almagor, J., McCann, M., Shaw, R., & Angione, C. (2021). Situating agent-based modeling in population health research. *Emerging Themes in Epidemiology, 18*(1), 10. https://doi.org/10.1186/s12982-021-00102-7

Singer, M. (2000). A dose of drugs, a touch of violence, a case of AIDS: Conceptualizing the Sava Syndemic. *Free Inquiry in Creative Sociology, 28*(1), Article 1.

Singer, M. (2011). Toward a critical biosocial model of ecohealth in Southern Africa: The hiv/aids and nutrition insecurity syndemic. *Annals of Anthropological Practice, 35*(1), 8–27. https://doi.org/10.1111/j.2153-9588.2011.01064.x

Stall, R., Mills, T. C., Williamson, J., Hart, T., Greenwood, G., Paul, J., Pollack, L., Binson, D., Osmond, D., & Catania, J. A. (2003). Association of cooccurring psychosocial health problems and increased vulnerability to HIV/AIDS among urban men who have sex with men. *American Journal of Public Health, 93*(6), 939–942. https://ajph.aphapublications.org. https://doi.org/10.2105/AJPH.93.6.939

Tatara, E., Gutfraind, A., Collier, N. T., Cotler, S. J., Major, M., Boodram, B., Ozik, J., & Dahari, H. (2019). Agent-based modeling of persons who inject drugs in metropolitan Chicago suggests that retreatment with antivirals of persons who are reinfected with Hep C is critical to achieve the WHO incidence reduction objective by 2030 [Preprint]. *Epidemiology.* https://doi.org/10.1101/653196

Tempalski, B., & McQuie, H. (2009). Drugscapes and the role of place and space in injection drug use-related HIV risk environments. *International Journal of Drug Policy, 20*(1), 4–13. https://doi.org/10.1016/j.drugpo.2008.02.002

Tempalski, B., Williams, L. D., Kolak, M., Ompad, D. C., Koschinsky, J., & McLafferty, S. L. (2022). Conceptualizing the socio-built environment: An expanded theoretical framework to promote a better understanding of risk for nonmedical opioid overdose outcomes in urban and non-urban settings. *Journal of Urban Health, 99*(4), 701–716. https://doi.org/10.1007/s11524-022-00645-3

Tsai, A. C. (2018). Syndemics: A theory in search of data or data in search of a theory? *Social Science & Medicine, 206*, 117–122. https://doi.org/10.1016/j.socscimed.2018.03.040

Tsai, A. C., Mendenhall, E., Trostle, J. A., & Kawachi, I. (2017). Co-occurring epidemics, syndemics, and population health. *The Lancet, 389*(10072), 978–982. https://doi.org/10.1016/S0140-6736(17)30403-8

Tsai, A. C., & Venkataramani, A. S. (2016). Syndemics and health disparities: A methodological note. *AIDS and Behavior, 20*(2), 423–430. https://doi.org/10.1007/s10461-015-1260-2

Tsing, A. L. (2012). On nonscalability: The living world is not amenable to precision-nested scales. *Common Knowledge, 18*(3), 505–524.

USGCRP. (2016). In A. Crimmins, J. Balbus, J. L. Gamble, C. B. Beard, J. E. Bell, D. Dodgen, R. J. Eisen, N. Fann, M. D. Hawkins, S. C. Herring, L. Jantarasami, D. M. Mills, S. Saha, M. C. Sarofim, J. Trtanj, & L. Ziska (Eds.), *The Impacts of Climate Change on Human Health in the United States: A Scientific Assessment* (p. 312). U.S. Global Change Research Program. https://doi.org/10.7930/J0R49NQX

VanderWeele, T. J. (2013). A three-way decomposition of a total effect into direct, indirect, and interactive effects. *Epidemiology (Cambridge, Mass.), 24*(2), 224–232. https://doi.org/10.1097/EDE.0b013e318281a64e

VanderWeele, T. J. (2014). A unification of mediation and interaction: A 4-way decomposition. *Epidemiology (Cambridge, Mass.), 25*(5), 749–761. https://doi.org/10.1097/EDE.0000000000000121

Verloo, M. (2013). Intersectional and cross-movement politics and policies: Reflections on current practices and debates. *Signs, 38*(4), 893–915. https://doi.org/10.1086/669572

World Health Organization. (2010). *A conceptual framework for action on the social determinants of health*. World Health Organization. Retrieved from https://apps.who.int/iris/handle/10665/44489

Wylie, J., Shah, L., & Jolly, A. (2007). Incorporating geographic settings into a social network analysis of injection drug use and bloodborne pathogen prevalence. *Health & Place, 13*(3), 617–628. https://doi.org/10.1016/j.healthplace.2006.09.002

Xu, R. (2018). Alternative estimation methods for identifying contagion effects in dynamic social networks: A latent-space adjusted approach. *Social Networks, 54*, 101–117. https://doi.org/10.1016/j.socnet.2018.01.002

Xu, R. (2020). Statistical methods for the estimation of contagion effects in human disease and health networks. *Computational and Structural Biotechnology Journal, 18*, 1754–1760. https://doi.org/10.1016/j.csbj.2020.06.027

Youm, Y., Mackesy-Amiti, M. E., Williams, C. T., & Ouellet, L. J. (2009). Identifying hidden sexual bridging communities in Chicago. *Journal of Urban Health, 86*(1), 107–120. https://doi.org/10.1007/s11524-009-9371-6

Young, A. M., Jonas, A. B., Mullins, U. L., Halgin, D. S., & Havens, J. R. (2013). Network structure and the risk for HIV transmission among rural drug users. *AIDS and Behavior, 17*(7), 2341–2351. https://doi.org/10.1007/s10461-012-0371-2

Chapter 16
Making Space: Walking as Qualitative Research with People Who Use Drugs and Experience Homelessness

Praveena K. Fernes and Tim Rhodes

Introduction

In the first 10 minutes of our walk around her neighborhood, Pooja [pseudonym] takes us[1] on a candid journey through the relationships that have left a mark on her path. From the bench where she and her brother share cherished moments to the stairwells under a block of flats where she and her ex-partner used to smoke crack, each location incorporates a piece of her life story. As we stroll, Pooja opens up about her past, emplacing her accounts of social experience in relation to the material aspects of the built environment we pass through. Our 45-minute walk weaves together various threads of time, objects, places, people, and affects that have disabled and enabled her lived experiences of care.

This walking interview method is, we will go on to consider, simultaneously a way into tracing the material emplacements of health and care experience in relation to space and a way of enacting space in the moment. In this chapter, we reflect on the walking interview as used in research looking at the intersections of illicit drug use and risk environments. We explore the walking interview as a method that not only attends to how space "shapes" or "determines" health (Macintyre *et al.*, 2002; Augustin *et al.*, 2023) but also acts as a practice that "makes space" through the materiality and unfolding movements of the research event itself (Michael, 2021). Through a focus on the walking interview, our aim is to move beyond an overly "structurally determined" account of the spatialization of health and harm (Galea

[1] While the first author, PKF, conducted all of the one-on-one walking interviews with interlocutors, we use "we" throughout to reflect our collaborative analysis.

P. K. Fernes (✉) · T. Rhodes
London School of Hygiene and Tropical Medicine, London, UK
e-mail: praveena.fernes@lshtm.ac.uk

© The Author(s) 2026
M. A. Kolak, I. K. Moise (eds.), *Place and the Social-Spatial Determinants of Health*, Global Perspectives on Health Geography,
https://doi.org/10.1007/978-3-031-88463-4_16

et al., 2009; O'Campo and Dunn, 2012; Cooper and Tempalski, 2014) to an account which emphasizes space as an emergent causation in the unfolding materiality of health and care, in which nonhuman elements "become-with" human actions (Connolly, 2004; Massey, 2005; Cummins *et al.*, 2007; Anderson, 2009; Duff, 2016; Rhodes and Lancaster, 2019), including the methods of research (Law, 2004; Michael, 2021).

There is increasing emphasis in social research on how space shapes health (Massey, 2005; Cummins *et al.*, 2007; Thrift, 2007; Neely and Nading, 2017), including in relation to illicit drug use and associated harms (Duff, 2007; Cooper *et al.*, 2009; Cooper and Tempalski, 2014). Through the walking interview, we can appreciate how lived experiences of health emerge in relation to their material contexts, in which nonhuman elements—from built environments to affects—come into play with human agency to shape risk environments (Duff, 2007; Rhodes, 2009; Müller-Mahn, 2013; Müller-Mahn *et al.*, 2018).

Space as Relational

Investigating health as an effect of space has historically tended to envisage space as a physical "determinant" of health in epidemiological models of "cause and effect," and this is largely the case in illicit drugs research (Thomas et al., 2008; Cooper *et al.*, 2009; Tempalski and McQuie, 2009; Strathdee *et al.*, 2010). Increasingly, such causative relations are recognized as dynamic and recursive in social-ecological epidemiological models (Krieger, 2001; Diez Roux, 2022; Galea, 2022). This shift begins to caution against an "overdeterminism" in the social determinants of health research (Galea, 2022), wherein environmental determinants are seen to act as constraints on action with insufficient attention to how action also impacts context in an unfolding relationship (Bourdieu, 1977; Giddens, 2003). Once contexts and environments themselves are treated as dynamic, as always in the process of being made through interaction, health and environment come to be seen as entangled effects of their emergent "relations" rather than as separates which "determine" the other (Massey, 2005; Andrews and Duff, 2019; Rhodes and Lancaster, 2019).

This relational shift is important for at least two reasons. First, it cautions against the tendency to artificially delineate 'types' of risk environments—such as predetermined dimensions of physical, social, economic, and policy environments—as if these operate separately from the other and without dynamic feedback (Rhodes, 2009). While emphasizing that health is shaped by environments, this emphasis may "overdetermine" a particular imagined causative and linear pathway which reproduces a hinterland of causative assumption. Yates-Doerr (2020), for example, notes that "the desire to address the roots of a pre-given problem imagined beginning at a measurable point and to then advance to a predictable (i.e., determinant) place sets us on a path toward prescriptive solutions that often do not result in the deep structural transformation they claim to inspire (p.380)." Second, thinking of health and environment as relational cautions against an overly "human-centered" depiction of agency, instead encouraging an understanding of agency that also incorporates the nonhuman and the material (Andrews and Duff, 2019; Rhodes and Lancaster, 2019). Here, then, we move to a relational understanding of agency and of health, one that is "socio-material" (Law, 2004; Massey, 2005; Andrews and Duff, 2019; Michael, 2021).

Engaging with such materialist thinking—to understand health as a material effect of networks or assemblages in which various human and nonhuman elements interact—offers new ways of tracing how socio-political forces mediate drug harms in the everyday of local risk environments (Fraser and Moore, 2011; Fraser, 2020). Drug use practices are locally materialized effects of their unfolding situations and environments, which entangle across multiple scales, from the local to the global (Bourgois, 2003; Duff, 2007; Rhodes, 2009). There is then, within the field of ethnographic and qualitative drugs research, a shift from political-economic and discursive understandings of risk environments (Bourgois, 1998; Singer, 1994; Keane, 2002; Rhodes, 2002, 2009) to socio-materialist understandings that trace the materiality of drugs, drug effects, drug harms, and drug settings (Fraser, 2006, 2020; Bøhling, 2014; Duff, 2014; Duncan et al., 2017; Malins, 2017; Dennis, 2020). Massey's (2005, p. 9) relational view of space and place is particularly helpful. Massey (2005) outlines space as "(1) the product of interrelations; (2) a sphere of the possibility for the existence of multiplicity; and (3) always under construction." In our analysis, we explore health related to drug use as a material effect of space, drawing attention to how methods of research are also actors in the making of space (Law, 2004; Michael, 2021).

What Walks Do

Renewed attention to walking and go-along interviews in geography and anthropology (e.g., Edensor, 2010; Vergunst & Ingold, 2016) is situated within a broader trend centering place and mobilities in the social sciences (Thrift, 1999; Urry, 2012). Since this wider 'mobility turn', a paradigm that foregrounds issues of movement, (Sheller and Urry, 2006), terms for go-along interview and walking interview have multiplied to include 'narrative walk in real-time' (Miaux et al., 2010), 'walking field-work approach,' 'mobile methods,' 'wheeling interview' (Parent, 2016), and 'docent interviews' (Chang 2017). Embedded in a range of human mobility practices—from promenading to protesting—walking is often acknowledged as an everyday part of the human experience (Lorimer, 2011). The power of the walk has been embraced by many traditions as a qualitative methodology, including the "go-along" (Kusenbach, 2003; Carpiano, 2009) and walking interviews in ethnographic research (Evans and Jones, 2011). Carpiano (2009) describes how the "go-along" interview method enables researchers to be "walked through" people's experiences within their local residential context while also exploring the interplay between structural conditions and individual agency for shaping action.

Much of walkability research assumes a deterministic relationship, where environmental factors directly dictate people's health outcomes. Such research also tends to take on the predominant use of a positivist epistemology, which prioritizes objective measures over residents' perceptions and experiences (Andrews *et al.*, 2012). In the drugs field, researchers deploying rapid assessment methods have incorporated walking in studies of the geospatial "determinants" of care (e.g., Singer *et al.*, 2000; Collins *et al.*, 2022). While the walking interview has afforded particular promises in the study of space as a relational effect of health, there are methodological and theoretical underpinnings that separate social epidemiological and sociological/anthropological approaches to understanding space (Andrews *et al.*, 2012). Springgay and Truman (2018) extend the major concepts that arise in walking research—place, sensory inquiry, embodiment, and rhythm—to bring a more-than-human approach by focusing on land and geos, affect, transmaterial, and movement. Thus, a convergence between new materialist thought and walking research may elicit live encounters with places that map more-than-human elements.

Approach

For this research, walks[2] aid in understanding how lived experiences and place-making are constructed through spatial practices of sociality and positionality. We extend this convergence between new materialist methodologies and walking research as we—participants and researchers—subject ourselves to the more-than-human agency of the elements that surround us on a walk. Collecting narrative-laden data on the move allows a more dynamic account than sit-down interviews because elements of the environment are incorporated into the materiality of the research event.

Furthermore, the theme of movement resonates with individuals experiencing homelessness and using substances. In qualitative and ethnographic research with individuals grappling with homelessness and substance use, movement emerges as a theme (e.g., Fast and Cunningham, 2018). The relentless quest for basic necessities compels many to be in perpetual motion, embedding movement as a form of agency and survival into many interlocutors' daily lives. Without stable transportation or funds, walking becomes a main form of transit, serving as a means of navigation and access to resources, as well as providing a sense of relative freedom from everyday situational and spatial constraints (Radley *et al.*, 2010). The use of walks in service providers' outreach strategies also underscores the relevance of place and movement in the lives of the people they serve (Bond *et al.*, 2022; Stambe *et al.*, 2023). The walking method has situational fit given how it intersects with the

---

[2] Importantly, this sentiment does not accommodate the diversity of body forms and abilities in society. In this study, "walking interviews" were expanded to understand movement as achieved through a variety of means, including wheelchairs, walking frames, and slowly in a 'non-normal' style.

everyday practices of those "on the move." We can consider movement as a matter of ontology; that is, as a way of thinking about how space and relations are in the "making." Here, we take up Dennis *et al.*'s (2020) ontological concern of movement: "the way human and nonhuman processes work together to create new formations of space-time-bodies" (p. 2). Beyond a metaphor and conflating movement for physical locomotion (i.e., walking), we try to hone in on practices (e.g., hostel rules) and entities (e.g., walls) that restrict and enable movement away from the materiality of harm.

Several scholars have also critically examined go-along interviews (Merriman, 2014; Spinney, 2015; Warren, 2017; Porkertová *et al.*, 2024). Given ableist conceptions of walking, talking and space—and its epistemological consequences (Porkertova *et al.*, 2024), we used the recruitment process to understand participants' mobility preferences and select a frequently traveled route or area. Merriman (2014) warns against the over-promise and over-reliance on mobile methods, arguing that researchers often conflate methods for mobilities research (Sheller and Urry, 2006, p. 217) with mobile methods (Büscher, Urry, and Witchger, 2010), leading them to mistakenly assume more 'accurate' ways of knowing and overlooking passive practices, engagements, and affective relations surrounding movement. Following his critique, our use of go-along interviews is not an attempt to more accurately know or represent practices, contexts and events, but we are interested in what go-along interviews do. Further, PKF's use of ethnography, of which the analysis here is a part, included documenting "other ways of experiencing mobilities" (Bissell, 2010, 58) such as stillness, waiting, slowness, tiredness, and boredom (Bissell, 2007, 2008; Bissell and Fuller, 2013).

We bring Michael's (2021) understanding of the research event as fluid and unfolding. Thus, the walking method is not merely a procedural tool but is also an active process of doing and making. This conceptual shift underscores the significance of the walking interview as an "evidence-making intervention" shaping and crafting space as an integral component of the research event itself (Rhodes and Lancaster, 2019; Dennis *et al.*, 2020). This distinction matters: It transcends the conventional notion of method as a means of "emplacing" or "locating" experiences within a given spatial context. Instead, it positions the method as part of a dynamic enactment whereby space also emerges as an element of relations. We bring these dual facets of method, one as a means of "finding out" and the other as a mode of "making," into our case study.

The walking interview drawn on in this chapter is part of a larger ethnographic doctoral study, which explores how people who use drugs and are homeless access and navigate social care services in London—and how place shapes such encounters. Parts of the writing in this chapter also appear in PKF's doctoral thesis. To contextualize the aims of these interviews, we describe how they fit into the wider study, which began in 2022 and is ongoing.[3]

---

[3] Writing this monograph during a study in motion parallels the ways in which interlocutors have also been documenting places in time.

Wider Study Context

Drug-related deaths and health harms for people who use drugs in the United Kingdom have increased dramatically in the last decade, particularly among those who experience homelessness. These harms have been exacerbated by rising living costs (Francis-Devine *et al.*, 2022), COVID-19, and cuts to social and health services (Stokes *et al.*, 2022). People who are homeless and use drugs are dealing with unprecedented hostility and everyday structural violence (Aldridge *et al.*, 2018).

Since 2012, community-based services in the United Kingdom are increasingly strapped for resources and must bid for short-term contracts by proving their impact via tick-box targets (Health and Social Care Act, 2012). Additionally, environmental features that impact well-being and belongingness for people who use drugs and are homeless—such as housing, employment, and access to harm reduction resources—are often spatially stratified. Community-based services therefore "must innovate to reach those most at risk" (Harris *et al.*, 2020, p. 3) in a manner that centers place and its social–spatial dimensions.

The study collaborates with a community-based "outreach" service that, like most services of this kind, is tasked to "fill in the cracks" and reach out to those who cannot access the 'front door' of the drug treatment service (which is located in a different building). Funded by the local council, the outreach team is based at a day center that serves people who are rough sleeping and provides guidance on safer substance use, harm reduction, access to treatment, community outreach, and referrals to external services. They operate both indoors, through groups and one-on-one support, and outdoors, through street outreach. This study has thus far investigated how the outreach service seeks to "reach out," including by following staff and practices ethnographically as they move beyond the day center's gate into the community and as they connect with other networks of organizations in the area.

Methods

The overall study currently uses ethnographic observation, go-along walking interviews, community-engaged visual data collection, and focus groups to map how bodies, places, and materials interact with and make socio-spatial conditions. The study nests the processes from the *Our Voice* Citizen Science method, developed within Stanford University's Healthy Aging Technology Solutions Lab, within the larger ethnographic study. Importantly, such processes use multiple participatory strategies to engage interlocutors in individual and group data collection and interpretation (King *et al.*, 2019; Pedersen *et al.*, 2022). This research not only traces how multiple realities are assembled but also interrogates how the methods assemblage—the combination and coordination of various research methods and techniques employed—creates that knowledge (Law, 2004).

For the one-on-one walking interviews, PKF meets each participant at a usual walking route of their choice, based on their level of comfort. Participants lead a walk to collect photographs of neighborhood features that they perceive as affecting their well-being and healthy living in good, bad, and complicated ways. Simultaneously, participants answer follow-up questions on and after the walk during a less structured interview to expand on topics of interest that arise and capture the "small talk" (Driessen and Jansen, 2013) that inevitably happens between stopping to record photos. Since encapsulating one's life stories into static images and a short audio narrative caption may flatten people's much more complex experiences, the go-along interview provides a means for deeper discussion.

At the conclusion of the walk, participants are invited to attend one or more focus groups to review and discuss their findings with other participants. While interview data feeds into our qualitative analysis, it is not used in the focus groups in which participants analyze their collective photographic data. Participants are aware of this distinction between data shared in the focus group (photos they take via a project device that interlocutors hold) and go-along interviews from our walk (recorded on a separate audio recorder that PKF holds).

As a volunteer at the service, PKF often sits in the Welcome Area and is tasked to check people in as they come in for a range of services such as a meal token, hot shower, meeting with a key worker, or a quick needle exchange. In addition to walks "beyond the gate," service–site interactions also serve as spaces by which stories are performed and knowledge is created.

PKF's ethnographic field notes trace (in)formal spaces of care, with a focus on reaching out and reaching in. The primary focus on service user experiences ("reaching in") is facilitated by walks and focus groups. By layering fieldwork while volunteering at the service, we have been able to observe practices of "reaching out" that complement and contradict service users' experiences of care. Charting these socio-material practices of reaching in and out has the potential to open new assemblages of care with their own plots, rhythms, and locations.

Out of the 21 interlocutors, nine are women and 12 are men. We do not share extensive demographic data due to the potential for indirect identification of individuals. Participants were recruited via a drug and alcohol outreach service housed at a day center in London to identify people who have lived experience with drug use and housing insecurity. The interviews lasted approximately 45 min (ranging from 30 min to 1.5 h) and followed a loose topic guide covering drug and alcohol use, housing history, and access to services. The interview transcripts were analyzed using grounded thematic techniques. Ethical approval for this study was granted by the London School of Hygiene and Tropical Medicine.

In the following analysis, we take you on one walk to show how space is eventuated through the doing of the walking interview and highlight what these stories can tell us about the spatial relations of care. We selected a single walk to showcase the powerful dynamism of a 30-min walking interview—a method that not only generates rich data but also enacts new meanings of space. Eric is originally from Rwanda and is living in the United Kingdom without legal documentation. He is in his early forties and was described a few times by service providers as someone who has been living off "the goodness of others" and "luck." While graciously accounting for a few positive relations of care, Eric also elucidates systemic failures that have disabled care for him in the long term. Eric does this all while speaking in ontologies of movement and stasis. On our walk, he moves through different stories of precarity shaped by waiting and exclusion.

Waiting as Slow Violence: Anti-Movement

We began our walk outside Eric's hostel accommodation and made our way toward a park he enjoys. While still close to his hostel, Eric takes a photo of the building and reflects on aspects of his accommodation that would help time pass:

> *Me and my friends, when we are bored, we cannot stay there a long time. Cuz, first of all, there's no TVs in the room. So, you just keep staring on the four walls. No phone to watch YouTube or something like that. Other people are fortunate, they have that. I don't, so that's why I'm mostly out, outdoors. Walking, you know.*

His feelings of boredom and escapism draw our attention to a mundane, incremental kind of violence that plays out across temporal scales. Coined by Nixon (2013) in relation to environmental degradation, "slow violence occurs gradually and out of sight, a violence of delayed destruction that is dispersed across time and space, an attritional violence that is typically not viewed as violence at all." Others have noted how seeking asylum in the United Kingdom sustains excluded individuals in a state of injury (Saunders and Al-Om, 2022; Mayblin *et al.*, 2020). Eric emphasizes how the daily "chore" of waiting including sitting still, feeling stuck, and experiencing sameness weighs on his mood and affects his mental health. He is describing here a situation of anti-movement. This he resists through locomotion, through walking away, by creating space from spaces that embody stasis.

In this example, we can see the materiality of the *mundane* at work. Feeling stuck is materialized by "four walls." In a similar way, the walking interview engages with both the mundane and material, not only the spectacular. Unlike much social epidemiological work on the determinants of health—which investigates population-level indicators of macro inequalities related to income, education, healthcare, and housing—we see how inequalities become embedded and dispersed into the everyday, so much so that they may have become out of sight (Nixon, 2013).

Eric further talks of how the space of his accommodation materializes his sense of restricted movement and freedom. He emphasizes how the everyday policies of the hostels sustain a gulf between his outside and inside worlds. One such rule is that residents are encouraged not to be outside after 10 pm and that visitors are prohibited after this time. Eric remarks that these rules create as well as exacerbate a sense of stress. He describes, for example, that many hostel residents "can't sleep." This is because "they're in a stressful situation—moneywise, housing, and things like that. So, you can't expect to tell adults to go to bed at 10 o'clock." Eric's feelings of isolation inside are facilitated by paternalistic rules. He explains, "It is like we are in a boarding school. It's weird. People cannot have visitors." Eric's hostel, in many ways, offers respite in the form of a material space of shelter and safety from the immediate threat of homelessness. Yet within its walls, Eric paints an image of lying awake, sleepless in a confined space shaped by a web of policies and physical boundaries. The hostel is both a place of care and a stark reminder of the very inequalities he seeks to escape, even as they are etched into its structure.

Slow violence—where structural inequalities become normalized through daily routines and where attempts to navigate these conditions can deepen harm—materializes in the space of the hostel, which is an assemblage of waiting, walls, and rules. Here, the built environment itself enacts and perpetuates a cycle of slow violence, where the same space that offers refuge also exacerbates feelings of stress and isolation. In this example, a walk *outside* with Eric draws attention to and enacts in the moment, Eric's sense of inside–outside relations in which space materializes his capacity to move and escape from harm in different ways. His story of being isolated and entrapped inside, and the relative freedom of the surrounding park, was prompted by being in and moving through this space. We move from the material, including the everyday and mundane, to stories of wider structural violence. Go-along walks can offer rich insights, perhaps especially in highly politicized environments (Anderson, 2004).

Keeping Moving: Taking on the Burden of Complex Systems

Eric associated being inside his rule-ridden accommodation with feeling stuck, while walking outside is described as enabling "peace and tranquility." He goes on to share that Mondays and Tuesdays are better days for him to participate in this walk because of his appointment-packed schedule, "It's just something to keep busy. Something to move out of the [accommodation]. So, I love being out. Yeah, I don't like being inside. So, as many appointments as I can get, I'll take it just to keep busy." While movement becomes a means of coping with his circumstances, a key point emerges: "keeping busy" by constantly scheduling appointments creates a semblance of freedom, yet simultaneously reinforces dependency on complex health and social care systems. Here, "keeping busy" can also be understood as a response to a systemic environment that perpetuates homelessness. The concept of "slow violence" (Nixon, 2013) finds resonance in Eric's experience of "keeping busy," as the attritional harm inflicted by systems' inefficiencies erodes his well-being over time. On the surface, this movement of keeping busy carries the charge of agency: an effort to improve his circumstances, reach safer spaces, or connect with sources of care. Yet as we walk, it becomes clear that this perpetual motion is also imposed under constraints. Amidst precarious environments and fractured care landscapes, his movement is shaped by the need to carefully navigate hostile or unsafe spaces. This tension between movement as action and movement as survival reveals how slow violence plays out over time as a relentless demand to stay in motion.

As Eric takes on the burden of structural violence and complex systems, movement again emerges as a central theme. Engaging extensively with interlinked housing, legal, benefit, and immigration systems over time often leads to feelings of overwhelm and a focus on individual responsibility (Guise, Burrows, and Marshall, 2022). Eric's narrative also attests to this phenomenon and frames service users' motivation to engage in services as crucial. He remarks,

> *You can take a donkey or a horse to the river, but you can't force it to drink the water. Yeah. So, the [harm] reduction services are good. But the person willing to partake in those is the most important thing. Yeah, if you're not ready, you're not ready.*

In the realm of movement, this metaphor vividly captures the idea that facilitating access or service providers reaching out (akin to leading the horse to the river) does not always result in enacting care. By Eric's account, the responsibility lies in the individual to have an intrinsic readiness to use harm reduction services (symbolized by drinking the water). Effective relations of care must align with the individual's own desire for movement and change, reflecting an understanding of their unique context and aspirations, much like the choice of a horse to quench their thirst at their own pace. This metaphor conveys a powerful internalization of structural violence that burdens the individual with a responsibility to change their circumstances within a harmful system.

In the context of ontologies of movement, "keeping busy" becomes a manifestation of how individuals like Eric negotiate space and "reach in" for care. Eric's continuous involvement in the interconnected yet fragmented systems related to housing, legal, benefits, and immigration exposes how the burden of navigating these systems falls onto those seeking help. As he immerses himself in these systems, his movements are constrained by bureaucratic entanglements, inhibiting his ability to progress toward stable housing and employment.

Making Memories of Harm and Care

When we finally get to the park, Eric describes positive experiences, ranging from picnicking with friends to enjoying the sunshine and tranquility. Throughout the wider study, interlocutors describe parks both positively (as an escape from judgment and a space for community) and negatively (encountering police and street violence). While urban green spaces are often perceived as sanctuaries offering privacy and refuge (Speer and Goldfischer, 2020), the wider study revealed how parks act as transitional, in-between spaces where people who use drugs and are homeless navigate negotiations of inclusion, exclusion, and danger. Eric's positive reflections are suddenly interrupted by the park-side view of high rises in Canary Wharf. His shoulders slump and his body language changes as if a weight has settled upon him. He recalls the breakdown of his romantic relationships that recently foreclosed connections to care and housing. He shares, "From 2010, I was in Canary Wharf… Before that I was in Manchester. And then I met my partner, my ex-partner. And then I moved from Manchester to London to be with her. We were together for 12 years. And then the relationship died. So, I ended up on the streets for two years. And then I got this place."

Here, the high rises of Canary Wharf serve as an impromptu interviewer. In a departure from the "conventional interrogative encounter" in research interviews, walking interviews are "a collage of collaboration: an unstructured dialogue where all actors participate in a conversational, geographical, and informational pathway creation" (Anderson, 2004, p. 260). Eric's shift in mood and decision to switch topics are intricately connected to the evolving environment he encountered.

The network enabling Eric's care in relation to his housing and mental health shifted after his relationship breakup. Having lost his home, Eric turned to the street for shelter for the first time in his life during the COVID-19 pandemic. We walk further, toward the local canal path, and we bump into a place where he used to rough sleep. After capturing a photo, Eric adds, "the security [guards] over here were very nice. In the morning, they would bring breakfast." He describes how he navigates seeking care in unofficial spaces and describes this particular area as a site of care, for now. He explains,

> Sometimes other places, they [residents] don't want you there… As long as it is tidy and don't have no cans of beer there, the food lying there… People are okay…some of them are coming out, giving me water. Sitting with me. Talking to me. Another one gave me, you know, them small portable TV? Just to watch.

His description of informal care is characterized by individuals extending support through entertainment or activities to pass the time, with formal outreach falling short of meeting these needs. This mode of "reaching out" contrasts the relations of care he depicts at formal care sites in the area, such as hostel accommodations and day centers. Eric puts it plainly, "[day center], it's programmed to make you fail." He emphasizes the narrow open hours such services offer, drawing a connection between this lack of access and people moving or retreating elsewhere:

> But it is when people have nowhere to go, that's why they end up in the park drinking because they have nothing to do, no activities. If the [day center which closes at 2pm] was doing some more activities, it would be better. Bring people in at least three times aweek. Open until 6pm, 7pm.

Eric describes how activities have been especially sparse since COVID-19. As we return and approach his hostel site, we segue into a conversation about Eric's legal journey of seeking asylum in the United Kingdom. To describe it as long and challenging is an understatement. He remarks, "I'm a professional in waiting [laughs sarcastically]. I've been waiting for almost 23 years just to get some papers in the country to be able to work and get housing." He goes on to describe the pain he feels from being prevented from working, through the waiting, having become "stuck." As Eric walks away from the sky rises and towering economic success in Canary Wharf, the contrast between the opulence these structures represent and his legal barriers to work is stark. Mapping out various sites and objects of anti-movement and exclusion, through the act of moving through space itself, helps to situate Eric's experience and story of material inequality. Eric's walk enables us to discern patterns of movement as he waits, keeps busy, and escapes the messy web of social care. Eric's story of co-produced movement helps make connections and meanings of space, a space that unfolds as an embodiment of "slow violence" in the past and every day.

Discussion

We conclude here by accentuating the potential of an approach that treats health relationally as a material effect of its spatial environment, and of the "walking interview" as a means to understanding, as well as making, space.

From Determinants to Relations

In tension with most depictions of the "social determinants of health" (see Yates-Doerr, 2020 for critique), we have accentuated the materiality of health and space as always in the making, as matters of *indeterminacy* and *becoming* (Andrews and Duff, 2019). Hyper-focusing on the "social" may elide the intricate interplay between the material and spatial. We saw, for example, that Eric's walk gave notice to various material objects along the way, which helped him, and us, tell a story: TVs, bedroom walls, high rises. Treating the social, and indeed, "context," as fixed, stable, and measurable artificially delineates material entanglements of the human and nonhuman in a given situation as if they were separate elements, apart from the other (Andrews and Duff, 2019; Rhodes and Lancaster, 2019). We have seen how the walking interview can invite an appreciation of indeterminacy that is made in the materiality of the event as part of the effects of the research process itself. For instance, Eric's walk was initially a tour of places of health and care but acted to situate and entangle such sites of health in relation with economics and migration. Bringing movement and space to the forefront of our analysis—with a focus on material entanglements—underscores how infrastructures like transportation networks, healthcare facilities, and housing arrangements are inseparable from relations of care and health.

Understanding health and care as relational encourages an approach that seeks to trace how health and care are "made to work" given their situation (Mol, Moser and Pols, 2010; Rhodes and Lancaster, 2019). In the wider study, observing the service enables PKF to trace how actors in the caregiving process "tinker" and craft unique and meaningful approaches for "reaching out" to people who have been historically *harmed* by social care. At the same time, service users must adapt, squeeze, and navigate to "reach in" to "make" care possible. As we saw in tracing the "relational extensions" within Eric's walk, there is movement, "reaching in" and "reaching out," in how care and support come to be in the moment. Rather than following a linear trajectory, as implied by some "determinants" thinking, service users "reach in" and make opportunities for meaningful care possible, which the walking interview, itself a movement, allows us to see (Rhodes *et al.*, 2019).

Material Risk Environments

Frameworks of "risk environment" in health and drugs research highlight risk as an effect of dynamic interplay across multiple dimensions of physical, social, economic, and political environment (Rhodes, 2009; Strathdee *et al.*, 2010; Collins *et al.*, 2019). A useful aspect of thinking across environment in the recursive relations that shape drug harm is the recognition of the significance of "non-drug" and "non-health" elements in harm production and reduction. Yet, we are suggesting that articulations of risk environment, as commonly applied within drug research, are neither relational nor material enough. We arguably need a more ecological approach (Rhodes *et al.*, 2021, 2023), that is, an understanding of the risk environment that not only delineates human health as a structural effect of the social, political, and economic environment but that also draws attention to health as an emergent matter of ecology in how humans and environments "become-with" the other (Andrews and Duff, 2019; Rhodes and Lancaster, 2019). Despite an increasing focus on how environments shape drug harms, including through spatial relations (Cooper *et al.*, 2009; Cooper and Tempalski, 2014), there is insufficient attention to health as an emergent dynamic of materiality in local practices (Duff, 2013; Dennis, 2020). How environments *materialize* drugs and drug harms in their local situations of interaction thus becomes a key focus of research that seeks to encourage a "more dynamic" and "more material" understanding of the risk environment (Rhodes *et al.*, 2023). One such illustration, as we have considered here, is the walking interview. The walking interview offers a socio-material approach to understanding the risk environment, a means of "becoming-with" the environment that helps, at once, to appreciate its material effects.

The spontaneously assembled path we took with Eric can be seen as a form of a living environment that is constantly made up in multiple ontologies of movement and therefore always on the move. Eric's walk traces the small and the large, human and nonhuman, social and material. These inter-scalar twists and turns emphasize that the risk environment is made up of elements that are not rigid or separate but interconnected and in flux. The environment in which Eric navigated was not confined to a single aspect but was instead a composite of various elements that interacted to shape the assemblage of his journey.

Indeterminacy is embodied in Eric's walk. As such, the walking interview under-scores how health is marked by uncertainties and unexpected shifts. The adaptations Eric made along the way were shaped ecologically and guided iteratively in the interplay between his felt needs and affects and the evolving environment he encountered. Eric's journey thus not only illustrates but also materializes the fluid and contingent nature of care and health, emphasizing the value of situated, adapt-able approaches that engage and "become-with" the environment over those that stand back, from afar, to generate more detached accounts.

Materializing the Social

Incorporating the go-along walk with ethnographic observation has served as a mode for telling different kinds of stories. By bringing a relational understanding of both care and place, we have been able to not only observe people's material envi-ronments but also study people's processing and navigation of them (Carpiano, 2009)—in the moment. This approach creates a space for embodied and marginal-ized experiences[4] to be expressed in ways that are not so easily spoken. As Eric's walk illustrates, such stories may have the potential to open up new pathways for combating stigma because they reveal how structural violence is located, often out of sight, in the mundane and every day.

A relational conception of space, following Massey (2005), has afforded an understanding of space as a "multiplicity of stories-so-far" (ibid, p. 100). Eric's health, as we have seen, is in the process of becoming in relation to place and space, which is afforded by the moments of the walking interview method. By embarking on walks that immerse us in affected spaces, landscapes of "slow violence" and "ontologies of movement" are materialized. While the method enables storytelling, it also serves to enact the material (Law and Urry, 2004). Circling back to Eric's walk, we see him shift his analysis from one that fixates on the individual-level intrinsic motivation of service to specific spatial forms (walls) and functions (hours) of the community-level service site that disabled care. This analytical move gives us a glimpse into how walks have the power to enact new imaginaries of place and space that contribute to the political project of harm reduction.

Acknowledgments We gratefully acknowledge interlocutors who agreed to take PKF on a walk, as well as service users and providers who have participated in the wider study. We thank PKF's supervisors, Simon Cohn and Magdalena Harris, for their support and guidance alongside TR. We also thank Sara Alavi, who provided a thoughtful review and editorial comments. Finally, our thanks to the editors of this book, Imelda Moise and Marynia A. Kolak.

---

[4] Given how the drug-using body is often treated as an object to dissect and study (Shilling, 2012), walking purposefully places the embodied knowledge of such stigmatized bodies at the fore for analysis.

References

Aldridge, R. W., Story, A., Hwang, S. W., Nordentoft, M., Luchenski, S. A., Hartwell, G., Tweed, E. J., Lewer, D., Vittal Katikireddi, S., & Hayward, A. C. (2018). Morbidity and mortality in homeless individuals, prisoners, sex workers, and individuals with substance use disorders in high-income countries: A systematic review and meta-analysis. *The Lancet, 391*(10117), 241–250. https://doi.org/10.1016/S0140-6736(17)31869-X

Anderson, J. (2004). Talking whilst walking: A geographical archaeology of knowledge. *Area, 36*(3), 254–261. https://doi.org/10.1111/j.0004-0894.2004.00222.x

Anderson, B. (2009). Affective atmospheres. *Emotion, Space and Society, 2*(2), 77–81. https://doi.org/10.1016/j.emospa.2009.08.005

Andrews, G. J., & Duff, C. (2019). Matter beginning to matter: On posthumanist understandings of the vital emergence of health. *Social Science & Medicine, 226*, 123–134. https://doi.org/10.1016/j.socscimed.2019.02.045

Andrews, G. J., Hall, E., Evans, B., & Colls, R. (2012). Moving beyond walkability: On the potential of health geography. *Social Science & Medicine, 75*(11), 1925–1932. https://doi.org/10.1016/j.socscimed.2012.08.013

Augustin, J., Andrees, V., Walsh, D., Reintjes, R., & Koller, D. (2023). Spatial aspects of health—Developing a conceptual framework. *International Journal of Environmental Research and Public Health, 20*(3), 1817. https://doi.org/10.3390/ijerph20031817

Bissell, D. (2007). Animating suspension: Waiting for mobilities. *Mobilities, 2*(2), 277–298. Available at: https://doi.org/10.1080/17450100701381581

Bissell, D. (2008). Comfortable bodies: Sedentary affects. *Environment and Planning A: Economy and Space, 40*(7), 1697–1712. Available at: https://doi.org/10.1068/a39380

Bissell, D. (2010). Narrating mobile methodologies: Active and passive empiricisms. In B. Fincham, M. McGuinness, & L. Murray (Eds.), *Mobile Methodologies*. London: Palgrave Macmillan UK, pp. 53–68. Available at: https://doi.org/10.1057/9780230281172_5

Bissell, D., & Fuller, G. (Eds.). (2013). *Stillness in a mobile world*. First issued in paperback. London New York: Routledge (International library of sociology).

Bøhling, F. (2014). Crowded contexts: On the affective dynamics of alcohol and other drug use in nightlife spaces. *Contemporary Drug Problems, 41*(3), 361–392. https://doi.org/10.1177/009145091404100305

Bourdieu, P. (1977). *Outline of a theory of practice* 1st edn. Translated by R. Nice. Cambridge University Press. Available at. https://doi.org/10.1017/CBO9780511812507

Bond, L., Wusinich, C., & Padgett, D. (2022). Weighing the options: Service user perspectives on homeless outreach services. *Qualitative Social Work, 21*(1), 177–193. https://doi.org/10.1177/1473325021990861

Bourgois, P. (1998). Just another night in a shooting gallery. *Theory, Culture & Society, 15*(2), 37–66. https://doi.org/10.1177/026327698015002002

Bourgois, P. (2003). Crack and the political economy of social suffering. *Addiction Research & Theory, 11*(1), 31–37. https://doi.org/10.1080/1606635021000021322

Büscher, M., Urry, J., & Witchger, K. (Eds.). (2010). *Mobile methods* 0 edn. Routledge. Available at: https://doi.org/10.4324/9780203879900

Carpiano, R. M. (2009). Come take a walk with me: The "go-along" interview as a novel method for studying the implications of place for health and Well-being. *Health & Place, 15*(1), 263–272. https://doi.org/10.1016/j.healthplace.2008.05.003

Chang, J. S. (2017). The docent method: A grounded theory approach for researching place and health. *Qualitative Health Research, 27*(4), 609–619. Available at: https://doi.org/10.1177/1049732316667055

Collins, A. B., Boyd, J., Cooper, H. L. F., & McNeil, R. (2019). The intersectional risk environment of people who use drugs. *Social Science & Medicine, 234*(1982), 234., 112384. https://doi.org/10.1016/j.socscimed.2019.112384

Collins, A. B., Edwards, S., McNeil, R., Goldman, J., Hallowell, B. D., Scagos, R. P., & Marshall, B. D. L. (2022). A rapid ethnographic study of risk negotiation during the COVID-19 pandemic among unstably housed people who use drugs in Rhode Island. *The International Journal on Drug Policy, 103*, 103626. https://doi.org/10.1016/j.drugpo.2022.103626

Connolly, W. E. (2004). 15 Method, problem, faith. In *Problems and methods in the study of politics* (p. 332).

Cooper, H. L., Bossak, B., Tempalski, B., Des Jarlais, D. C., & Friedman, S. R. (2009). Geographic approaches to quantifying the risk environment: A focus on syringe exchange program site access and drug-related law enforcement activities. *The International Journal on Drug Policy, 20*(3), 217–226. https://doi.org/10.1016/j.drugpo.2008.08.008

Cooper, H. L. F., & Tempalski, B. (2014). Integrating place into research on drug use, drug users' health, and drug policy. *The International Journal on Drug Policy, 25*(3), 503–507. https://doi.org/10.1016/j.drugpo.2014.03.004

Cummins, S., Curtis, S., Diez-Roux, A. V., & Macintyre, S. (2007). Understanding and representing 'place' in health research: A relational approach. *Social Science & Medicine, 65*(9), 1825–1838. https://doi.org/10.1016/j.socscimed.2007.05.036

Dennis, F. (2020). Mapping the drugged body: Telling different kinds of drug-using stories. *Body & Society, 26*(3), 61–93. https://doi.org/10.1177/1357034X20925530

Dennis, F., Rhodes, T., & Harris, M. (2020). More-than-harm reduction: Engaging with alternative ontologies of 'movement' in UK drug services. *International Journal of Drug Policy, 82*, 102771. https://doi.org/10.1016/j.drugpo.2020.102771

Diez Roux, A. V. (2022). Social epidemiology: Past, present, and future. *Annual Review of Public Health, 43*(1), 79–98. https://doi.org/10.1146/annurev-publhealth-060220-042648

Driessen, H., & Jansen, W. (2013). The hard work of small talk in ethnographic fieldwork. *Journal of Anthropological Research, 69*(2), 249–263. https://doi.org/10.3998/jar.0521004.0069.205

Duff, C. (2007). Towards a theory of drug use contexts: Space, embodiment and practice. *Addiction Research & Theory, 15*(5), 503–519. https://doi.org/10.1080/16066350601165448

Duff, C. (2013). The social life of drugs. *The International Journal on Drug Policy, 24*(3), 167–172. https://doi.org/10.1016/j.drugpo.2012.12.009

Duff, C. (2014). The place and time of drugs. *International Journal of Drug Policy, 25*(3), 633–639. https://doi.org/10.1016/j.drugpo.2013.10.014

Duff, C. (2016). Atmospheres of recovery: Assemblages of health. *Environment and Planning A: Economy and Space, 48*(1), 58–74. https://doi.org/10.1177/0308518X15603222

Duncan, T., Duff, C., Sebar, B., & Lee, J. (2017). "Enjoying the kick": Locating pleasure within the drug consumption room. *The International Journal on Drug Policy, 49*, 92–101. https://doi.org/10.1016/j.drugpo.2017.07.005

Edensor, T. (2010). Walking in rhythms: Place, regulation, style and the flow of experience. *Visual Studies, 25*, 69. https://doi.org/10.1080/14725861003606902

Evans, J., & Jones, P. (2011). The walking interview: Methodology, mobility and place. *Applied Geography, 31*(2), 849–858. Available at: https://doi.org/10.1016/j.apgeog.2010.09.005

Fast, D., & Cunningham, D. (2018). "We don't belong there": New geographies of homelessness, addiction, and social control in vancouver's inner city. *City & Society, 30*(2), 237–262. https://doi.org/10.1111/ciso.12177

Francis-Devine, B., Barton, C., Harari, D., Keep, M., Bolton, P., & Harker, R. (2022). Rising cost of living in the UK. Retrieved from https://commonslibrary.parliament.uk/research-briefings/cbp-9428/.

Fraser, S. (2006). The chronotope of the queue: Methadone maintenance treatment and the production of time, space and subjects. *International Journal of Drug Policy, 17*(3), 192–202. https://doi.org/10.1016/j.drugpo.2006.02.010

Fraser, S. (2020). Doing ontopolitically-oriented research: Synthesising concepts from the ontological turn for alcohol and other drug research and other social sciences. *International Journal of Drug Policy, 82*, 102610. https://doi.org/10.1016/j.drugpo.2019.102610

Fraser, S., & Moore, D. (Eds.). (2011). *The drug effect: Health, crime and society*. Cambridge: Cambridge University Press. Available at: https://doi.org/10.1017/CBO9781139162142

Galea, S. (2022). Moving beyond the social determinants of health. *International Journal of Health Services, 52*(4), 423–427. https://doi.org/10.1177/00207314221119425

Galea, S., Hall, C., & Kaplan, G. A. (2009). Social epidemiology and complex system dynamic modelling as applied to health behaviour and drug use research. *International Journal of Drug Policy, 20*(3), 209–216. https://doi.org/10.1016/j.drugpo.2008.08.005

Giddens, A. (2003). *Modernity and self-identity: Self and society in the late modern age*. Stanford, Calif: Stanford Univ. Press.

Guise, A., Burrows, M., & Marshall, A. (2022). A participatory evaluation of legal support in the context of health-focused peer advocacy with people who are homeless in London, UK. *Health & Social Care in the Community, 30*(6), e6622–e6630. https://doi.org/10.1111/hsc.14111

Harris, M., Scott, J., Hope, V., Wright, T., McGowan, C., & Ciccarone, D. (2020). Navigating environmental constraints to injection preparation: The use of saliva and other alternatives to sterile water among unstably housed PWID in London. *Harm Reduction Journal, 17*(1), 24. https://doi.org/10.1186/s12954-020-00369-0

Health and Social Care Act. 2012. Health and Social Care Act. Chapter 7. Retrieved from http://www.legislation.gov.uk/ukpga/2012/7/contents/enacted.

Keane, H. (2002). *What's wrong with addiction?* Melbourne University Press.

King, A. C., Winter, S. J., Chrisinger, B. W., Hua, J., & Banchoff, A. W. (2019). Maximizing the promise of citizen science to advance health and prevent disease. *Preventive Medicine, 119*, 44–47. https://doi.org/10.1016/j.ypmed.2018.12.016

Krieger, N. (2001). Theories for social epidemiology in the 21st century: An ecosocial perspective. *International Journal of Epidemiology, 30*(4), 668–677. https://doi.org/10.1093/ije/30.4.668

Kusenbach, M. (2003). Street phenomenology: The go-along as ethnographic research tool. *Ethnography, 4*(3), 455–485. https://doi.org/10.1177/146613810343007

Law, J. (2004). *After method: Mess in social science research*. Routledge.

Law, J., & Urry, J. (2004). Enacting the social. *Economy and Society, 33*(3), 390–410. https://doi.org/10.1080/0308514042000225716

Lorimer, H. (2011). Walking: New forms and spaces for studies of pedestrianism. In *Geographies of mobilities: Practices, spaces, subjects*. Routledge.

Macintyre, S., Ellaway, A., & Cummins, S. (2002). Place effects on health: How can we conceptualise, operationalise and measure them? *Social Science & Medicine, 55*(1), 125–139. https://doi.org/10.1016/S0277-9536(01)00214-3

Malins, P. (2017). Desiring assemblages: A case for desire over pleasure in critical drug studies. *International Journal of Drug Policy, 49*, 126–132. https://doi.org/10.1016/j.drugpo.2017.07.018

Massey, D. B. (2005). *For space*. SAGE.

Mayblin, L., Wake, M., & Kazemi, M. (2020). Necropolitics and the slow violence of the everyday: Asylum seeker welfare in the postcolonial present. *Sociology, 54*(1), 107–123. https://doi.org/10.1177/0038038519862124

Merriman, P. (2014). Rethinking Mobile methods. *Mobilities, 9*(2), 167–187. Available at: https://doi.org/10.1080/17450101.2013.784540

Miaux, S., et al. (2010). Making the narrative walk-in-real-time methodology relevant for public health intervention: Towards an integrative approach. *Health & Place, 16*(6), 1166–1173. Available at: https://doi.org/10.1016/j.healthplace.2010.08.002

Michael, M. (2021). *The research event: Towards prospective methodologies in sociology*. London: Routledge. Available at: https://doi.org/10.4324/9781351133555.

Mol, A., Moser, I., & Pols, J. (2010). *Care in practice: On tinkering in clinics, homes and farms*. Transcript.

Müller-Mahn, H.-D. (Ed.). (2013). *The spatial dimension of risk: How geography shapes the emergence of riskscapes*. New York.

Müller-Mahn, D., Everts, J., & Stephan, C. (2018). Riskscapes revisited—Exploring the relationship between risk, space and practice. *Erdkunde, 72*(3), 197–213. https://doi.org/10.3112/erdkunde.2018.02.09

Neely, A. H., & Nading, A. M. (2017). Global health from the outside: The promise of place-based research. *Health & Place, 45*, 55–63. https://doi.org/10.1016/j.healthplace.2017.03.001

Nixon, R. (2013). *Slow violence and the environmentalism of the poor*. (First Harvard University Press paperback edition). Harvard University Press.

O'Campo, P., & Dunn, J. R. (Eds.). (2012). *Rethinking social epidemiology: Towards a science of change*. Springer.

Parent, L. (2016). The wheeling interview: Mobile methods and disability. *Mobilities, 11*(4), 521–532. Available at: https://doi.org/10.1080/17450101.2016.1211820

Pedersen, M., Wood, G. E. R., Fernes, P. K., Goldman Rosas, L., Banchoff, A., & King, A. C. (2022). The "our voice" method: Participatory action citizen science research to advance behavioral health and health equity outcomes. *International Journal of Environmental Research and Public Health, 19*(22), Article 22. https://doi.org/10.3390/ijerph192214773

Porkertová, H., et al. (2024). "Wait, really, stop, stop!": Go-along interviews with visually disabled people and the pitfalls of ableist methodologies. *Qualitative Research, 24*(5), 1230–1252. Available at: https://doi.org/10.1177/14687941231224595

Radley, A., Chamberlain, K., Hodgetts, D., Stolte, O., & Groot, S. (2010). From means to occasion: Walking in the life of homeless people. *Visual Studies, 25*(1), 36–45. https://doi.org/10.1080/14725861003606845

Rhodes, T. (2002). The 'risk environment': A framework for understanding and reducing drug-related harm. *International Journal of Drug Policy, 13*(2), 85–94. https://doi.org/10.1016/S0955-3959(02)00007-5

Rhodes, T. (2009). Risk environments and drug harms: A social science for harm reduction approach. *The International Journal on Drug Policy, 20*(3), 193–201. https://doi.org/10.1016/j.drugpo.2008.10.003

Rhodes, T., Egede, S., Grenfell, P., Paparini, S., & Duff, C. (2019). The social life of HIV care: On the making of 'care beyond the virus'. *BioSocieties, 14*(3), 321–344. https://doi.org/10.1057/s41292-018-0129-9

Rhodes, T., Harris, M., Sanín, F. G., & Lancaster, K. (2021). Ecologies of drug war and more-than-human health: The case of a chemical at war with a plant. *International Journal of Drug Policy, 89*, 103067. https://doi.org/10.1016/j.drugpo.2020.103067

Rhodes, T., & Lancaster, K. (2019). Evidence-making interventions in health: A conceptual framing. *Social Science & Medicine, 238*, 112488. https://doi.org/10.1016/j.socscimed.2019.112488

Rhodes, T., Ordoñez, L. S., Acero, C., Harris, M., Holland, A., & Sanín, F. G. (2023). Caring for coca, living with chemicals: Towards ecological harm reduction. *The International Journal on Drug Policy, 120*, 104179. https://doi.org/10.1016/j.drugpo.2023.104179

Saunders, N., & Al-Om, T. (2022). Slow resistance: Resisting the slow violence of asylum. *Millennium, 50*(2), 524–547. https://doi.org/10.1177/03058298211066339

Sheller, M., & Urry, J. (2006). The new Mobilities paradigm. *Environment and Planning A: Economy and Space, 38*(2), 207–226. https://doi.org/10.1068/a37268

Shilling, C. (2012). *The body and social theory* (3rd ed.). Sage.

Singer, M. (1994). Aids and the health crisis of the U.S. urban poor; the perspective of critical medical anthropology. *Social Science & Medicine, 39*(7), 931–948. Available at: https://doi.org/10.1016/0277-9536(94)90205-4

Singer, M., Stopka, T., Siano, C., Springer, K., Barton, G., Khoshnood, K., Gorry de Puga, A., & Heimer, R. (2000). The social geography of AIDS and hepatitis risk: Qualitative approaches for assessing local differences in sterile-syringe access among injection drug users. *American Journal of Public Health, 90*(7), 1049–1056.

Speer, J. & Goldfischer, E. (2020) 'The city is not innocent: Homelessness and the value of urban parks', *Capitalism Nature Socialism*, 31(3), pp. 24–41. Available at: https://doi.org/10.1080/10455752.2019.1640756

Spinney, J. (2015). Close encounters? Mobile methods, (post) phenomenology and affect. *Cultural Geographies, 22*(2), 231–246. Available at: https://doi.org/10.1177/1474474014558988

Springgay, S., & Truman, S. E. (2018). *Walking methodologies in a more-than-human world: Walking lab.* Milton Park, Abingdon, Oxon ; New York, NY: Routledge. (Routledge advances in research methods).

Stambe, R., Kuskoff, E., Parsell, C., Plage, S., Ablaza, C., & Perales, F. (2023). Seeing, sharing and supporting: Assertive outreach as a partial solution to rough sleeping. *The British Journal of Social Work*, bcad251. https://doi.org/10.1093/bjsw/bcad251

Stokes, J., Bower, P., Guthrie, B., Mercer, S. W., Rice, N., Ryan, A. M., & Sutton, M. (2022). Cuts to local government spending, multimorbidity and health-related quality of life: A longitudinal ecological study in England. *The Lancet Regional Health—Europe, 19*, 100436. https://doi.org/10.1016/j.lanepe.2022.100436

Strathdee, S. A., et al. (2010). HIV and risk environment for injecting drug users: The past, present, and future. *The Lancet, 376*(9737), 268–284. Available at: https://doi.org/10.1016/S0140-6736(10)60743-X

Tempalski, B., & McQuie, H. (2009). Drugscapes and the role of place and space in injection drug use-related HIV risk environments. *International Journal of Drug Policy, 20*(1), 4–13. https://doi.org/10.1016/j.drugpo.2008.02.002

Thomas, Y., Richardson, D. and Cheung, I. (2008). 'Integrating Geography and Social Epidemiology in Drug Abuse Research', in Y.F. Thomas, D. Richardson, and I. Cheung (eds) Geography and Drug Addiction. Dordrecht: Springer Netherlands, pp. 17–26. Available at: https://doi.org/10.1007/978-1-4020-8509-3_2

Thrift, N. J. (1999). Steps to an ecology of place. In D. Massey, J. Allen, & P. Sarre (Eds.), *Human geography today* (pp. 295–323). Polity Press.

Thrift, N. J. (2007). *Non-representational theory: Space, politics, affect.* Routledge.

Urry, J. (2012). *Mobilities.* (Reprint). Polity Press.

Vergunst, J. L., & Ingold, T. (Eds.). (2016). *Ways of walking: Ethnography and practice on foot.* (First issued in paperback). Routledge, Taylor & Francis Group.

Warren, S. (2017). Pluralising the walking interview: Researching (im)mobilities with Muslim women. *Social & Cultural Geography, 18*(6), 786–807. Available at: https://doi.org/10.1080/14649365.2016.1228113

Yates-Doerr, E. (2020). Reworking the social determinants of health: Responding to material-semiotic indeterminacy in public health interventions. *Medical Anthropology Quarterly, 34*(3), 378–397. https://doi.org/10.1111/maq.12586

Part V
Empirical Illustrations of Social-Spatial Determinants of Health

Chapter 17
Applications of GIS to Spatial Patterns of Disease and Health

Joseph R. Oppong and Katherine A. Lester

Introduction

GIS is a powerful tool for geographic analysis of disease and health. It allows us to unleash the power of geographic analysis to interrogate the complex relationships between humans (their biological makeup, behavior, and practices that hinder or facilitate disease spread and exposure), their environment (social, physical, and political), and disease-causing pathogens to produce geographic patterns of sickness and death. GIS has long been a favored tool for exploring the contributing factors of death and disease, including an array of social-spatial determinants of health (SDoH). In fact, it hardly seems possible that anyone could use SDoH to effectively target communities without GIS.

While the expanding use of GIS in health and disease is exciting, considerable caution is necessary to avoid erroneous and misleading conclusions. This paper examines the potential and promise of GIS for the spatial analysis of health and disease. It also highlights some common methodological challenges of GIS, including spatial autocorrelation, spatial scale, and small data. We argue that to realize the full potential of GIS, robust surveillance data on the geography of environmental conditions, disease agents, and health outcomes over time are essential. Moreover, the role of power and vulnerability in explaining the spatial patterns of health and disease must become more prominent.

J. R. Oppong (✉)
Department of Geography and The Environment, University of North Texas,
Denton, TX, USA
e-mail: oppong@unt.edu

K. A. Lester
Spatial Sciences Institute, University of Southern California, Los Angeles, CA, USA

© The Author(s) 2026
M. A. Kolak, I. K. Moise (eds.), *Place and the Social-Spatial Determinants of Health*, Global Perspectives on Health Geography,
https://doi.org/10.1007/978-3-031-88463-4_17

The Promise of GIS and Health

Diseases vary across geographic space. Medical geography seeks to explain these geographic patterns—who gets what diseases where and why? This requires a thorough examination of the complex spatial variations between places, including the physical environment—temperature, vegetation, fauna, built environment, and the hydrosphere—as well as the human response to these. Human behavior may facilitate or hinder exposure, sickness, and death from a disease, but the place of residence influences and constrains human behavior. For example, when the absence of neighborhood walkways and green spaces limits opportunities for daily exercise, the obesity and related diseases of the residents must be seen in their broader context—the lived environment, not just in the behavior of the occupants (King, 2010). Therefore, the lived environment may shape the behavior of occupants by limiting access to resources that facilitate regular exercise. The same could be said for education, economic stability, healthcare quality, social context, and other aspects of the built environment.

Location is the most effective way to organize and analyze these disparate sources of data. GIS allows us to assemble, organize, and query thousands of pieces of locational data to better target interventions and control disease outbreaks. For example, the WHO effort to control onchocerciasis, or river blindness, employs GIS. Mapping the geographic patterns of onchocerciasis now allows the WHO ONCHO program to target the 217.5 million people who live in these endemic areas (WHO, 2019) with mass drug administration of ivermectin for treatment. Additionally, it reveals the best areas to target pesticides to eliminate Simulium flies, the main vector of onchocerciasis.

Moonan et al. (2004, 2006) present another fascinating example of how targeting interventions using GIS can facilitate disease control. Despite being preventable, treatable, and curable, tuberculosis (TB) has emerged as the leading killer of infectious disease, surpassing COVID in 2022. Although mostly controlled in high-income countries, low-income countries continue to struggle with high TB disease burdens, low vaccination rates, and poor treatment outcomes. In contrast, in high-income countries where low TB rates are the norm, the challenge is to detect the hidden pockets of TB infection. In such areas, the recommended intervention strategy is targeted at testing and treatment of people most at risk and stopping further spread. Because such people are usually found in pockets or small populations, no clear path to implementing such a strategy was available. However, by linking GIS technology with molecular surveillance data, the authors identified geographical areas of ongoing tuberculosis transmission. The research used data collected on people newly diagnosed with culture-positive tuberculosis at the Tarrant County Health Department (TCHD) between January 1, 1993, and December 31, 2000. Clinical isolates were molecularly characterized using IS6110-based RFLP analysis and spoligotyping methods. Patient residential addresses at the time of diagnosis were geocoded and mapped according to strain characterization (Moonan et al., 2004).

Evaluating the spatial distribution of cases within zip-code boundaries revealed distinct geographical clusters of the same strain of disease. Concluding that these geographical areas had an increased likelihood of ongoing transmission, they became the target of geographically based screening and treatment programs. These enhanced targeted screening and control efforts improved case discovery and led to the interruption of disease transmission and incidence reduction. Additionally, expanded surveillance in geographical areas of increased incidence with a high percentage of unique strains helped to locate cases in hard-to-reach foreign-born populations with treatment, interrupting further transmission. The study described above demonstrates that using existing health data, GIS can identify previously undetected TB transmission. These results were used to design new targeted screening efforts (Moonan et al., 2006).

GIS is also a critical tool for examining potential associations between environmental exposure and negative health outcomes. Boakye et al. (2022) provide a fascinating example of using GIS to interrogate the association between on-road air pollution and adverse health. The study uses census tract-level cancer and noncancer risk estimates from the National-Scale Air Toxics Assessment (NATA) and sociodemographic variables from the US Census Bureau. GIS analysis indicated that census tracts with the highest cancer and noncancer risks clustered in the major urban areas occupied by high percentages of Black, Indigenous, and People of Color (BIPOC). Similarly, cancer and noncancer risks from on-road air pollution were high in nonmetropolitan census tracts occupied by minority populations. The results suggest that densely concentrated minority ethnic groups are more likely to experience toxins from on-road air pollutants. Thus, census tracts with a high percentage of minority ethnic groups (African American, Hispanic, Asian) and a high proportion of people with low (socioeconomic status) SES had higher cancer and noncancer risks from on-road pollution. Their findings are consistent with those of other studies (Chakraborty et al., 2017; Garcia, 2018). Such studies provide insights that can guide interventions for addressing the socio-spatial disparities in air pollution exposure.

Despite the benefits of such studies, it is important to recognize one vital limitation—humans are continually moving within their environments, thus changing their exposure patterns to disease-causing pathogens and environmental pollutants. Consequently, the spatial and temporal patterns of disease and their causal relationships are not simple, static, or fixed, but complex, dynamic, multifactorial, and multidirectional. As Kwan (2012) argues, geographies of health are far too complex to explain with any single perspective or group of factors (e.g., individual attributes, environmental features, social relations, institutional processes, and cultural systems). Physical, social, environmental, and biological factors must be considered facets of the whole. Accordingly, place-based analysis of health behaviors and outcomes must address numerous processes and contexts that interact in an overly complex manner (King, 2010). This is vital because human exposure to disease pathogens occurs in disparate places and environments and at various times during our lifetimes, not simply in our place of usual residence or place of work. We need to move beyond the traditional focus on static locations or places, such as residential addresses, and embrace the role of human mobility and movement on health (Kwan, 2012). Mobility is fundamental to spatiotemporal experiences, and these complex experiences, exposures, and risks transcend where people live.

Figure 17.1 provides a sampling of relevant methods in GIS. Applications range from relatively simple, such as choropleth mapping, to complex (i.e., space-time

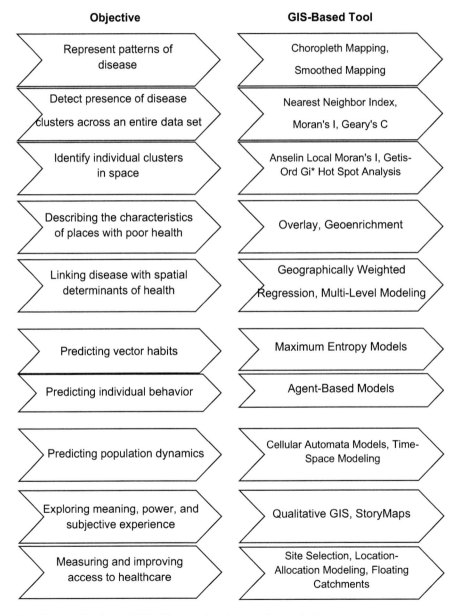

Fig. 17.1 Applications of GIS with examples of appropriate methods

modeling, etc.). Spatial science permeates all areas of health science, including cluster detection, vector habitat modeling, epidemic spread, and access to health-care. Additionally, qualitative GIS is a burgeoning field which can incorporate lived, subjective experience within the traditionally hyper-quantitative family of GIS tools (Taylor et al., 2020). In this section, we have introduced the promise and potential of GIS applications across the health domain. While GIS is uniquely suited to addressing these complexities, much care is necessary to address its inherent limita-tions and avoid misleading conclusions. The next section details some of these potential pitfalls.

Methodological Limitations of GIS

The ability of GIS to handle complex data structures, statistical analysis, and model-ing also comes with a high risk of misapplication, misinterpretation, and propagated error. Figure 17.2 shows several of the most pressing challenges facing GIS analysis today, including data availability, rate instability, non-stationarity, and autocorrela-tion. This section details these challenges and the potential for erroneous results and conclusions as well as methods to address them.

Data Availability and Quality

One of the most basic challenges to health analysis using GIS is the availability and reliability of spatial data. Creating primary datasets can be costly and fraught with privacy challenges. Therefore, most geographers rely on secondary datasets avail-able from disease surveillance programs and previously published research. Hence, researchers rarely have complete control over the geographic scale of their studies and are frequently limited to counties, states, and zip codes, which may be inap-propriate for specific applications. Furthermore, this data may be subject to collec-tion or sampling issues that are undisclosed in the metadata. Matching the scale of analysis to the questions being asked needs special attention. As described below, patterns at one scale may not be visible at another, and rate stability standards may vary.

Available data often oversimplifies the complex reality of morbidity and expo-sure. For instance, comorbidity data are rarely available. Also, from a spatial per-spective, issues can arise from assigning one location to a health outcome. Mei-Po Kwan (2018) challenges this neighborhood effect averaging problem by arguing that most people do not spend all their time in one neighborhood. They occupy resi-dential, professional, commercial, and recreational spaces. When researchers assign an individual to only their residential neighborhood, they have an elevated risk of missing relationships among all these other spaces. Researchers must be mindful of these issues in their analysis and interpretations.

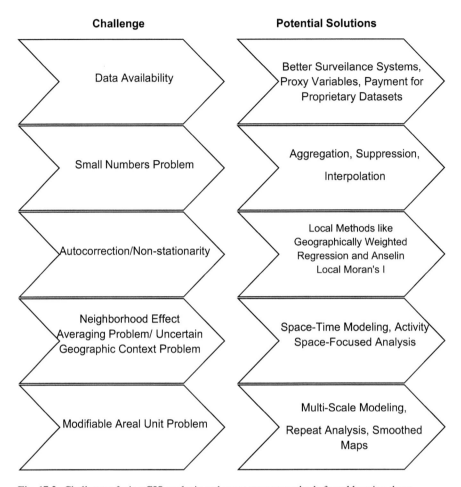

Fig. 17.2 Challenges facing GIS analysis and some current methods for addressing them

Autocorrelation

Another unique challenge of spatial data is autocorrelation. This relates to the first law of geography coined by Waldo Tobler (1970)—"everything is related to everything else, but near things are more related than distant things" (p. 236). In some ways, this rule is intuitive; we would expect the conditions in Los Angeles County to be more like neighboring Orange County than the conditions in New York City. However, this creates a statistical issue—we cannot automatically consider spatial samples as independent. Therefore, to use nonspatial statistical methods on spatial data, researchers should evaluate the data for autocorrelation using a tool such as Moran's I. If no autocorrelation is detected, nonspatial methods are adequate, although unit-level results such as regression residuals should be double checked for autocorrelation as well. However, most spatial data will contain some autocorrelation, and spatial statistics such as geographically weighted regression are typically more appropriate. Analysis of spatial data that ignores spatial autocorrelation is bound to produce erroneous results and conclusions.

Small Numbers

When choosing data, health geographers must balance the desire for detailed analysis with rate stability. The small numbers problem (SNP) refers to instability in the data caused by small observations in spatial units (Waller & Gotway, 2004). For instance, consider the rate calculations in Table 17.1 from three Texas counties. In a large county such as Harris County (containing the City of Houston), an increase of one death in the numerator changes the rate little. However, as the number of cases and the size of the population decrease, one additional death increases the rate by a substantial number. Thus, in Nacogdoches, one additional suicide increases the rate by 7.71%, while in tiny Carson County, one additional death increases the county rate by 50%.

This effect raises additional concerns. Rate instability in smaller counties with few deaths makes comparisons among them meaningless and trends over time impossible to capture. Is there anything useful to learn from a county where one person's suicide impacts the rate so greatly? Most likely, not. What does it mean for our models if one person's death is inconsequential on one side of the county line but inflates the county rate on the other?

There are multiple ways to work around the SNP, but they require clear and justified decisions from the researcher. First, researchers must determine the appropriate threshold at which they consider the numbers sufficiently stable. This is no easy task. For example, the Centers for Disease Control and Prevention classifies rates into three categories: suppressed, unreliable, and reliable. Reliable rates are calculated with at least 20 cases, unreliable rates are calculated with 10–19 cases, and less than 10 cases are suppressed. For some health outcomes, this classification may be adequate, but even in twenty cases, one additional case increases the rate by 5%. Other government statistical agencies use different thresholds. For example, the

Table 17.1 Comparing suicide in three Texas counties, 2021

County	Harris	Nacogdoches	Carson
Population	4,728,030	64,668	5746
Suicide cases	566	13	2
Crude rate	11.97 per 100,000	20.10 per 100,000	34.81 per 100,000
Crude rate, with one additional death	11.99 per 100,000	21.65 per 100,000	52.21 per 100,000
Percent change	0.17%	7.71%	50.00%
CDC rate classification	Reliable	Unreliable	Suppressed

National Household Survey on Drug Abuse only considers a stable rate if it contains at least 25 cases (Klein et al., 2002). Rate stability may also be determined based on population instead of incidence/mortality counts.

Once a researcher has settled on a suitable threshold, they must choose a strategy for addressing the SNP. The most common solution is to simply drop the smaller, unstable units from the analysis. However, this strategy is rarely unbiased, and rural areas are likely to become underrepresented. A second way to address this is by increasing the spatial scale, since larger units may have more cases and thus more stable rates. This can be done by choosing a larger unit of spatial support (such as states instead of counties) or through selective aggregation using protocols such as Sun and Wong (2017), which combine smaller units while leaving larger units intact. Finally, researchers can choose to expand the time scale. For example, in Table 17.1, Nacogdoches only has 13 cases of suicide in 2021, but when the dates are expanded to 2018–2021, Nacogdoches has 57 cases. Fifty-seven cases are adequate to assume a stable rate for comparison purposes. In any case, failing to accommodate the SNP is likely to lead to unstable trends and misleading interpretations and conclusions.

Modifiable Areal Unit Problem

The SNP requires researchers to think critically about spatial support and time periods. However, this introduces a new caveat. The modifiable areal unit problem (MAUP) refers to the issues that arise from changing definitions of spatial support. The MAUP can be further broken down into the scale problem and the zoning problem (Ye & Rogerson, 2022). The scale problem references the granularity of the spatial units under study. Individual-level data are rarely available; most geographic studies utilize aggregated data. However, these units may reveal different patterns at different scales.

To address and explore the MAUP, geographers commonly repeat their analysis at multiple scales. Ideally, this type of research has individual data to compare to the aggregated units. One study in Scotland, Lee et al. (2020), investigates the relationship between air pollution and respiratory and circulatory diseases. They found that at a finely aggregated scale, the associations between air pollution and disease only deviated from the effect size found in individuals by approximately 10%. However, as aggregation increased, accuracy decreased. In larger units (analogous to US counties), the relationship diminished to 80% of the initial effect size. One of the biggest problems with the scale effect is clear in air pollution studies. Within a spatial unit, not everyone is exposed the same way, especially when pollution is from point sources.

Rare disease outcomes often suffer from both the SNP and the MAUP. Because the case numbers are so low, aggregation of large areas may be necessary to derive stable disease rates. Sánchez-Díaz et al. (2020) designed a study to examine

Huntington's disease in Spain at three administrative levels: province, district, and municipality. Their data showed that the province level was too large and revealed extraordinarily little in terms of patterns. Municipalities had too few cases to justify confidence in the stability of their rates. In this situation, they decided on the district scale as the most stable with the highest resolution. Unfortunately, this only solves the issue for one disease outcome in one geographic context. There can be no concrete guidelines mandating which scale is best; every situation is unique.

Examples of the scale problem abound in every corner of medical geography. Araujo Navas et al. (2020) investigated the impact of the MAUP on the distribution of Schistosoma japonicum, the vector for schistosomiasis. The authors created an assortment of buffer scales from 30 meters to one kilometer. They found that once the study area reaches one kilometer, significant associations are lost, and analysis results become useless, or in some cases, dangerously misleading. This study highlights the profound consequences of mis-specifying the unit of study. Readers unfamiliar with the MAUP are likely to accept the results of analysis at any scale as true, even though it may just be noise.

The consequences of misspecification are a concern for any scientific results, but they may be dangerous when applied to humans. Javanmard et al. (2023) examined public transit reliability and equity in Winnipeg. Assessments were conducted at the level of individual bus stops, neighborhoods, and routes. The route-level analysis shows equity between minority and nonminority neighborhoods. However, the individual stop and neighborhood analyses show significant social inequities in minority neighborhoods. The authors express concern that political agencies may misinterpret results or cherry-pick the scale they choose to report. Acting on the results from the individual stop analysis may increase transportation parity between neighborhoods, while reporting the route-level results could communicate that there are no problems.

The scale problem is complemented by the zoning problem. The zoning problem refers to changes in perception caused by changing the boundaries of spatial units. Gerrymandering congressional districts is an excellent example of how the zoning problem has been weaponized for political purposes. In health, the zoning problem can interfere with the way disease rates appear across space in aggregated units. Neighborhood studies are notoriously plagued by the zoning problem. Geographers often use the term neighborhood to describe a unit that is smaller than a city and may or may not have some similar characteristics. Some cities, such as New York, have clearly delimited civic neighborhoods, but others, such as Houston, are not so clearly partitioned, so researchers must define their own criteria. In a systematic review of the literature on food environments, Chen et al. (2022) conclude that unless neighborhood zonation aligns well with natural socioeconomic partitions, statistical analysis of socioeconomic associations will be poor. Postal codes are not a sufficient substitute for a well-defined neighborhood. Kwan (2018) has suggested that the neighborhood might be defined as an individual region based on activity space. Everyone's neighborhood may be different depending on where they live and work.

Where neighborhood boundaries are drawn can have a substantial impact on how healthcare and sociodemographic variables align with indicators of poor health. Jakobsen (2021) recently demonstrated this effect in Denmark, comparing psychiatric medication use and socioeconomic deprivation across multiple redistricted neighborhood scales. While there was a slight relationship between psychiatric medication use and socioeconomic deprivation at the microscale, it disappeared in parishes and postal codes.

Both scale and zoning problems always act on spatially aggregated data. Wang and Di (2020) explored the effect of the MAUP on associations between COVID-19 mortality and atmospheric NO_2 levels. Wang and Di reaggregated the data in four separate ways: city level and provincial level (scale) and two alternate aggregation techniques (zoning). The four aggregation strategies yielded four extremely different regression models, including a positive linear relationship at the city level, a null relationship at the provincial level, and exponential positive and linear negative relationships in the two alternate aggregation strategies. Based on only four aggregation schemes, the same data yielded four results that would be interpreted very differently if only one scale was independently investigated.

The MAUP has a relative called the modifiable temporal unit problem (MTUP) (Ye & Rogerson, 2022). In the same way that the scale and zone of spatial data can change results, so can the period. In Western Australia, Yap et al. (2021) examined the effect of drought on mental health by comparing rainfall and mental health emergencies across various scales of both time and space. The relationships were quite different across spatial scales. However, they found positive associations between mental health emergencies when rain was measured in the summer and negative associations when rainfall was measured in the winter.

The challenge presented by the MAUP is not easy to overcome, but some remedies are available in the research. First, like some of the studies highlighted above, it is becoming increasingly popular to conduct analysis at multiple scales and/or zonation strategies. This study design produces multiple analysis results that researchers and readers can compare and may be more desirable than "black box" operations such as hierarchical modeling, which combines multiple scales within the analysis before producing a result.

A primary drawback of these studies is data availability. If data are only available on one scale and zonation, then the only course of action is to aggregate into increasingly larger units, which may not make any sense for the question. For example, exposure studies have clearly shown that individual-level data work best (Lee et al., 2020; Wang & Di, 2020). These authors also concede that if aggregation is necessary, then the finest scale possible is the best choice. Unless an author was trying to make a point about the MAUP, it would make no sense to investigate air pollution exposure at the state level just for the sake of a multiscale study design.

Smoothing techniques are another potential means to overcome the MAUP in a GIS environment. However, not all smoothing techniques are the same. Smoothing based on single-aggregation maps may overcome issues of visual perception but remain tied to one level of aggregation. In recent years, Tucson et al. (2020) proposed an overlay aggregation method that borrows data from multiple scales and zones to build a smooth surface for disease maps. This raises new questions about the theoretical compatibility of data collected at multiple scales but offers a new methodological path to explore. Similarly, Zhang et al. (2022) encourage spatial epidemiologists to engage more with raster techniques that create a continuous surface across space instead of polygons and points.

The simplest solution is to choose the appropriate scale and zonation for the question under study. In the methods, the authors must acknowledge the MAUP and why they believe the scale they have chosen is appropriate (Chen et al., 2022). Should analysis be conducted on a single scale, limitations, and conclusions should reflect that the results may be completely different on a smaller/larger scale or with a varied zonation strategy.

Geographic Information Systems and Social-Spatial Determinants of Health Research: The Promise and Challenge

Our review so far indicates that GIS can offer powerful tools for SDoH research. A patient's social determinants of health—factors outside of the traditional healthcare setting that impact an individual's health (e.g., food insecurity, transportation access, job training, and housing)—contribute more to their health than their race/ethnicity, gender, or medical care. Consequently, SDoH data can be used to predict exposure and health risk and target appropriate interventions (Bazemore et al., 2016) . In this regard, GIS can provide powerful tools for visualizing the spatial patterns of SDoH, spatially targeting, and evaluation to improve health outcomes. For example, using GIS, we can identify spatial variations in exposure to environmental pollution and the associated risk of disease. However, the assumption that all individuals who live in specified spatial units have similar characteristics or exposure is untenable.

GIS provides powerful tools for identifying and visualizing SDoH, but its limitations must be clearly understood, particularly in terms of spatiotemporal dimensions. Specifically, human exposure is neither fixed nor static. Because humans are continually moving within their environments, SDoH cannot be simple, static, or fixed. They must be complex and dynamic. Moreover, within a fixed environment, such as a census tract or zip code, individual exposures may vary significantly in a way that GIS is currently unable to capture (Kwan, 2012).

Similarly, while social determinants of health must be grounded in spatial terms, spatial variations in exposure to disease pathogens and environmental pollutants frequently transcend the social determinants or at least do not align neatly with the social boundaries (Kwan, 2018). Mobility is fundamental to exposure and risk. For example, most people do not spend all their time in one location or neighborhood.

Additionally, while research on neighborhood food environments can benefit from using GIS tools to determine distance impedance, the GIS is unable to capture how individual or neighborhood conditions may modify geographic distance and distort geographic access. Moreover, mobility is an important element of people's spatiotemporal experiences, and these complex experiences, exposures, and risks cannot simply be replaced by place of residence (Kwan, Richardson & Chenghu, 2015). Such temporal variations in exposure must be addressed.

Data limitations are another problem. Typically, the level of data available does not directly translate to solving the problem at hand. Missing or mismatching data can make determining a community's needs or level of exposure more difficult, particularly when traditional geographic units fail to adequately capture the community's boundaries. The larger the spatial unit, the more variability exists in it. But while smaller geographies represent residents more realistically, the risk of breaching confidentiality increases. One zip code may cover two towns, which makes estimates for one town difficult. Additionally, the distribution of social determinants of health can vary drastically even within a single zip code.

Thus, while GIS can shed light on differences in built environments that may contribute to disease and health outcomes, and especially social determinants of health, its focus on space and inherent assumption of uniformity of exposures and characteristics of residents in spatial units limits this utility.

Theoretical Considerations and Future Directions

While GIS may be a new tool in some disciplines, it is fundamental to the practice of human geography. GIS is the gateway to effective visualization, modeling, data organization, and spatial statistical analysis. Advances in information technology, the availability of large datasets (overtime and space), and increased capacity to manage, integrate, model, and visualize complex data in (near) real time offer extraordinary opportunities to integrate sophisticated space-time analysis and models in the study of complex environmental, social, and biological health dynamics (Song & Wu, 2021).

However, the health geography literature is teeming with challenges to the effective implementation of GIS methods. First, high-quality data is still costly and time-consuming to collect. Health data, including SDoH indicators, is limited by privacy concerns, suppression, and measurement error. Additionally, reliably linking the space-time data of people's movements to other relevant attributes (e.g., activity type, real-time exposure) is extremely difficult. Moreover, modeling human movements in space and time also brings complex issues of uncertainty (Chun et al., 2019). Thus, notwithstanding the increased capacity to manage, integrate, model, and visualize complex data that GIS provides, much remains to be done in future research to more fully exploit the potential of big data and other recent technologies in the spatial analysis of health and disease. For example, many fundamental notions in geographic research still tend to be conceptualized in static spatial terms, ignoring the importance of time and human mobility. Mobility is an essential element of people's spatiotemporal experiences, and these complex experiences, exposures, and risks cannot simply be replaced by place of residence (Kwan et al., 2015).

People's exposure to contextual influences could also vary widely over time. Consequently, studies on people's exposure to health risk factors must consider their residential history (Kwan, 2012). For instance, exposure to traffic-related air pollution varies as people move through the polluted environment over time during the day. A person may have several occupations throughout their lifetime. Chronic poverty affects the body differently than short-term poverty. A single snapshot cannot adequately capture the effects of environmental exposure or SDoH. Thus, the assessment of people's exposure to environmental influences must reflect dynamic notions of context that include both time and human mobility (Kwan, 2012).

Thus far, this paper has introduced many technical limitations, including data availability, spatial autocorrelation, the small numbers problem (SNP), the modifiable areal unit problem (MAUP), and its cousin the modifiable temporal unit problem (MTUP). Additionally, issues of geographic uncertainty and mobility raise important theoretical concerns for exposure risk assessment. However, this paper would be incomplete without briefly addressing the social, political, and historical limitations of quantitative modeling, especially in the areas of social vulnerability and power.

Socioeconomic and political status influence the risk and unequal impacts of disease and reveal structural inequality. The SDoH are not static, neutral, or independent. The poor are more likely to get sick and have worse outcomes because health problems and poverty interrupt and create complex causal chains with synergistic effects. Poverty produces and results from disease. This is what Farmer (Farmer, 2004) called structural violence—social structures and institutions undermine people's ability to meet basic needs or protect themselves from disease, and as a result, diseases disproportionately affect the poor and marginalized. Consequently, areas occupied by the poor and powerless are more likely to experience disease while lacking the resources to fight disease, whether at the global or local level.

Geographic analysis of disease must address both structural and politico-economic conditions alongside far less ordered processes reflecting complexity, uncertainty, contingency, and context specificity (Leach & Scoones, 2013). We need to understand how disease exploits social inclusions and exclusions and power relations in society. The vulnerability and consequences of disease are not merely a question of nature but one of politics and economy (Wisner et al., 1976).

However, too often, studies of global health issues ignore analyses of structural conditions—of power, politics, and economic relations—and understandings of local agency and mobilization—with all the attendant complexities, contingencies, and uncertainties of contexts, histories, and ecologies. We need to be wary of overly simplistic, linear, causal narratives of disease outbreaks and other health phenomena. GIS allows us to use the tools to engage in this complexity, and we must proceed along this path with enthusiasm and caution.

References

Araujo Navas, A. L., Osei, F., Soares Magalhães, R. J., Leonardo, L. R., & Stein, A. (2020). Modeling the impact of MAUP on environmental drivers for Schistosoma japonicum prevalence. *Parasites & Vectors, 13*(1), 112. https://doi.org/10.1186/s13071-020-3987-5

Bazemore, A. W., Cottrell, E. K., Gold, R., Hughes, L. S., Phillips, R. L., Angier, H., Burdick, T. E., Carrozza, M. A., & DeVoe, J. E. (2016). "Community vital signs": incorporating geocoded social determinants into electronic records to promote patient and population health. *Journal of the American Medical Informatics Association: JAMIA, 23*(2), 407–412. https://doi.org/10.1093/jamia/ocv088

Boakye, K. A., Iyanda, A. E., Oppong, J. R., & Lu, Y. (2022). A multiscale analysis of social and spatial determinants of cancer and noncancer hazards from on-road air pollution in Texas. *Spatial and Spatio-Temporal Epidemiology, 41*, 100484.

Chakraborty, J., Collins, T. W., & Grineski, S. E. (2017). Cancer risks from exposure to vehicular air pollution: A household level analysis of intraethnic heterogeneity in Miami, Florida. *Urban Geography, 38*(1), 112–136. https://doi.org/10.1080/02723638.2016.1150112

Chen, X., Ye, X., Widener, M. J., Delmelle, E., Kwan, M., Shannon, J., . . . Jia, P. (2022). A systematic review of the modifiable areal unit problem (MAUP) in community food environmental research. *Urban Informatics, 1*(1) doi:https://doi.org/10.1007/s44212-022-00021-1.

Chun, Y., Kwan, M., & Griffith, D. A. (2019). Uncertainty and context in GIScience and geography: Challenges in the era of geospatial big data. *International Journal of Geographical Information Science: IJGIS, 33*(6), 1131–1134. https://doi.org/10.1080/13658816.2019.1566552

Farmer, P. (2004). An Anthropology of Structural Violence. *Current Anthropology, 45*(3), 305–325.

Garcia, M. G. L. (2018). *Environmental injustice and racial/ethnic heterogeneity in Houston, Texas*. The University of Texas at El Paso.

Jakobsen, A. L. (2021). Neighborhood socioeconomic deprivation and psychiatric medication purchases. different neighborhood delineations, different results? A nationwide register-based multilevel study. *Health & Place, 72*, 102675. https://doi.org/10.1016/j.healthplace.2021.102675

Javanmard, R., Lee, J., Kim, J., Liu, L., & Diab, E. (2023). The impacts of the modifiable areal unit problem (MAUP) on social equity analysis of public transit reliability. *Journal of Transport Geography, 106*, 103500. https://doi.org/10.1016/j.jtrangeo.2022.103500

King, B. (2010). Political ecologies of health. *Progress in Human Geography, 34*(1), 38–55. https://doi.org/10.1177/0309132509338642

Klein, R. J., Proctor, S. E., Boudreault, M. A., & Turczyn, K. M. (2002). *Healthy people 2010 criteria for data suppression*. Washington, DC. Retrieved from. https://www.cdc.gov/nchs/data/statnt/statnt24.pdf

Kwan, M. (2012). Geographies of health. *Annals of the Association of American Geographers, 102*(5), 891–892.

Kwan, M. P. (2015). Beyond space (as we knew it): toward temporally integrated geographies of segregation, health, and accessibility. In M. P. Kwan, D. Richardson, D. Wang, & C. Zhou (Eds.), *Space-time integration in geography and GIScience*. Springer. https://doi.org/10.1007/978-94-017-9205-9_4

Kwan, M. (2018). The limits of the neighborhood effect: Contextual uncertainties in geographic, environmental health, and social science research. *Annals of the American Association of Geographers, 108*(6), 1482–1490. https://doi.org/10.1080/24694452.2018.1453777

Kwan, M., Richardson, D., Wang, D., & Zhou, C. (2015). *Space-time integration in geography and GIScience. Springer*.

Leach, M., & Scoones, I. (2013). The social and political lives of zoonotic disease models: Narratives, science and policy. *Social Science & Medicine, 1982(88)*, 10–17. https://doi.org/10.1016/j.socscimed.2013.03.017

Lee, D., Robertson, C., Ramsay, C., & Pyper, K. (2020). Quantifying the impact of the modifiable areal unit problem when estimating the health effects of air pollution. *Environmetrics (London, Ont.), 31*(8), n/a. https://doi.org/10.1002/env.2643

Moonan, P. K., Bayona, M., Quitugua, T. N., Oppong, J., Dunbar, D., Jost, K. C., Jr., et al. (2004). Using GIS technology to identify areas of tuberculosis transmission and incidence. *International Journal of Health Geographics, 3*(1), 23. https://doi.org/10.1186/1476-072X-3-23

Moonan, P. K., Oppong, J., Sahbazian, B., Singh, K. P., Sandhu, R., Drewyer, G., et al. (2006). What is the outcome of targeted tuberculosis screening based on universal genotyping and location? *American Journal of Respiratory and Critical Care Medicine, 174*(5), 599–604. https://doi.org/10.1164/rccm.200512-1977OC

O'Keefe, P., Westgate, K., & Wisner, B. (1976). Taking the naturalness out of natural disasters. *Nature, 260*.

Sánchez-Díaz, G., Alonso-Ferreira, V., de la Paz, M. P., & Escobar, F. (2020). New insights around the modifiable areal unit problem (MAUP) in its relation to cartographic representation of rare diseases. *Investigaciones Geográficas, 74*, 71–84.

Song, W., & Wu, C. (2021). Introduction to advancements of GIS in the new IT era. *Annals of GIS, 27*(1), 1–4. https://doi.org/10.1080/19475683.2021.1890920

Sun, M., & Wong, D. W. (2017). Spatial aggregation as a means to improve attribute reliability. *Computers, Environment and Urban Systems, 65*, 15–27.

Taylor, F. E., Millington, J. D. A., Jacob, E., Malamud, B. D., & Pelling, M. (2020). Messy maps: Qualitative GIS representations of resilience. *Landscape and Urban Planning, 198*, 1–11.

Tobler, W. R. (1970). A computer movie simulating urban growth in the Detroit region. *Economic Geography, 46*(2), 234–240. https://doi.org/10.2307/143141

Tuson, M., Yap, M., Kok, M. R., Boruff, B., Murray, K., Vickery, A., . . . Whyatt, D. (2020). Overcoming inefficiencies arising due to the impact of the modifiable areal unit problem on single-aggregation disease maps. *International Journal of Health Geographics, 19*(1), 40. doi:https://doi.org/10.1186/s12942-020-00236-y.

Waller, L. A., Gotway, C. A., & NetLibrary, I. (2004). *Applied spatial statistics for public health data*. John Wiley & Sons.

Wang, Y., & Di, Q. (2020). Modifiable areal unit problem and environmental factors of COVID-19 outbreak. *The Science of the Total Environment, 740*, 139984. https://doi.org/10.1016/j.scitotenv.2020.139984

Wisner, B., O'Keefe, P., & Westgate, K. (1976). Taking naturalness out of natural disasters. *Nature, 260*(5552), 566–567.

World Health Organization. (2019). Elimination of human onchocerciasis: Progress report, 2018–2019. *Weekly Epidemiological Record, 94*(45), 513–523.

Yap, M., Tuson, M., Turlach, B., Boruff, B., & Whyatt, D. (2021). Modeling the relationship between rainfall and mental health using different spatial and temporal units. *International Journal of Environmental Research and Public Health, 18*(3), 1–15. https://doi.org/10.3390/ijerph18031312

Ye, X., & Rogerson, P. (2022). The impacts of the modifiable areal unit problem (MAUP) on omission error. *Geographical Analysis, 54*(1), 32–57. https://doi.org/10.1111/gean.12269

Zhang, S., Wang, M., Yang, Z., & Zhang, B. (2022). Do spatiotemporal units matter for exploring the microgeographies of epidemics? *Applied Geography, 142*, 102692. https://doi.org/10.1016/j.apgeog.2022.102692

Chapter 18
Understanding Malaria Transmission and Control within and Between Regions in Zambia Using a Socio-Spatial Determinants of Health Framework

Jailos Lubinda and Oliver Mweemba

Introduction

Malaria is a febrile illness caused by Plasmodium parasites that spread to people through the bite of infected female Anopheles mosquitoes. It affects nearly half (≈ 3 billion people) of the global population, living in approximately 85 countries, causing about 250 million cases and over 600,000 deaths annually (World Health Organization [WHO], 2022). Historically, the progressive contraction in the distribution of malaria transmission is a consequence of control efforts and, largely, development and population growth (Hay et al., 2004).

Geographically, the disease is more prevalent among poor nations and communities. The disease affects groups on sociodemographic lines such as age, gender, ethnicity, education level, and income, depending on "place." Globally, children under five account for nearly 80% of all malaria.

deaths (WHO, 2022). Similarly, pregnant women, migrants, and those in rural areas carry a disproportionate share of the malaria burden. In terms of approach, malaria epidemiological studies have historically focused on the prevalence of infection in populations, associating the observed infection levels with various parasitologic, climatologic, and entomologic parameters (Institute of Medicine, 1991). A modified epidemiological approach was recommended by an expert committee of the WHO in the late 1980s, emphasizing the need to follow systematic identification and ranking of principal determinants of malaria that would point to specific control strategies that should be considered within given paradigms. The eight determinants

J. Lubinda (✉)
Telethon Kids Institute, Perth Children's Hospital, Perth, Australia
e-mail: jailos.lubinda@telethonkids.org.au

O. Mweemba
Department of Health Promotion and Education, School of Public Health, University of Zambia, Lusaka, Zambia

© The Author(s) 2026
M. A. Kolak, I. K. Moise (eds.), *Place and the Social-Spatial Determinants of Health*, Global Perspectives on Health Geography,
https://doi.org/10.1007/978-3-031-88463-4_18

identified included understanding the level of endemicity, parasite species, mosquito vectors, human population characteristics, health infrastructure, social, behavioral, and economic considerations, availability and effectiveness of antimalarial drugs, and the influence of development projects (Institute of Medicine, 1991). There are relatively large volumes of studies and publications from the first four determinants compared to the latter ones, suggesting a persisting emphasis on the epidemiological triad.

The distribution of malaria is disproportionately concentrated in sub-Saharan Africa (SSA), bearing 95% of cases and 96% of deaths of the global burden annually (World Health Organization, 2022). Since the 1990s, the population at risk has increased, and the number of cases and deaths has nearly tripled, mostly driven by population growth, urbanization, and enhanced reporting through improved health information systems. However, the population-adjusted indicators are relatively reduced. For example, the global prevalence, incidence, and mortality rates showed tremendous declines between 1990 and 2015 (World Health Organization, 2022). Nonetheless, these declining trends showed stagnation thereafter before slowly rebounding between 2015 and 2019. The estimated annual percentage change in incidence rates showed that during this period, most SSA experienced increases of up to 1000%, with only a few countries showing marginal decreases >50% (Liu et al., 2021).

While the burden of malaria can be influenced by global or regional factors, as demonstrated by Caminade et al. (2014), sustained progress in the fight against malaria primarily depends on local factors of a place (Birkholtz et al., 2012). Place represents the primary environmental characteristics, directly and indirectly affecting the prevalence of malaria by enhancing or hindering various prerequisites for the malaria parasites and vectors to thrive. For example, variations in the suitability of malaria transmission in lowlands vs. highlands, dryer vs. wetlands, farming lands vs. industrial areas, and hot vs. cool places differentiate the parasite replication, vector density, and human and vector behaviors. These local factors specific to place are usually complemented by those often encapsulated within the social determinants of the health framework and intricately linked to human development indices. Together, these emphasize the need for a nuanced understanding of the diverse contexts in which malaria prevention and control strategies must be implemented or explain the observed significant sub-national or geographical heterogeneities. For example, communities with the poorest populations tend to bear the highest malaria burden, and the risk of becoming infected with malaria more frequently is higher than that of richer people (Nawa, 2019; Mwangu et al., 2022; Roll Back Malaria, 2001).

Furthermore, mortality rates remain higher in poorer households and more so among children, especially those from poorer families (Roll Back Malaria, 2001). As recently reported, SSA's observed pace of poverty reduction is unchanged, so it is no coincidence that the top five poverty-contributing countries are also among the top malarious countries in the world, four of which are in SSA (World Bank 2022). With an average of nearly 60% of people living in rural areas, malaria incidence in SSA is generally higher among rural populations, characterized by the lowest sociodemographic indices, most marginalized populations on the planet, and the most susceptible to disease and death from malaria (Nawa, 2019).

However, a narrative that solely emphasizes climatic, environmental, or ecological factors as explanations for observed malaria patterns overlooks the intricate interplay between these factors and the neglected social determinants that play a role in the distribution and persistence of malaria in some communities (Ricci, 2012). This is because the environmental narrative mainly captures the direct influence on vector transmission mechanisms, such as mosquito breeding, density, and high entomological inoculation rates (Nissan et al., 2021). However, it regrettably fails to recognize important social determinants that significantly influence the success or failure of intervention efforts or could explain the psychosocial behaviors contributing to the high exposure risk of communities in these areas. Failure to account for these social determinants may help explain why, even in Zambia, where most universal coverage targets have been achieved, and similar interventions and coverage thresholds are applied countrywide, regions other than Lusaka and Southern have struggled to achieve meaningful, substantial impact, particularly those currently classified as high-transmission regions.

Thus, this chapter delves into the intricate interplay of biophysical, geospatial, and social determinants of health (SDoH) in influencing malaria transmission and control in Zambia. The focus is on how these factors contribute to the differential exposure and effects of malaria, emphasizing the need for comprehensive health promotion interventions that address both social and ecological dimensions.

The chapter is structured to provide a holistic understanding of malaria transmission dynamics in Zambia. Key sections include: (i) Introduction, providing an overview of malaria's impact globally and specifically in Zambia, detailing historical and current epidemiological trends; (ii) Country Context, giving an analysis of Zambia's spatial-temporal patterns of malaria risk, focusing on the geographical and socioeconomic disparities that influence disease distribution; (iii) Current Interventions and Coverage, which reviews the primary malaria interventions implemented in Zambia, their targets, and the observed outcomes. (iv) Social-Spatial Determinants of Malaria in Zambia, providing an in-depth exploration of how social and environmental factors intersect to shape malaria transmission and control efforts in Zambia. This section discusses not only the geospatial determinants of malaria in Zambia but also some SDoH, such as behavioral and psychosocial factors, health systems factors, and factors related to material circumstances and socioeconomic position. (v) Application of the SSDoH Framework, providing practical examples and suggestions for integrating social-spatial determinants into malaria intervention strategies using the Rainbow model. (vi) Conclusion, which emphasizes a call for a multipronged approach to malaria control that includes improving healthcare access, addressing behavioral dimensions and socioeconomic disparities, and implementing adaptable mosquito control measures. By the end of this chapter, the reader should gain a better appreciation of the multifaceted determinants of malaria in Zambia and the importance of integrating social and spatial considerations into the control intervention strategies to enhance their effectiveness.

Country Context: Zambia's Spatial-Temporal Patterns of Malaria Risk

Like many other SSA countries, Zambia is highly endemic to malaria, with transmission cycles and distribution varying over time and space. Malaria transmission is very seasonal in the Southern half of the country, with the peak transmission occurring during the rainy season from November to April. The Northern and Eastern regions generally have longer and more intense malaria transmission seasons than the other regions. Nonetheless, the timing and intensity of transmission vary depending on the location within the country (Lubinda et al., 2021a, b).

Transmission is highest in five of Zambia's provinces: Luapula, Northern, Muchinga, Northwestern, and Western. These five provinces also have the highest proportion of rural populations, have the lowest socioeconomic status, and are the most deprived when assessed against multidimensional overlapping deprivation techniques for deriving multidimensional poverty indices. Safe to say, however, that these provinces also exhibit higher malaria suitability indices. Figure 18.1a–c show multidimensional poverty levels, while Fig. 18.1d–f show malaria prevalence in respective provinces yearly. The provinces with higher malaria prevalence also have the highest levels of socioeconomic deprivation (i.e., nutrition, education, access to information, housing, water, sanitation, and health), computed as multidimensional poverty (Oxford Poverty and Human Development Initiative [OPHI], 2022). These provinces are furthest from the capital city or prominent regional capitals, and most

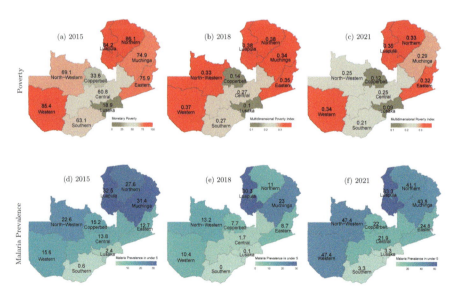

Fig. 18.1 Maps **a**, **b**, and **c** (top row) show poverty indices, and **d**, **e**, and **f** (bottom row) show malaria prevalence at the province levels. Darker shades of red or blue denote the highest poverty or prevalence, while lighter colors indicate lower values. The author created these maps based on the most recent World Bank Poverty reports, and malaria indicator survey reports from 2015 to 2021

of their populations often suffer from limited access to healthcare services due to long distances to public health facilities and the scarce availability of private healthcare facilities (De la Fuente et al., 2015; Shifa et al., 2017). The Southern, Lusaka, and Copperbelt provinces, notably urban-dominated, enjoy lower malaria prevalence and low multidimensional poverty, largely due to a combination of relatively high socioeconomic status, easy access to care services, and a drier climate.

Often based on local environmental and socioeconomic factors, there are significant variations in malaria transmission within regions, largely attributed to their favorable geographic and climatic conditions, which allow mosquitoes to breed easily and for malaria cases to spread (Bennett et al., 2016; Shimaponda-Mataa et al., 2017a, b; Lubinda et al., 2021a, b). For example, areas with higher rainfall (e.g., Luapula, Northern, Muchinga, and Copperbelt provinces) and standing water, such as floodplains, marshlands, swamps, or large protected forest lands (e.g., Western and Northwestern or Eastern provinces), are particularly more prone to malaria transmission than those without. Further, communities around Zambian lakes, such as Kariba in the Southern province, Mweru, Mweru-wantipa, and Bangweulu in the Luapula province, and Tanganyika in the Northern province, experience higher incidences than elsewhere. The same applies to communities or villages near slow-flowing rivers, streams, and water reservoirs (Kamanga et al., 2010). Meanwhile, the Western province is, by extension, located in the heart of Zambia's floodplains, known for small-scale migratory fishing and farming, but also provides mass mosquito breeding habitats common to both humans and vector mosquitoes. On the other hand, low malaria cases in the Southern, Copperbelt, and Lusaka regions have been attributed to unfavorable climatic and geographical conditions and, sometimes, intervention effectiveness.

Zambia: Current Interventions and Coverage

Primary Malaria Interventions

Historically, Zambia has had at least five-year malaria strategic plans that progressively set ambitious goals to scale up malaria interventions and to enhance control, consolidation, and elimination efforts (Chanda et al., 2012; MoH., 2011, 2017, 2022a, b). Increasingly, the fight against malaria through prevention, control, and case management was broadly involved:

- Increasing indoor residual spraying (IRS) coverage.
- Promoting ownership and usage of insecticide-treated nets (ITNs).
- Improving malaria case management with effective diagnostics and drugs.
- Preventing and controlling malaria in pregnancy and infants via intermittent presumptive treatments (IPT).
- And information, education, and communication (IEC)/behavioral change communication (BCC) strategies (Chanda et al., 2012).

Furthermore, entomological surveillance through larval source management (spraying the water bodies in communities to reduce the density of mosquitoes) and mass drug administration was implemented to reduce the malaria burden before the peak seasons (i.e., September and October), and thereafter ITNs, as well as IRS, were rolled out.

Intervention Scale-Up, Targets, and Effect

Between 2000 and 2015, Zambia made great strides in the fight against malaria on all fronts, transforming its epidemiological profile from a ubiquitously high countrywide endemicity to a stratified one (MoH, 2010a, b, c). These strides follow an aggressive use of a combination of new or improved tools for case management (e.g., rapid diagnostics' tools) and the various primary interventions and strategies discussed earlier (MoH., 2022b).

The target for core preventive interventions, including ITNs, was to achieve universal coverage for all sleeping spaces (Miller et al., 2008; MoH, 2008a, b; MoH., 2022b). IRS aimed to achieve at least 80% of the targeted numbers of sleeping structures in eligible districts (MoH, 2009) or, in the recent case, given precedence over ITNs as a primary vector control intervention between 2020 and 2022 (MoH., 2022b). Meanwhile, diagnostic testing and treatment sought to reach a minimum of 80% of patients receiving prompt (within 24 h of symptom onset) and ensure that 80% of pregnant women had access to IPT interventions (MoH., 2022b). Furthermore, malaria case investigation was also implemented in low malaria burden areas. Community health workers take immediate action when a malaria case is reported. They follow up with the index households and conduct tests on all their members and neighboring households. Subsequently, they provide ITNs to reduce parasite reservoirs and prevent the spread of malaria further.

For ITN, the main distribution channels are nationwide mass ITN distribution campaigns conducted every 3 years (Masaninga et al., 2018) as well as more recently through routine health facility-based antenatal care distribution to pregnant women and children via the expanded program for immunization (Miller et al., 2022). Meanwhile, IRS, the mode of distribution is annual, often done in predetermined areas with targeted coverage thresholds. Initially, however, areas of economic importance, such as irrigation schemes, mines, and tourism centers, were a priority for IRS due to high population densities and economic viability or costs of malaria. However, the private sector often advanced such investments, including mining companies. This is because the populations in these areas usually comprise a mix of local populations, often with higher socioeconomic status, living in modern housing with less exposure to malaria risk, and expatriate migrant workers with low immunity to malaria. The justification was that, though these populations had a low risk of exposure to infection, the risk of severe disease and death in the event of a malaria infection was high. The suitability for IRS in other areas included, among others, being urban and having walls suitable for spraying. Until recently, this criterion remained primary for household IRS eligibility. However, with new technological improvements, more rural housing is increasingly sprayed with IRS yearly (MoH., 2022b).

With consistent scale-up and targeted universal coverage of these interventions, reductions in malaria burden were accompanied by changing spatial and temporal patterns into clusters of microgeographic regions or areas. These changes affected distinct demographic subpopulations, primarily characterized by shared gender, age, social, behavioral, economic, environmental, or geographical risk factors. These shifts in populations at risk of malaria raise new questions, particularly for communities experiencing various epidemiological changes, such as Zambia, where primary control interventions are no longer as effective in many areas.

Gaps in the Interventions and Importance of Integrating Social-Spatial Determinants of Health Framework for Policy and Programming

Despite nationwide mass ITN distribution campaigns and yearly IRS in targeted areas, studies in northern Zambia reported only modest declines in malaria prevalence in the rainy season and no declines during the dry season following 3 years of IRS (Chanda et al., 2012; Hast et al., 2019). The key questions now include:

- Why does there seem to be a stalled effect, or even rebounding malaria, in some communities around the country, even in areas where reductions were significant?
- Are the interventions no longer efficacious?
- Has there been a change in the biological or molecular response of vectors to interventions?
- Is dynamic human behavior hindering the effectiveness of interventions in most areas where little to no progress has been recorded?
- Are interventions emphasizing tackling the spatial/environmental and biological determinants of the observed epidemiology, while neglecting the strong influence of some social determinants of health, able to explain this?

Based on Fig. 18.2, it is evident that current malaria risk mapping relies on environmental risk factors and prevailing epidemiological profiles. Preventive interventions prioritize achieving targeted coverage thresholds, while curative interventions primarily address physical access to healthcare services and care accessibility, often overlooking factors related to acceptability and service quality. By and large, these factors still need to ensure high utility or uptake patterns, are mediated by local socioeconomic pressure to engage in risk-prone activities, and rarely engage in resolving cultural bottlenecks hindering the effectiveness of highly efficacious malaria interventions.

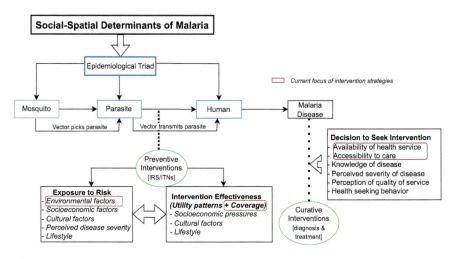

Fig. 18.2 Social-spatial determinants of malaria

The Social-Spatial Determinants of Malaria in the Zambian Context

Epidemiological Approaches to Malaria Control

Most Zambian studies on malaria still stress the epidemiologic approach, emphasizing local variability in the distribution of malaria problems and calling for the design of appropriate and suitable control strategies and monitoring and evaluation in different ecological areas (Chanda et al., 2012; Masaninga et al., 2013; Lubinda et al., 2021a, b). They emphasize environmental determinants of malaria risk by modeling the role of temperature, rainfall, humidity, vegetation indices, elevation, distance from pools, rivers, and lakes, length of the wet season, travel time to cities, and persistence of breeding pools in the prevalence of malaria (Weiss et al., 2015; Endo & Eltahir, 2016; Pfeffer et al., 2018). These variables are often used to help predict vector bionomics, transmission dynamics, and malaria risk (Weiss et al., 2015; Nissan et al., 2021).

By implication, environmental changes due to climate change or human-induced ecological disturbances may directly and indirectly exacerbate or reduce the spread of malaria and are often associated with prevalence, incidence, and mortality. Ironically, social determinants associated with the indirect effects of climate variability and environmental change on the distribution of malaria are understudied and often unquantified or deemed as results or consequences of the environmental determinants of the disease and rarely as contributors (Nissan et al., 2021). Many studies treat them as only associated and not causally linked to malaria, possibly because studies tend to use sociodemographic characteristics (gender, age, education level of parents, ethnicity, marital status, pregnancy), which are non-modifiable (Bayode & Siegmund, 2022).

The social-spatial determinants of health (SSDoH) framework extends the analysis and offers a comprehensive understanding of both the geospatial and social determinants of malaria, which are crucial to understanding the risk of exposure to malaria and the suitable interventions against it. It, therefore, includes the social, economic, political, cultural, ethnic/racial, psychological, and behavioral circumstances that influence the occurrence of health problems and their risk factors in places where people are born, live, work, and age in the context of their everyday life (Pell et al., 2011; Shayo et al., 2015; Thornton et al., 2016). These contexts influence the individual and community's socioeconomic position, which sets the stage for the transitional determinants, which include material circumstances, psychosocial circumstances, behavioral and/or biological factors, and the health system. These determine individuals' and communities' health situation or status (Thornton et al., 2016).

Therefore, social-spatial determinants of malaria refer to how social and environmental factors intersect to influence the prevailing epidemiology, control of malaria, and well-being. The concept recognizes that health outcomes are not solely a product of disease etiology or genetics, but also shaped by larger social and environmental factors. Fewer studies on malaria have combined the two dimensions in analyzing the distribution of malaria and informed the effectiveness of interventions.

Mostly, researchers choose to focus on one approach (i.e., set of factors) or the other. For instance, geospatial factors, comprised of physical environmental characteristics, have been studied on their own, while the social sciences and fields of health promotion separately try to link the SDoH and malaria prevalence, distribution, or the success of malaria interventions, instead of building an integrated approach (Pell et al., 2011; Shayo et al., 2015).

Behavior and Psychosocial Factors

Among the identified social determinants of malaria are the behavior and psychosocial factors of the affected people and communities. These include health-seeking behaviors in which affected members delay seeking health services for various reasons, including lack of correct knowledge of the signs and symptoms of malaria and their perceived risk of fatality. Other behavioral and social factors involve the inadequate appreciation of available interventions, leading to the misuse of interventions such as ITNs, lack of adherence to the treatment guidelines, and preference for alternative medicines (Cardona-Arias, 2023; Mwangu et al., 2022). The current intervention strategies, i.e., ITNs, IRS, IPT, and facility-based case management, do not address the list above, which is common in rural communities that also bear the most significant burden of malaria.

Evening outdoor lifestyles, such as cooking, drinking beer, and overnight church and funeral practices, which expose people to mosquito bites, are also common in these settings (Janko et al., 2018). The effectiveness of ITNs depends on going to bed early, sleeping under a bed net, fewer outdoor evening activities, and increased distance from the animal kraals, which tend to attract more vector mosquitoes. However, the list above is highly embedded in the community's cultural norms and rarely considered in global or local intervention planning strategies.

These behaviors, which undermine the effectiveness of interventions, are predominantly observed and reported in high-prevalence areas of Zambia. Take Luapula province, for instance, where abundant water bodies and livelihoods based on fishing and farming often lead to the misappropriation of ITNs. These nets are frequently diverted for use in fishing, garden fencing, or resale, resulting in a scarcity of ITNs for malaria prevention. While ownership figures may increase on paper, the actual use of ITNs may lag unless they are correctly deployed upon distribution (Masaninga et al., 2018; Mwangu et al., 2022). Even then, there is no guarantee that the nets will not be tampered with, removed, or used improperly.

Moreover, even though rural households in Zambia frequently report similar or higher ownership and use of ITNs (Shimaponda-Mataa et al., 2017a, b; MoH, 2022b), less-educated and poor rural households may not proactively replace their ITNs when they wear out before the next mass distribution campaigns, primarily due to a lack of interest in personal procurement (Shimaponda-Mataa et al., 2017a, b). ITNs are subject to relatively higher wear and tear in these regions, more prone to accumulating dirt from ubiquitous dust, and susceptible to damage from leaking thatched roofs. Consequently, they require more frequent washing to maintain their integrity, making them more susceptible to developing holes.

Besides, low education levels and higher illiteracy rates among women in rural Zambia have been associated with higher malaria prevalence than those with secondary education (Mutale & Mbewe, 2017). The caregiver's education level also hinders the reception of medical instructions, drug dosage, perception of symptoms, and disease severity, affecting care-seeking behavior. Unfortunately, the effects of low literacy levels are also knitted into the gender constructs of most Zambian communities, where even though men are more likely to be more educated than women, women remain the primary caregivers of most households, thereby directly influencing most health-related household decisions.

Health Systems

Within the framework of SDoH and the right to health, health system factors contributing to delayed care-seeking include extended travel times to health facilities in the absence of motorized transport options for long journeys and are associated with a higher economic burden for transport. Limited availability and retention of qualified healthcare professionals, difficulties in procuring and stocking essential malaria medications, and the perceptions of healthcare quality may be associated with the high prevalence of malaria in Zambia, as noted in other studies (Aberese-Ako et al., 2019; Hill et al., 2013). In some areas, individuals either delay seeking formal health services for malaria prevention and treatment or resort to local alternative remedies and traditional practitioners, often as the primary or initial choice (Zingani et al., 2017; Aberese-Ako et al., 2019; Mwangu et al., 2022; Cardona-Arias, 2023). Unsurprisingly, the provinces with the highest malaria prevalence also show the worst compliance with regulatory quality of health services standards, which are proxies for quality. All the above breach the expected operational effectiveness and efficacy of curative interventions for malaria.

Additionally, hard-to-reach areas with poor road networks or nonexistent roads, widely dispersed housing, and physical barriers, such as rivers and mountains, are major obstacles for most malaria intervention distribution, including IRS (Pell et al., 2011). Many areas in Luapula, Muchinga, and Western provinces are isolated, hard to reach, and mostly cut off during the rainy season, which makes it difficult to implement timely, effective interventions in these areas (Shimaponda-Mataa et al., 2017a; Zingani et al., 2017).

Intertwined with infrastructure challenges during heavy rainfall and tough climatic conditions, predominantly rural and remote, lacking proper road infrastructure, hard-to-reach areas exert further geographical isolation by creating significant barriers to establishing accessible health facilities, retaining qualified healthcare staff, delivering and maintaining sufficient stocks of essential malaria medications, consequently affecting the effective delivery of quality health services (Pinchoff et al., 2015; Nawa et al., 2019; Nissan et al., 2021).

With average distances to health facilities between 5 and 10 km, it takes 30 min and 2 h of cycling or walking to access care, respectively. In most areas where motorized transport is limited, bad health-seeking behavior is strongly influenced by the indirect costs of health services, such as out-of-pocket expenditure to access care and treatment. For instance, even though public health services in rural areas may be officially "free," patients still incur costs related to their transport, registration fees, and occasionally prescribed medications. These added financial burdens adversely impact decision-making processes concerning prompt and effective malaria case management. All these are directly linked to socioeconomic status and, broadly, multidimensional poverty, which is also strongly associated with provinces of high prevalence (MoH, 2022b; Human Development Initiative (OPHI), 2022).

Material Circumstances and Socioeconomic Position

Poor communities, especially in rural areas, struggle to have decent housing and sleeping spaces where mosquito exposure can be prevented (Ricci, 2012). Communities with poor housing and sanitation and limited access to preventive measures (such as ITNs and IRS) or exposure-prone cultures are at higher risk of malaria. It is common practice for farming and fishing communities to spend most of their time outdoors, migrating to the fields, forests, farming or fishing camps, or living in temporary housing structures (Aberese-Ako et al., 2019). These movements to temporary shelters for various economic-related activities such as planting, harvesting, cattle grazing, and fishing make them susceptible to malaria, take them away from health services, and make it more difficult to access preventive and curative services from health facilities (Zingani et al., 2017; Aberese-Ako et al., 2019). The behavior of migratory households is beyond what the current interventions can address, and any complementary strategy for ITN or IRS may not satisfy both residences either (Zingani et al., 2017).

In places of high malaria prevalence, such as Luapula, Western, Northern, and Northwestern provinces, communities are largely agrarian and fishing with limited resources to spend on malaria interventions (their health) or their livelihood (Mwangu et al., 2022). Most households here have large families and must make hard decisions about who should sleep under the few bed nets available (freely obtained or purchased). Their housing structures are mostly not up to standard, making it difficult to employ interventions such as IRS and nets (Chanda et al., 2012; Nawa et al., 2019). Even where rural households receive IRS, traditionally built rural housing often requires frequent replastering, which may affect the desired chemical half-life effect. Hence, most IRS are easy to implement in urban and peri-urban areas where modern houses are common (Chanda et al., 2012).

Application of Social-Spatial Determinants of Health Framework to Improve Effectiveness of Malaria Intervention Using the Rainbow Model

According to Dahlgren and Whitehead (2021), the Rainbow model, or the Dahlgren and Whitehead model of SDoH approaches, suggests four modifiable layers to achieve greater impact through (i) proximal (individual's lifestyle), (ii) social and community networks, (iii) living and working conditions, and (iv) distal (general socioeconomic, cultural, and environmental conditions) influence. These influences are depicted in concentric rings, illustrating domains of influence and the corresponding actors at each level. The innermost ring, labeled "individual," represents those most directly affected. The subsequent rings depict individuals with direct contacts who wield influence over attitudes, beliefs, and actions, shaping community norms and access to resources. The outermost ring encompasses those who indirectly impact those most affected, constituting a favorable environment. While this model considers individual behavior because of multiple influences, it aims at driving individual changes that shape the broader social context.

In application to the determinants of malaria and the cascade as shown in Fig. 18.2, the Rainbow model's rings would translate to (i) the individual affected (due to their lifestyle and interactions with exposure risk, preventive interventions, and curative interventions), (ii) and (iii) the immediate family/household and friends (i.e., parents, siblings, friends, and school or religious peers), (iv) teachers, religious and community leaders, respected elders, and employers. Here, we try to explore and apply the various levels of influence to pinpoint modifiable critical factors, encompassing individual motivation, knowledge, social and gender norms, skills, and supportive environments.

While grappling with higher illiteracy rates and aligning with the "living and working conditions" layer, relevant to the socioeconomic and sociocultural factors discussed earlier, teachers and peers (rings ii and iii), especially in primary education, can promote behavior change by imparting malaria knowledge to students, who, in turn, can reinforce and influence each other and the wider community by sharing this information. Integrating malaria awareness into annual school events (such as World Malaria Day commemorations) and teaching its prevention in primary schools can bolster community knowledge among school-going children, creating growing generations of malaria ambassadors and vigilant little whistle-blowers from the most affected age group. However, while school-level interventions like school-based net distribution are a promising way to involve the most affected, they overlook households without school-going children, especially in rural areas where education is one of their areas of deprivation. Nonetheless, this can be bridged by antenatal care and expanded programs for immunization distribution that capture pregnant mothers or their under five pre-educational children.

Outside school and facility-based distribution or messaging, social and behavior change communication (SBCC), facilitated through information sharing within "social and community networks" based on the Dahlgren and Whitehead model, can influence education and individual or collective lifestyles. Factors like exposure to local radio messages (which capture the older age groups) and sports activities (that capture the youths) can enhance a community's perception of risk and improve an individual's response to exposure and self-efficacy. Enhanced knowledge of malaria transmission symptoms and severity can drive preventive behaviors. Besides, we know that where community engagement is strong, malaria control and elimination programs thrive because of enhanced community trust and involvement, high acceptability of advocated health practices, high intervention uptake, and good community buy-in (Arroz, 2017). Conversely, SBCC investments may prove less effective in communities where strategic engagement is inadequate or lacking.

In high-pressure economic contexts, multidimensional poverty diminishes the impact of vector control interventions unless coordinated with socioeconomic programs. Co-creation and direct community involvement in finding acceptable solutions to reduce the risk of exposure and increase knowledge and intervention uptake would help, even for highly risk-prone groups. For instance, where suitable, seasonal malaria chemoprevention and wide use of potable insect or spatial repellents can be explored to help such mobile populations in farming and fishing communities (Stevenson et al., 2018; Syafruddin et al., 2020).

Finally, a network of community volunteers, such as community health workers, health committees, religious volunteers, political activists, and traditional health practitioners involved in malaria prevention and SBCC, or made to preside over local and village-specific programs, ensure enhanced uptake and utility of malaria interventions.

While our intention was not to provide an exhaustive list of examples and solutions, we highlighted the feasibility and importance of leveraging spatial-SDoH to innovate in conveying malaria prevention approaches, driving behavior change, implementing exposure-reduction strategies, and determining suitable interventions. Doing so is essential for optimizing the synergistic potential of current malaria interventions, particularly in Zambia and other communities across Africa, where their effectiveness has waned. We call for developing integrated approaches within this framework, capable of using environmental and socioeconomic covariates in malaria modeling for risk or epidemiological profiles. Perhaps the challenge is the need to generate high-resolution socioeconomic covariates for use alongside well-established high-resolution environmental covariates. Nonetheless, with many similar works that have been developed during the last decade, e.g., high-resolution rasters (100 × 100 m) for population density, urbanity, etc., extending similar methodology to generate publicly available SDoH datasets would go a long way.

Conclusion

Using the SSDoH framework to discuss the distribution of malaria in Zambia, we argue that the distribution of malaria cases is determined not just by climate and environmental factors but also by SDoH influence. We showed that many communities where malaria prevalence remains high have suitable environmental drivers for the parasite and vector but also struggle with multidimensional poverty, socioeconomic deprivation, and high sociocultural malaria exposure risk factors. We equally highlighted how the current intervention strategies are skewed toward the epidemiological approach and how environmental factors define all exposure to the risk of malaria instead of integrated approaches that include the SDoH.

This creates a complex set of SDoH that undermine the impact of well-intended interventions by causing low intervention utility, misuse, poor health-seeking behavior, and poor sense of personal or community practice of preventive approaches to malaria exposure risk. Addressing these social-spatial determinants of malaria requires a multipronged approach that includes improving access to healthcare services, addressing poverty and education disparities, implementing locally adaptable and effective mosquito control measures, and understanding and harnessing the role of gender and age in promoting behavioral change through education and outreach programs. Therefore, the proposed comprehensive approach encompassing SSDoH offers more than a holistic understanding; it offers an opportunity for strategic adaptation and the necessary rethinking of ongoing malaria interventions. This directly responds to the uniqueness of various community risks and strengths considered in malaria control. This way, tailored and sustainable approaches will be ensured, and the investment will be impactful because it addresses the root of differential causes and effects of malaria sustained by inequalities in multidimensional poverty and other social indices but amplified by inherent susceptibility to environmental risk exposures.

References

Aberese-Ako, M., Magnussen, P., Ampofo, G. D., & Tagbor, H. (2019). Health system, sociocultural, economic, environmental, and individual factors influencing bed net use in the prevention of malaria in pregnancy in two Ghanaian regions. *Malaria Journal, 18*(1), 1–13.

Arroz, J. A. H. (2017). Social and behavior change communication in the fight against malaria in Mozambique. *Revista de Saúde Pública, 51*, 18.

Bayode, T., & Siegmund, A. (2022). Social determinants of malaria prevalence among children under five years: A cross-sectional analysis of Akure, Nigeria. Scientific African, 16, e01196.

Bennett, A., Yukich, J., Miller, J. M., Keating, J., Moonga, H., Hamainza, B., et al. (2016). The relative contribution of climate variability and vector control coverage to changes in malaria parasite prevalence in Zambia 2006–2012. *Parasites & Vectors, 9*(1), 1–12.

Birkholtz, L. M., Bornman, R., Focke, W., Mutero, C., & De Jager, C. (2012). Sustainable malaria control: Transdisciplinary approaches for translational applications. *Malaria Journal, 11*, 1–11.

Caminade, C., Kovats, S., Rocklov, J., Tompkins, A. M., Morse, A. P., Colón-González, F. J., et al. (2014). Impact of climate change on global malaria distribution. *Proceedings of the National Academy of Sciences, 111*(9), 3286–3291.

Cardona-Arias, J. A. (2023). Synthesis of qualitative evidence on malaria in pregnancy, 2005–2022: A systematic review. *Tropical Medicine and Infectious Disease, 8*(4), 235.

Chanda, E., Kamuliwo, M., Steketee, R. W., Macdonald, M. B., Babaniyi, O., & Mukonka, V. M. (2012). An overview of the malaria control programme in Zambia. *ISRN Prev Med, 2013*, 495037. https://doi.org/10.5402/2013/495037

Dahlgren, G., & Whitehead, M. (2021). The Dahlgren-Whitehead model of health determinants: 30 years on and still chasing rainbows. *Public Health, 199*, 20–24. https://doi.org/10.1016/j.puhe.2021.08.009

Endo, N., Eltahir, E.A.B. (2016). Environmental determinants of malaria transmission in African villages. Malar J 15, 578. https://doi.org/10.1186/s12936-016-1633-7.

de la Fuente, A., Murr, A., & Rascón, E. (2015). Mapping Subnational Poverty in Zambia. Dotter, C., & Klasen, S. (2017). The multidimensional poverty index: achievements, conceptual and empirical issues (No. 233). Courant Research Centre: Poverty, Equity and Growth-Discussion Papers.

Hast, M. A., Chaponda, M., Muleba, M., Kabuya, J. B., Lupiya, J., Kobayashi, T., Shields, T., Lessler, J., Mulenga, M., Stevenson, J. C., Norris, D. E., & Moss, W. J. (2019). The Impact of 3 Years of Targeted Indoor Residual Spraying With Pirimiphos-Methyl on Malaria Parasite Prevalence in a High-Transmission Area of Northern Zambia. American journal of epidemiology, 188(12), 2120–2130. https://doi.org/10.1093/aje/kwz107.

Hay, S. I., Guerra, C. A., Tatem, A. J., Noor, A. M., & Snow, R. W. (2004). The global distribution and population at risk of malaria: Past, present, and future. (2004). *The Lancet Infectious Diseases, 4*(6), 327–336. https://doi.org/10.1016/S1473-3099(04)01043-6

Hill, J., Hoyt, J., van Eijk, A. M., D'Mello-Guyett, L., Ter Kuile, F. O., Steketee, R., et al. (2013). Factors affecting the delivery, access, and use of interventions to prevent malaria in pregnancy in sub-Saharan Africa: A systematic review and meta-analysis. *PLoS Medicine, 10*(7), e1001488.

Janko, M. M., Irish, S. R., Reich, B. J., et al. (2018). The links between agriculture, anopheles mosquitoes, and malaria risk in children younger than 5 years in The Democratic Republic of the Congo: A population-based, cross-sectional, spatial study. *Lancet Planet Health, 2*(2), e74–e82. https://doi.org/10.1016/s2542-5196(18)30009-3

Kamanga, A., Moono, P., Stresman, G., et al. (2010). Rural health centres, communities and malaria case detection in Zambia using mobile telephones: A means to detect potential reservoirs of infection in unstable transmission conditions. *Malar Journal, 9*, 96. https://doi.org/10.1186/1475-2875-9-96

Liu Q, Jing W, Kang L, Liu J, Liu M. Trends of the global, regional and national incidence of malaria in 204 countries from 1990 to 2019 and implications for malaria prevention. J Travel

Med. 2021 Jul 7;28(5):taab046. https://doi.org/10.1093/jtm/taab046. PMID: 33763689; PMCID: PMC8271200.

Lubinda, J., Bi, Y., Hamainza, B., Haque, U., & Moore, A. J. (2021a). Modelling of malaria risk, rates, and trends: A spatiotemporal approach for identifying and targeting sub-national areas of high and low burden. *PLoS Computational Biology, 17*(3), e1008669.

Lubinda, J., Haque, U., Bi, Y., Hamainza, B., & Moore, A. J. (2021b). Near-term climate change impacts sub-national malaria transmission. *Scientific Reports, 11*(1), 1–13.

Masaninga, F., Chanda, E., Chanda-Kapata, P., Hamainza, B., Masendu, H. T., Kamuliwo, M., Kapelwa, W., Chimumbwa, J., Govere, J., Otten, M., Fall, I. S., & Babaniyi, O. (2013). Review of the malaria epidemiology and trends in Zambia. *Asian Pac J Trop Biomed, 3*(2), 89–94. https://doi.org/10.1016/S2221-1691(13)60030-1

Masaninga, F., Mukumbuta, N., Ndhlovu, K., Hamainza, B., Wamulume, P., Chanda, E., ... & Kawesha-Chizema, E. (2018). Insecticide-treated nets mass distribution campaign: benefits and lessons in Zambia. Malaria Journal, 17, 1–12.

Miller, J. E., Malm, K., Serge, A. A., Ateba, M. J., Gitanya, P., Sene, D., et al. (2022). Multi-country review of ITN routine distribution data are ANC and EPI channels achieving their potential. *Malaria Journal, 21*(1), 1–11.

Miller J. M., Robinson A. L., Seiber E., Chanda P., Chizema E., Mukuka C., Mohamed A., and Steketee R. W. (2008). Household coverage of insectide-treated mosquito nets associated with a reductioncin febrile episodes, malaria, and anemia in Zambian children: results of a national malaria indicator survey, The American Journal of Tropical Medicine and Hygiene. 79.

MoH (2008a). Guidelines on the Distribution and Utilization of Insecticide Treated Nets for Malaria Prevention and Control, Ministry of Health, Lusaka, Zambia.

MoH (2008b). Zambia National Malaria Indicator Survey Report, 2008, Ministry of Health, Lusaka, Zambia.

MoH (2009). National Guidelines for Indoor Residual Spraying in Zambia, Ministry of Health, Lusaka, Zambia.

MoH. (2010a). *Achievements in malaria control: The Zambian story 2000–2010.* Ministry of Health.

MoH. (2010b). *Zambia national malaria indicator survey report.* PATH, MACEPA, CDC, WHO, Ministry of Health. Retrieved from http://www.nmcc.org.zm/files/FullReport/ ZambiaMIS2010_000pdf

MoH. (2010c). *Zambia national malaria programme performance review* (p. 2010). Ministry of Health.

MoH. (2011). *National Malaria Control Programme Strategic Plan for FY 2011–2015: "consolidating malaria gains for impact".* National Malaria Control Programme, Ministry of Health.

MoH. (2017). *National Malaria Elimination Strategic Plan 2017–2021: Moving from accelerated burden reduction to malaria elimination in Zambia.* National Malaria Elimination Programme, Ministry of Health.

MoH. (2022a). *National Malaria Elimination Strategic Plan 2022–2026: A strengthened, tailored approach to accelerate burden reduction and establish malaria-free communities.* National Malaria Elimination Programme, Ministry of Health.

MoH. (2022b). *Zambia national malaria indicator survey 2021.* National Malaria Elimination Programme, Ministry of Health.

Mutale, P., & Mbewe, P. (2017). *Socioeconomic determinants of malaria among children in Zambia.* GRIN Verlag.

Mwangu, L. M., Mapuroma, R., & Ibisomi, L. (2022). Factors associated with non-use of insecticide-treated bed nets among pregnant women in Zambia. *Malaria Journal, 21*(1), 1–9.

Nawa, M. (2019). Influence of history, geography, and economics on the elimination of malaria: A perspective on disease persistence in rural areas of Zambia. *International Journal of Travel Medicine and Global Health, 7*(4), 113–117.

Nawa, M., Hangoma, P., Morse, A. P., & Michelo, C. (2019). Investigating the upsurge of malaria prevalence in Zambia between 2010 and 2015: a decomposition of determinants. *Malar Journal, 18*, 61.

Nissan, H., Ukawuba, I., & Thomson, M. (2021). Climate-proofing a malaria eradication strategy. *Malaria Journal, 20*, 1–16.

Oxford Poverty and Human Development Initiative (OPHI). (2022). *Global MPI Country Briefing 2022: Zambia (Sub-Saharan Africa).* Oxford. OPHI.

Pell, C., Straus, L., Andrew, E. V., Menaca, A., & Pool, R. (2011). Social and cultural factors affecting uptake of interventions for malaria in pregnancy in Africa: A systematic review of the qualitative research. *PLoS One, 6*(7), e22452.

Pfeffer, D. A., Lucas, T. C. D., May, D., et al. (2018). malariaAtlas: An R interface to global malariometric data hosted by the malaria atlas project. *Malar Journal, 17*, 352. https://doi.org/10.1186/s12936-018-2500-5

Pinchoff, J., Hamapumbu, H., Kobayashi, T., Simubali, L., Stevenson, J., Norris, D., et al. (2015). Factors associated with sustained use of long-lasting insecticide-treated nets following a reduction in malaria transmission in southern Zambia. *Am J Trop Med Hyg, 93*, 954–960.

Ricci, F. (2012). Social implications of malaria and their relationships with poverty. *Mediterranean Journal of Hematology and Infectious Diseases, 4*(1), e2012048.

Roll Back Malaria (2001) Country Strategies and Resource Requirements. WHO/CDS/RBM/2001.34. https://iris.who.int/bitstream/handle/10665/66951/WHO_CDS_RBM_2001.34.pdf?sequence=1.

Shayo, E. H., Rumisha, S. F., Mlozi, M. R., Bwana, V. M., Mayala, B. K., Malima, R. C., et al. (2015). Social determinants of malaria and health care seeking patterns among rice farming and pastoral communities in Kilosa District in Central Tanzania. *Acta Tropica, 144*, 41–49.

Shifa, Muna, and Leibbrandt, Murray (2017), Profiling Multidimensional Poverty and Inequality in Kenya and Zambia at Sub-National Levels, Consuming Urban Poverty Project Working Paper No. 3, African Centre for Cities, University of Cape Town.

Shimaponda-Mataa, N. M., Tembo-Mwase, E., Gebreslasie, M., Achia, T. N., & Mukaratirwa, S. (2017a). Modelling the influence of temperature and rainfall on malaria incidence in four endemic provinces of Zambia using semiparametric Poisson regression. *Acta Tropica, 166*, 81–91.

Shimaponda-Mataa, N. M., Tembo-Mwase, E., Gebreslasie, M., & Mukaratirwa, S. (2017b). Knowledge, attitudes and practices in the control and prevention of malaria in four endemic provinces of Zambia. *Southern African Journal of Infectious Diseases, 32*(1), 29–39.

Stevenson, J. C., Simubali, L., Mudenda, T., et al. (2018). Controlled release spatial repellent devices (CRDs) as novel tools against malaria transmission: A semi-field study in Macha, Zambia. *Malar Journal, 17*, 437. https://doi.org/10.1186/s12936-018-2558-0

Syafruddin, D., Asih, P. B. S., Rozi, I. E., Permana, D. H., Nur Hidayati, A. P., Syahrani, L., Zubaidah, S., Sidik, D., Bangs, M. J., Bøgh, C., Liu, F., Eugenio, E. C., Hendrickson, J., Burton, T., Baird, J. K., Collins, F., Grieco, J. P., Lobo, N. F., & Achee, N. L. (2020). Efficacy of a spatial repellent for control of malaria in Indonesia: A cluster-randomized controlled trial. *The American Journal of Tropical Medicine and Hygiene, 103*(1), 344–358. https://doi.org/10.4269/ajtmh.19-0554

Thornton, R. L., Glover, C. M., Cené, C. W., Glik, D. C., Henderson, J. A., & Williams, D. R. (2016). Evaluating strategies for reducing health disparities by addressing the social determinants of health. *Health Affairs, 35*(8), 1416–1423.

Weiss, D. J., Mappin, B., Dalrymple, U., et al. (2015). Re-examining environmental correlates of Plasmodium falciparum malaria endemicity: A data-intensive variable selection approach. *Malar Journal, 14*, 68. (2015). https://doi.org/10.1186/s12936-015-0574-x

World Health Organization. (2022). *World malaria report 2022.* World Health Organization.

World Bank. (2022). *Poverty and shared prosperity 2022: Correcting Course.* World Bank. Retrieved October13, 2022, from https://openknowledge.worldbank.org/bitstream/handle/10986/37739/9781464818936.pdf.

Zingani, E., Kalungia, G. M. C., Mukosha, M., & Banda, A. (2017). Socioeconomic and sociocultural factors affecting malaria control interventions in Zambia. *Journal of Preventive and Rehabilitative Medicine. Medicine, 1*(1), 25–33.

Chapter 19
The Interstate Highway System as a Tool of Segregation and the Subsequent Impact on Community Health

Brooke Ury, Gregory N. Gibson, and Diana S. Grigsby-Toussaint

Introduction

The Interstate Highway System was touted as a demonstration of America's wealth and prosperity, a revolutionary idea to spur economic growth and development through the connection of America's largest cities. However, recent scholarship has challenged the altruistic vision of the Interstate Highway System after analyzing the impact of highways on minority communities throughout the USA. In many cases, the interstate highways cut through minority communities, displacing millions and disrupting thriving communities. There is a growing sentiment that portions of the Interstate Highway System do more harm than good; in fact, dozens of cities across the USA are planning or even committing to removing portions of the Interstate Highway System to reconnect minority communities (Popovich et al., 2021).

Using a social determinants of health lens, this chapter will explore the history of the Interstate Highway System, its impact on minority communities, and potential policy changes or considerations to improve health and living conditions. First, the chapter will provide an overview of the social determinants of the development of the Interstate Highway System in the USA and the implications for health and well-being. Then, we will explore how the interstate routes deliberately and directly targeted minority communities and the lasting impacts of highway construction on the health and well-being of these communities, drawing upon the cities of Atlanta and Detroit as case studies. Finally, the chapter will review potential policy solutions to mitigate and reduce the detrimental impact of highways on communities, including important considerations for current and future projects.

B. Ury · G. N. Gibson · D. S. Grigsby-Toussaint (✉)
School of Public Health, Brown University, Providence, RI, USA
e-mail: diana_grigsby-toussaint@brown.edu

© The Author(s) 2026   317
M. A. Kolak, I. K. Moise (eds.), *Place and the Social-Spatial Determinants of Health*, Global Perspectives on Health Geography,
https://doi.org/10.1007/978-3-031-88463-4_19

Social Determinants of the Development of the Interstate Highway System in the USA

Expositions of social determinants of health require an exploration of both historical and contemporary factors that lead to certain health-promoting or health-harming outcomes. To fully understand the reasoning for the spatial distribution of highways in the USA, it is important to delve into the role those policies in various spheres of life (e.g., racial residential segregation, education) played in determining which groups bore the brunt of the changes implemented by the government (Braveman, 2023). Laws such as Plessy v. Ferguson (1896) codified the notion of "separate but equal" accommodations for Blacks and Whites, greatly impacting the lived experience of Blacks. Clear differences existed in access to quality educational opportunities, which impact future earnings prospects, and cachet to take on leadership roles in one's community. Additionally, more recent policies related to racial covenants, exclusionary zoning, and redlining limited the capacity for racial and ethnic minority groups to build wealth, a precursor to leadership opportunities.

As shown in Fig. 19.1, upstream factors, such as the implementation of the Federal-Aid Highway Act of 1944, were not implemented using diverse and inclusive planners or community input to ensure that certain groups were not disadvantaged. Without diverse participation, many racially minoritized communities were seen as an afterthought as highways were being built (Archer, 2020; Dillon & Poston, 2021; Semuels, 2016). Consequently, this leads to limited access to resources, such as inadequate transportation and the destruction of homes. The lack of resources in one's surroundings may impact downstream factors, such as the increased likelihood of social isolation and chronic stress, and ultimately adverse health outcomes, such as poor sleep and cardiovascular outcomes, among others.

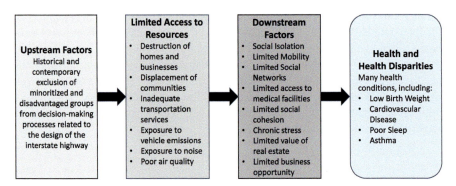

Fig. 19.1 Overview of upstream and downstream factors that influence health and health disparities due to interstate highway design: A framework to describe how upstream and downstream factors influence health and health disparities due to the design of the Interstate Highway System, including the historical and contemporary excursion of minority populations, limited access to resources, and the subsequent impact on the health and well-being of communities

The Interstate Highway System as a Tool of Urban Renewal and Segregation

Planning for the Interstate Highway System in the USA began in the 1930s and was largely a collaboration between federal, state, and local officials. An early feasibility study completed in 1938 recommended a toll-free interregional highway network that followed existing roads wherever possible, leading President Franklin D. Roosevelt to appoint a national highway committee in 1941 (Weingroff, 2017; Weingroff, n.d.). In 1944, the highway committee released a report entitled "Interregional Highways," including a tentative selection of rural routes (National Interregional Highway Committee, 1944). This report was particularly detailed in the planning of interstate highways in cities, demonstrating great concern over the "slums" and "blighted areas" that would persist without rehabilitation measures (National Interregional Highway Committee, 1944; Weingroff, n.d.). The report noted that "downtown areas" are "cramped, crowded, and depreciated," and that "so long... as the central areas of the cities are poor places in which to live and rear children, people will continue to move to the outskirts. Undoubtedly a factor that has facilitated this movement has been the improvement of highways" (National Interregional Highway Committee, 1944, p.54).

The "Interregional Highways" report recommended freeways that cut through the center of major cities, even though a disproportionate number of minorities lived in these areas. These ideas of redesigning cities were prevalent in urban planning at the time; it was commonplace for planners to call for the clearing and development of urban areas to "redeem" them (Semuels, 2016). One of the most influential post-World War II urban planners, Robert Moses, who oversaw all public works projects in New York City, noted in a 1959 speech that "our categorical imperative is action to clear the slums... We can't let minorities dictate that this century-old chore will be put off another generation or finally abandoned" (Naylor, 2016, para. 5). Black neighborhoods were considered to be blight and slums in need of eradication, a sentiment echoed in the language of the "Interregional Highways" report (Dillon & Poston, 2021).

Based on the tentative selection of routes included in the 1944 report, the Federal-Aid Highway Act of 1944 directed that the interstate highway routes would be finalized by joint efforts of state and local officials and then reviewed by the Public Roads Administration (PRA) and Department of Defense (Federal-Aid Highway Act, 1944). By 1955, the PRA published the "General Location of National System of Interstate Highways," which included plans for 37,000 miles of the overall highway system, as shown in Fig. 19.2 (U.S. Department of Commerce, 1955; Weingroff, n.d.).

Although the creation of this 1955 report and other subsequent plans did involve the collaboration of government at many levels, minority communities were unable to meaningfully contribute to the planning of the Interstate Highway System due to historical oppression, limited power in government, and socioeconomic inequalities. This was compounded by a lack of representation in government: At the time,

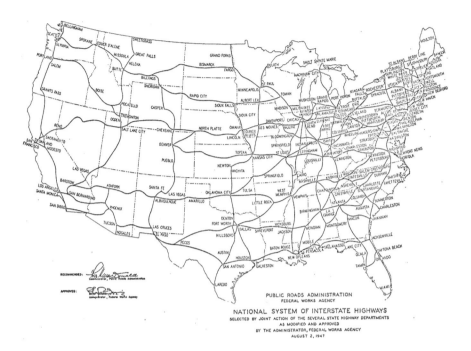

Fig. 19.2 1955 Proposed Routes for the Interstate Highway System: Map of the proposed network of interstate highways prepared by the Public Roads Administration and published in the General Location of the National System of Interstate Highways in 1955. *Source:* U.S. Department of Commerce 1955

lawmakers in federal, state, and local governments were mostly White men (Epperly et al., 2020). Minority community members were also excluded from participating in the decision-making process, leading White residents to be more informed and better able to advocate for themselves. For example, when the District of Columbia Health and Welfare Council interviewed Washington DC residents in 1963 to assess public opinion on highway proposals, White individuals were generally more informed of the proposals and were more empowered to impact highway planning (Kemp, 1965). A resident of the predominantly White, upper-class neighborhood believed they were "financially able to put up a strong fight against putting a freeway through their property" (Kemp, 1965, p. 96). In contrast, a resident of a predominantly Black, low-income neighborhood expressed her concern over her inability to impact highway planning, noting she "did not care when and if construction started because she could not do anything about it. 'You don't have the vote in Washington, so there's nothing you can say'" (Kemp, 1965, p. 96). Regardless of the opinion of the Black community, there was a widespread belief that the government would not consider these points of view.

Further evidence of the exclusion of minority residents from the interstate planning process is demonstrated by early protests against the construction of the Interstate Highway System. While well-connected community members were able to block the interstate planned to go through the French Quarter in New Orleans, then a mostly White neighborhood, the plan to create an interstate through the majority-minority Tremé neighborhood went ahead as planned—there was no public hearing, nor did officials bother to consult with minority residents (Gershon, 2021). Similarly, Nashville civil officials added a curve to the proposed interstate design to avoid a White neighborhood after protests, instead redirecting the interstate to cut directly through a Black neighborhood (Dillon & Poston, 2021). White residents had additional connections with planners and policymakers, were more informed of proposed interstate routes, and overall had much more political power, allowing them to influence and even change interstate routes, while minority communities were effectively excluded from the planning process.

The Interstate Highway System required seizing an unprecedented amount of land under eminent domain, effectively creating the opportunity for social engineering. While highway construction supported White Americans, Black Americans were uprooted and displaced—city, state, and the federal government effectively financed the containment of Black Americans in old urban centers (Mohl, 2002; Seligman, 2003). In some communities, highway development was a tool of the segregationist agenda, whereby highways were purposefully enacted to "protect" White people from Black migration through the creation of highways that separated Black and White communities (Archer, 2020).

The Interstate Highway System exacerbated segregation throughout the country, limiting access to resources and destroying communities. After the Interstate Highway Act of 1956 allocated the 26 billion dollars necessary to construct the Interstate Highway System, construction demolished 37,000 urban housing units each year in the 1960s, which were disproportionately owned by low-income and Black residents (Mohl, 2002; Weingroff, n.d.). After the expansion of a Miami interstate, for instance, only 8000 of an estimated population of 40,000 remained in the once-thriving "Harlem of the South" neighborhood (Evans, 2023). The destruction of neighborhoods limited social networks, destroyed businesses, and reduced real estate value, effectively reducing the wealth of these communities and limiting economic opportunities (Dillon & Poston, 2021). While White communities in New Orleans were unaffected after successfully blocking the construction through the French Quarter, the inability of minority residents to advocate for themselves resulted in a divided neighborhood and the destruction of 500 homes, yielding a subsequent drop in economic activity and quality of life (Gershon, 2021). By displacing so many people and businesses while creating physical barriers between neighborhoods, the Interstate Highway System severely restricted the ability of minority communities to thrive.

Throughout the development of the Interstate Highway Systems, planners targeted minority communities. Socioeconomic barriers and political repression prevented minority communities from any meaningful involvement in the development of the Interstate Highway System, while an emphasis on "urban renewal" led to the integration of interstates into the segregationist agenda. As a result, planners built the Interstate Highway System to bring economic development to White communities while deliberately demolishing minority neighborhoods and displacing minority residents.

Impact of Highway Development on Community Well-being

As aforementioned, highway planners presented demolition programs as part of an "urban renewal" campaign, ostensibly aimed at benefiting residents by replacing old, blighted infrastructures with new constructions (Sugrue, 2005). However, James Baldwin, a prominent voice in the American Civil Rights movement, proclaimed that "urban renewal means negro removal" (Graham, 2015). This is evident in major cities, such as Atlanta, Georgia, and Detroit, Michigan, where highways were used as a tool for segregation and destruction of Black neighborhoods. Ultimately, the construction of interstate highways in these cities resulted in the erasure of culture and the creation of unhealthy living conditions.

Atlanta Interstate Highway Project

During the 1950s, Atlanta was recognized as the "cradle of the modern Civil Rights Movement" (Atlanta History Center, n.d., para. 1). This recognition was partly attributed to the thriving Black community and the civic organizations situated on Sweet Auburn, such as the Southern Christian Leadership Conference (SCLC), that contributed to social and civil rights advancements in the South (Inwood, 2007). Following the post-World War II migration of Black Americans to Atlanta in pursuit of economic opportunities, city planners made efforts to enforce segregation through the creation of the Interstate Highway System (Bayor, 1996; Kruse, 2019).

In a 1960 report created by the Atlanta Bureau of Planning, the Interstate 20 (I-20) West was designed to "be the boundary between the White and Negro communities" (Bayor, 1996, p. 101). As depicted in Fig. 19.3, the creation of I-20 created a barrier between Black neighborhoods in the north and White neighborhoods in the south (Bayor, 1996). Construction of the interstate led to the removal of approximately 7500 people and the destruction of 2200 homes, disproportionately affecting the Black community (Bayor, 1996; Grimminger & Kenny, 2023).

Moreover, city planners deliberately converged Interstate 75 (I-75) and Interstate 85 (I-85) into Interstate 75/85 and routed the interstate through Sweet Auburn, as shown in Fig. 19.4 (Keating, 2001). Interstate 75/85, also known as the Downtown Connector, resulted in the displacement of 68,000 people, who were majority Black (Inwood, 2007; Keating, 2001). Subsequently, in the 1970s and 1980s, the Downtown connector caused the deterioration and decline of Sweet Auburn, once known as the "richest [Black] street in the world" (Grant, 1993, p. 543).

Today, it is estimated that Atlanta loses $6.4 million a year in property tax due to the creation of I-20 (Grimminger & Kenny, 2023). Moreover, because of the

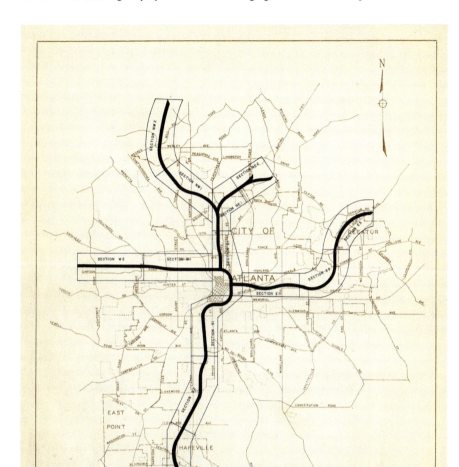

Fig. 19.3 Map of the 1946 Lochner Plan: The Lochner plan for the metro Atlanta expressway system created by the Georgia Department of Transportation showing Interstate 20 traveling east to west and Interstate 75 and Interstate 85 converging into the Downtown Connector. *Source*: H. W. Lochner and Company, and DeLeuw, Cather and Company. (1946)

emphasis on racial segregation instead of the functionality of Atlanta's interstate highways (converging I-75 and I-85 through Sweet Auburn to create the Downtown Connector and creating I-20 as a racial boundary), Atlanta's traffic is one of the worst in the United States (American Transportation Research Institute, 2022). The American Transportation Research Institute (ATRI) ranked the area outlined in yellow in Fig. 19.4, where the Downtown connector and I-20 merge, as the 4th most traffic-congested area in the nation due to the merging of three major interstates:

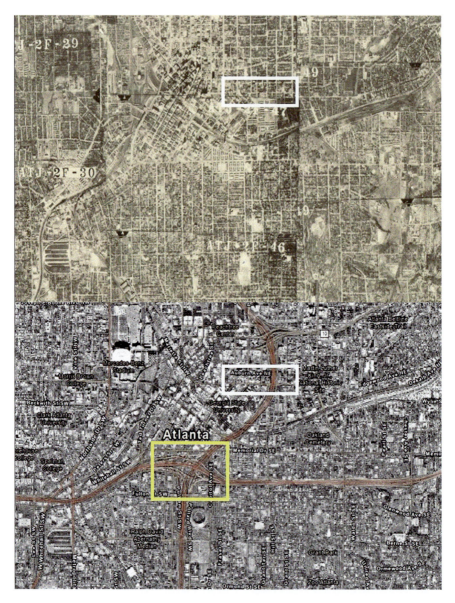

Fig. 19.4 Map of Atlanta before and after the construction of the Downtown Connector: Top—Aerial image of Atlanta, Georgia, in 1949 with a white box depicting the approximate location of Sweet Auburn. Bottom—Aerial image of Atlanta, Georgia, in 1992 after the construction of the Downtown Connector (illustrated as red lines, extending both north and south) with a white box depicting the approximate location of Sweet Auburn. The yellow box shows where the Downtown Connector and I-20 (illustrated as red lines, extending both east and west) merge. *Source:* U.S. Agricultural Stabilization and Conservation Service Aerial Photography Division, 1978; City of Johns Creek, GA, 1993. Compiled with ArcGIS Pro 3.1.0 (Esri, 2023). https://www.esri.com/en-us/arcgis/products/arcgis-pro/overview

I-75, I-85, and I-20 (ATRI, 2023). As a result, the traffic in Atlanta has adverse implications on individuals' health and has been shown to disproportionately affect Black individuals (Servadio et al., 2019). The mechanism between highway traffic and community well-being and health will be discussed later in the chapter.

Detroit Interstate Highway Project

The construction of the highway system as a mechanism for the erasure and marginalization of Black culture, wealth, and communities extended beyond the confines of the southern USA. This phenomenon manifested throughout the country, with implications evident in Detroit, Michigan. During the 1940s, Detroit highway planners constructed new expressway plans that razed through Black neighborhoods such as Paradise Valley and Black Bottom while minimally disrupting White middle-class neighborhoods (Sugrue, 2005).

Before the interstate highway project, Paradise Valley and Black Bottom were home to over 120,000 Black residents (Coleman, 2021). In 1942, these vibrant communities hosted over 300 Black-owned institutions, encompassing churches, nightclubs, bars, hair salons, hotels, restaurants, and medical offices (Coleman, 2021; Sugrue, 2005). In the late 1940s, city planners targeted these areas for highway construction. The Chrysler Expressway, as shown in Fig. 19.5, was placed through Paradise Valley and Black Bottom. To make way for the expressway, city officials condemned and destroyed numerous Black-owned institutions and displaced thousands of Black residents (Sugrue, 2005).

A Black business owner in the area described the aftermath as "no man's land of deterioration and abandonment" (Sugrue, 2005, p. 47). Figure 19.6 shows the loss of city infrastructure due to the Detroit Expressway project. Compounding the issue, ongoing housing discrimination in the city inhibited displaced Black residents from seeking new residences in other parts of the city (Coleman, 2021).

Today, the presence of freeways in Detroit has become a significant contributor to noise and air pollution (Brinkman & Lin, 2022). These "freeway disamenities" have contributed to a decline in the central city population (Brinkman & Lin, 2022). The most substantial burden of pollution is concentrated in areas adjacent to interstate highways (Martenies et al., 2017). Considering the proximity of these highways to minority neighborhoods, the adverse effects of pollution disproportionately impact the marginalized populations in Detroit (Data Driven Detroit, 2013; Martenies et al., 2017).

Ultimately, these two cases in Atlanta and Detroit underscore the deliberate placement of interstate highways, aimed at enforcing segregation between White and Black communities while concurrently erasing the cultural and economic affluence within Black communities. The upstream effects of excluding minoritized groups from highway planning and displacing these groups resulted in substantial economic losses and limited social mobility for minoritized groups (Grimminger & Kenny, 2023). In particular, the destruction of Black homes and institutions

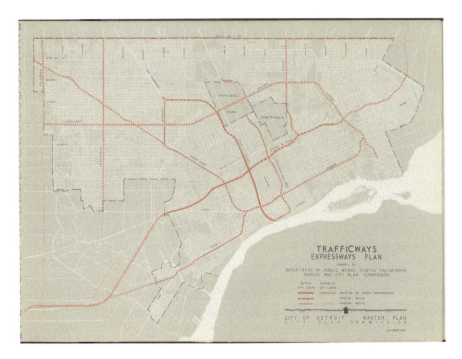

Fig. 19.5 Map of the 1959 Detroit Expressway System plan: The original plans for the Detroit Expressway System, which cut through Black neighborhoods such as Paradise Valley and the Black Bottom. *Source:* Walter P. Reuther Library Archives of Labor and Urban Affairs, Wayne State University, 1959

hindered Black residents from participating in modes of generational wealth building such as homeownership, causing lasting impacts on Black wealth. The forced displacement of Black residents to homes near highways resulted in a depreciation of their property values, exacerbating the overall negative impact on their wealth (Allen et al., 2015). Considering the profound influence of wealth on health outcomes, the racial wealth gap stemming from interstate highway construction could partially explain the observed downstream racial health disparities within urban cities across the USA (Braveman & Gottlieb, 2014). Moreover, the downstream effects of the placement of highways near minority communities have also exacerbated the health impacts that resulted from the construction of highway systems.

Direct Health Impacts of Living Near Highways

Residing near highways can have significant and varied effects on health, particularly concerning noise, air, and heat pollution. The proximity of residential areas to highways exposes residents to elevated levels of noise, air pollutants, and heat, leading to a range of adverse health outcomes.

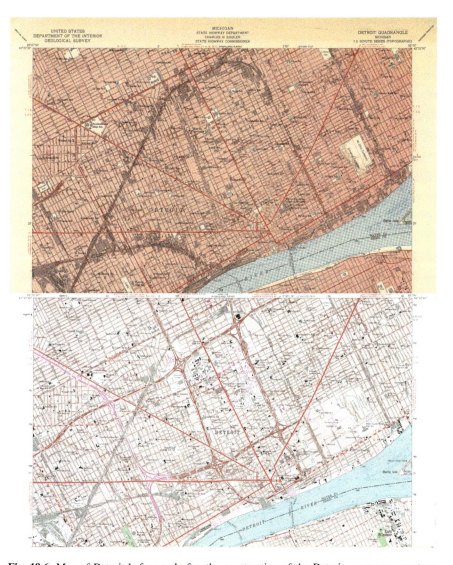

Fig. 19.6 Map of Detroit before and after the construction of the Detroit expressway system: Top—1952 Topographic map of Detroit, Michigan, before construction of expressways. Expressways are visually represented as red, dashed lines. Bottom—1968 Topographic map of Detroit, Michigan, after construction of expressways. Expressways are visually represented as red, dashed lines. *Source*: U.S. Geological Survey, 1952; U.S. Geological Survey, 1968. Compiled with ArcGIS Pro 3.1.0 (Esri, 2023). https://www.esri.com/en-us/arcgis/products/arcgis-pro/overview

Numerous studies highlight the detrimental impact of highway-related noise pollution on health. Living near highways often subjects residents to noise levels exceeding recommended guidelines (Yusoff & Ishak, 2005). A study conducted in Italy found that women with medium to high noise exposure are at higher risk for cardiovascular diseases (CVD) (Bustaffa et al., 2022). Similarly, in Amsterdam, Leijssen et al.'s, 2019 study found a positive association between high-road traffic noise exposure and depressed mood. Additionally, nocturnal road traffic noise in Norway is associated with reduced sleep duration for girls (Weyde et al., 2017). According to the World Health Organization (WHO) estimates, exposure to high traffic noise levels in Western Europe is estimated to result in the loss of at least one million disability-adjusted life years annually (World Health Organization, 2011). Furthermore, in the United States, areas with larger proportions of non-White and lower-socioeconomic status residents have higher noise levels (Casey et al., 2017). This suggests that there could be potential socioeconomic inequalities regarding the association between highway traffic noise exposure and poor health outcomes in the USA. These studies show that highway-related noise pollution is associated with conditions such as CVD, decreased sleep duration, and depressive symptoms, causing significant burden to individuals living near interstates.

Highways also contribute to traffic-related air pollution (TRAP), which has been shown to affect birth outcomes and cognitive, respiratory, and cardiovascular health. Specifically, there is an increased risk of low birth weight associated with proximity to major roads (Dadvand et al., 2014). This association is mediated by exposure to air pollution and heat (Ibid, 2014). Increased exposure to TRAP is also associated with increased odds of having a low birth weight and slower cognitive development among children (Wilhelm et al., 2012; Sunyer et al., 2015). Moreover, TRAP exposure is also associated with adverse respiratory outcomes. A study conducted in Harlem found that TRAP increased the risk of chronic respiratory symptoms in children (Oosterlee et al., 1996). Among children living near major roadways with asthma, diesel exhaust exposure was found to exacerbate symptoms of asthma such as coughing and wheezing (Spira-Cohen et al., 2011). Additionally, long-term exposure to TRAP is associated with increased chronic obstructive pulmonary disease (COPD) and COPD hospitalization (Andersen et al., 2011; Sugiri et al., 2006). Moreover, a recent study found associations between TRAP and lung cancer (Cheng et al., 2022).

Highways also contribute to the urban heat island effect by absorbing and re-emitting heat (Environmental Protection Agency, 2014). This effect results in higher temperatures in urban areas, increasing the risk of heat-related illnesses (Ibid, 2014). Studies have found that individuals' proximity to major roadways puts them at increased risk of heat-related illness (Sorensen & Hess, 2022).

Ultimately, as outlined in the Atlanta and Detroit case studies and Fig. 19.1, the planning and construction of interstate highways have had profound implications on the health of minority groups. As many Black Americans were displaced to areas near freeways, such as in Atlanta and Detroit, they bore disproportionate exposure to traffic-related pollution, raising their risk for health conditions such as noise-related stress, sleep deprivation, respiratory diseases, and CVD. These detrimental health impacts underscored the need for comprehensive policies and interventions to remedy the effects of interstate highways.

Policy Solutions and Considerations

In recognition of the detrimental impacts of the Interstate Highway System, highway removal policies are becoming increasingly common throughout the United States. Although a paradigm shift away from highways has yet to occur, transportation planning is increasingly turning away from highways: Currently, over 30 cities nationwide are discussing some form of highway removal (Popovich et al., 2021; Khalaj et al., 2020). The most common reasons cited for removing highways include economic factors (such as financial concerns and structural damage) but also demand from local communities and academics (Khalaj et al., 2020).

Part of this shift in transportation planning is evident in the Reconnecting Communities and Neighborhoods Grant Program, created by the Inflation Reduction Act of 2022 and spearheaded by the Department of Transportation. The act will distribute $20 billion in grants to "reconnect communities by removing, retrofitting, or mitigating highways or other transportation facilities that create barriers to community connectivity" (Department of Transportation [DOT], n.d., para. 2). The transportation plan acknowledges the disproportionately negative impact of the Interstate Highway System on minority communities, with the Secretary of Transportation, Pete Buttigieg, himself recognizing the racism physically built into the Interstate Highway System, noting in his confirmation hearing that "at their worst, misguided policies and missed opportunities in transportation can reinforce racial and economic inequality, by dividing or isolating neighborhoods and undermining government's basic role of empowering Americans to thrive" (Bagshaw et al., n.d., para.1; Testimony of U.S. Department of Transportation Secretary-Designate Pete Buttigieg, 2021).

Many highway redevelopment projects have already been widely successful. Projects include freeway lids, which are bridges built over highways. A freeway lid containing a park built over an interstate in Seattle in 1975 is lauded as connecting Seattle neighborhoods while bringing vital greenspace to the area (Freeway Park Association, n.d.). Other projects aim to remove sections of interstate entirely. For example, the Central Artery Tunnel Project in Boston (otherwise known as "The Big Dig") was a huge venture to redirect an interstate highway underground and was completed in 2007. The project reconnected communities and created more than 45 parks and public plazas, all while providing around $168 million in time and cost savings to travelers (Commonwealth of Massachusetts, n.d.). Demand from community members, as well as the success of projects like these in Seattle and Boston, has inspired proposals for highway reconstruction efforts.

However, careful planning is necessary to ensure that the benefits of future interstate projects go directly to the communities most impacted by the construction of the Interstate Highway System in the first place. One major unintended consequence of these projects is gentrification: Frequently, an unintended consequence of neighborhood improvements through the construction of new parks or housing is the rapid increase in home prices, driving lower-income communities from the area and thereby excluding the original communities from benefiting from the construction (Tajima, 2003).

The creation of parks where railroad tracks used to be, such as the High Line in New York City or the 606 in Chicago, are well-studied examples of how revisioning transportation can result in gentrification. In New York, the value of homes adjacent to the High Line, which were historically owned by the working-class, increased by 50.6% since June 2011, while comparison areas only appreciated 31.4% (Jo Black & Richards, 2020; Quintana, 2016). In Chicago, prices for buildings along the 606 have increased by nearly 244 percent since 2012, threatening the stability of the historically working-class and racially diverse neighborhood (Dudah et al., n.d.; Jo Black & Richards, 2020).

A similar phenomenon is seen in highway reconstruction projects, such as the rerouting of the Cypress Freeway and the construction of a street-level boulevard in West Oakland. Compared to the rest of West Oakland, there were larger increases in property values (184%) and decreases in the long-time Black population (−28%) (Patterson & Harley, 2019). Similarly, although the"Big Dig" project in Boston increased property values by nearly $1 billion, it is estimated that this improvement will negatively impact low-income minority groups who live in rental housing units, as many will be unable to meet rising housing costs (Tajima, 2003).

Future highway reconstruction efforts must consider existing communities to limit the impact of these projects on gentrification. Fortunately, the Reconnecting Communities and Neighborhoods Grant Program already has elements to ensure that the benefits of the projects go to minority communities. To receive funding, projects must include the voices of communities previously ignored, incorporate objectives such as reconnecting highway-separated communities, or build afford-able housing close to urban centers (Bagshaw et al., n.d.; DOT, n.d.). Furthermore, the grant program aims to prioritize projects with measures that would ensure affordable housing (DOT, n.d.; Popovich et al., 2021). But even smaller-scale proj-ects or projects not funded by this grant program should consider the potential impact on gentrification. By focusing on equity and community, planners can ensure the benefits of these projects go to the communities harmed by the development of the highway in the first place. Previous successful projects, such as the Milwaukee Park East Project in Wisconsin, demonstrate methodologies to improve interstate reconstruction projects.

Milwaukee Park East Project

The construction of Interstate 43 in Milwaukee, Wisconsin, devastated many com-munities. The creation of the interstate faced significant contention—construction was halted in 1972 due to environmental concerns, and further construction of con-necting highways was ultimately canceled in 1977 (Snyder, 2016). However, by the end of 1971, over 1500 homes—nearly all of the properties needed for the Park East project—were taken through eminent domain and demolished (Cutler, 2001). The interstate divided African American, German, Jewish, and White communities and quickly devalued land near the interstate, ultimately affecting as many as 17,300 homes and 1000 businesses (Federal Highway Administration [FHWA], n.d.-a).

The Wisconsin Park East Freeway Removal Project, completed in 2003, removed one mile of the elevated interstate, replacing it with a 6-land boulevard, and effectively opened 24 acres of downtown property for redevelopment, as shown in Fig. 19.7 (Snyder, 2016; City of Milwaukee, n.d.). Already, the project has yielded significant economic impact: since 2003, the Park East Corridor has seen over $1 billion in private investment and brought vitality, jobs, and a physical expansion downtown to the city (City of Milwaukee, n.d.).

One of the factors that made the project so impactful was the focus on producing benefits for the existing community. This included smaller measures, such as a development requirement for a public-private partnership of riverfront property, effectively granting public access to the waterfront (FHWA, n.d.-a). Perhaps most effective, however, was the creation of a Community Benefits Agreement (CBA), a legally binding contract that requires development in specific areas to meet thresholds for affordable housing, a living wage, or even hiring directly from a community (FHWA, n.d.-a; Office of Energy Justice and Equity, n.d.). In Milwaukee, a CBA entitled the "Park East Redevelopment Compact" (PERC) was leveraged and included provisions for affordable housing units, expanded transportation options, and equitable hiring practices for construction (FHWA, n.d.-a). Under the PERC, the county has agreed to fund at least 20% of affordable housing units on county-owned lands, helping to combat gentrification by increasing housing options for low-income community members (FHWA, n.d.-a). Furthermore, the PERC enabled the project to directly re-invest in minority communities, evidenced by provisions including that 25% of construction jobs must be from businesses designated as Disadvantaged Business Enterprises or Minority Business Enterprises, and developers must provide training and apprenticeships to low-skilled and low-income residents (Milwaukee County Board of Supervisors, 2004; Office of Policy Development and Research, n.d.).

Ultimately, the Park East Redevelopment project brought significant industry and investment to Milwaukee, as numerous businesses opened storefronts and offices alongside the development of additional housing (FHWA, n.d.-a). Although redevelopment projects will never undo the original consequences of building the interstate, maintaining an emphasis on community impact ensures the benefits of the reconstruction projects are equitably distributed. By incorporating community involvement and maintaining a diversity focus through hiring, the resulting Park East Boulevard in Milwaukee has been lauded as a place of "authentic diversity," with multiple architects, investors, and developers coming together to create a unique district design (Steuteville, 2020). Although more research is needed to fully understand the impact of the PERC on community redevelopment, this Community Benefits Agreement supported low-income and minority communities through the interstate development process.

As interstate development projects are becoming more common, other projects are also emphasizing equity in the planning processes, such as the removal of the

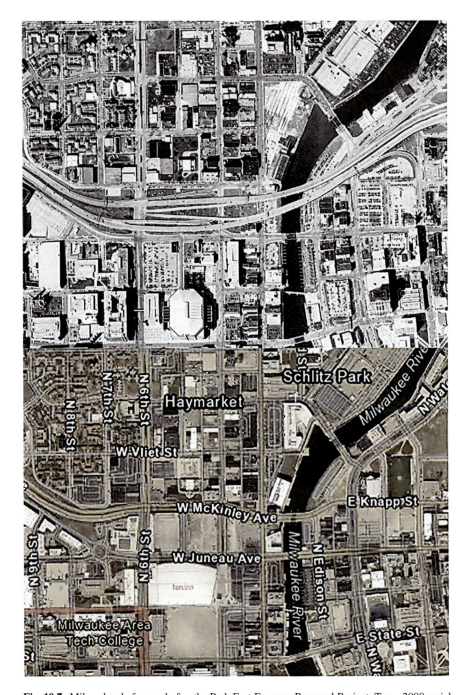

Fig. 19.7 Milwaukee before and after the Park East Freeway Removal Project: Top—2000 aerial image of Milwaukee before the Park East Freeway Removal Project, centered on Interstate 43. Bottom—2020 aerial image of Milwaukee after the Park East Freeway was removed and replaced with the boulevard McKinley Avenue, as well as numerous business and housing developments. *Source:* The Milwaukee County GIS & Land Information Office, 2000; The Milwaukee County GIS & Land Information Office 2020. Compiled with ArcGIS Pro 3.1.0 (Esri, 2023). https://www. esri.com/en-us/arcgis/products/arcgis-pro/overview

Inner East Loop in Rochester, New York. The project raised $229 in economic investment through the construction of new business offices while ensuring that 60% of the over 500 housing units developed were for community members earning less than the area median income (Popovich et al., 2021; City of Rochester, NY, n.d.). Moreover, by removing the highway, both walking and biking increased by 50%, showcasing how highway reconstruction projects can increase mobility and transportation options, which can directly impact community health (FHWA, n.d.-b). There is also growing discourse about how different gentrification solutions can be adapted to fit interstate demolition projects, such as rent-to-own models, community land trusts, and rent control (Popovich et al., 2021; National Low Income Housing Coalition, 2019).

The success of projects such as the Milwaukee Park East Project or even the Rochester Inner East Loop project demonstrates how highway redevelopment projects can employ methodologies to reverse some of the detrimental impacts of the initial development of the Interstate Highway System. By reconnecting communities, supporting local economies, and creating affordable housing, interstate reconstruction projects can strengthen communities and reconnect neighborhoods to vital resources.

Conclusion

The history of the Interstate Highway System in the United States reveals a complex narrative that transcends its initial promise of economic growth and urban development. Despite its portrayal as a symbol of national prosperity, recent literature underscores the adverse impact of interstate highways on minority communities. The chapter highlighted the deliberate targeting of minority neighborhoods during the construction of the Interstate Highway System. The intersection of racial inequality, historical oppression, and lack of minority representation in interstate planning resulted in disproportionate harm toward minority communities.

Highlighting the Atlanta and Detroit interstate highway projects, this chapter demonstrates the lasting consequences of highway development in Black communities. The deliberate placement of highways as a tool for segregation resulted in the erasure of cultural and economic affluence within Black communities. Ultimately, the economic losses and forced displacement continue to disproportionately affect Black wealth and contribute to significant health disparities. The deliberate placement of minority neighborhoods near highways adversely affects the health of residents, mediated by traffic-related noise, air, and heat pollution. These detrimental health impacts underscore the need for comprehensive highway removal policies.

Currently, there is an increasing number of highway removal policies across the USA. However, in some cases, the removal of these highways to construct new parks and improved homes has exacerbated gentrification, threatening the stability of historically working-class and racially diverse neighborhoods. The Reconnecting Communities and Neighborhoods Grant Program represents a significant step in the right direction, emphasizing community involvement, affordable housing, and measures to prevent gentrification. Drawing on successful equity-focused projects like the Wisconsin Park East Project, this chapter advocates for policies like Community

Benefits Agreements which ensure development in specific areas meet thresholds for affordable housing and living wages, ensuring that the benefits of highway reconstruction directly benefit communities most affected.

Ultimately, this chapter demonstrates the imperative need for a paradigm shift in transportation planning, moving toward more equitable and community-centered solutions. By learning from the history of interstate highway development and prioritizing equity-focused interventions, future highway redevelopment projects have the potential to foster community health, address historical injustices, and contribute to the overall well-being of marginalized communities.

References

Allen, M. T., Austin, G. W., & Swaleheen, M. (2015). Measuring highway impacts on house prices using spatial regression. *Journal of Sustainable Real Estate, 7*(1), 83–98. https://doi.org/10.108 0/10835547.2015.12091876

American Transportation Research Institute. (2022). *The nation's top truck bottlenecks 2022.* ATRI. Retrieved from https://truckingresearch.org/wp-content/uploads/2022/02/ATRI-2022-Top-Truck-Bottlenecks-Executive-Summary.pdf

American Transportation Research Institute. (2023). *Atlanta, GA: I-285 at I-85 (North).* ATRI. Retrieved from https://truckingresearch.org/wp-content/uploads/2023/01/bn004-2023.pdf

Andersen, Z. J., Hvidberg, M., Jensen, S. S., Ketzel, M., Loft, S., Sørensen, M., Tjønneland, A., Overvad, K., & Raaschou-Nielsen, O. (2011). Chronic obstructive pulmonary disease and long-term exposure to traffic-related air pollution: A cohort study. *American Journal of Respiratory and Critical Care Medicine, 183*(4), 455–461. https://doi.org/10.1164/rccm.201006-0937OC

Archer, D. N. (2020). "White men's roads through Black men's homes": Advancing racial equity through highway reconstruction. *Vanderbilt Law Review, 73*(5), 1259–1330.

Atlanta History Center. (n.d.). *Civil rights activism | Atlanta in 50 objects | exhibitions.* Atlanta History Center. Retrieved from https://www.atlantahistorycenter.com/exhibitions/atlanta-in-50-objects/civil-rights-activism/

Bagshaw, S., Bonjukian, S., & Feit, J. (n.d. December 15). *Reconnecting what freeways severed: Addressing the historical toll on communities split by highways.* Harvard Advanced Leadership Initiative Social Impact review. Retrieved from https://www.sir.advancedleadership.harvard.edu/articles/reconnecting-what-freeways-severed-addressing-the-historical-toll-on-communities-split-by-highways

Bayor, R. H. (1996). *Race and the shaping of twentieth-century Atlanta.* The University of North Carolina Press.

Braveman, P. (2023). *The social determinants of health and health disparities.* Oxford University Press.

Braveman, P., & Gottlieb, L. (2014). The social determinants of health: It's time to consider the causes of the causes. *Public Health Reports, 129*(Suppl 2), 19–31.

Brinkman, J., & Lin, J. (2022). *Freeway revolts! The quality of life effects of highways (working paper)* (pp. 22–24). Federal Reserve Bank of Philadelphia. https://doi.org/10.21799/frbp.wp.2022.24

Bustaffa, E., Curzio, O., Donzelli, G., Gorini, F., Linzalone, N., Redini, M., Bianchi, F., & Minichilli, F. (2022). Risk associations between vehicular traffic noise exposure and cardiovascular diseases: A residential retrospective cohort study. *International Journal of Environmental Research and Public Health, 19*(16), 10034. https://doi.org/10.3390/ijerph191610034

Casey, J. A., Morello-Frosch, R., Mennitt, D. J., Fristrup, K., Ogburn, E. L., & James, P. (2017). Race/ethnicity, socioeconomic status, residential segregation, and spatial variation in noise exposure in the contiguous United States. *Environmental Health Perspectives, 125*(7), 077017. https://doi.org/10.1289/EHP898

Cheng, I., Yang, J., Tseng, C., Wu, J., Shariff-Marco, S., Park, S.-S. L., Conroy, S. M., Inamdar, P. P., Fruin, S., Larson, T., Setiawan, V. W., DeRouen, M. C., Gomez, S. L., Wilkens, L. R., Le Marchand, L., Stram, D. O., Samet, J., Ritz, B., & Wu, A. H. (2022). Traffic-related air pollution and lung cancer incidence: The California multiethnic cohort study. *American Journal of Respiratory and Critical Care Medicine, 206*(8), 1008–1018. https://doi.org/10.1164/rccm.202107-1770OC

City of Johns Creek, GA. (1993). *Imagery 1993 [Map]*. Retrieved from https://www.arcgis.com/home/item.html?id=db02bee319344f0e86d527152dc25b9e

City of Milwaukee. (n.d.). *Park East Corridor*. City of Milwaukee. Retrieved December 3, 2023, from https://city.milwaukee.gov/DCD/Projects/ParkEastredevelopment

City of Rochester, NY. (n.d.). *Inner loop east project*. City of Rochester, NY. Retrieved December 4, 2023, from https://www.cityofrochester.gov/InnerLoopEast/

National Low Income Housing Coalition. (2019, April 5). *Gentrification and neighborhood revitalization: What's the difference?*. National Low Income Housing Coalition. Retrieved from https://nlihc.org/resource/gentrification-and-neighborhood-revitalization-whats-difference

Coleman, K. (2021). *The people and places of black bottom, Detroit*. The National Endowment for the Humanities. Retrieved from https://www.neh.gov/article/people-and-places-black-bottom-detroit

Commonwealth of Massachusetts. (n.d.). *The big dig: Project background*. Commonwealth of Massachusetts. Retrieved January 9, 2024, from https://www.mass.gov/info-details/the-big-dig-project-background

Cutler, R. W. (2001). *Greater Milwaukee's growing pains, 1950–2000: an insider's view*. Milwaukee County Historical.

Dadvand, P., Ostro, B., Figueras, F., Foraster, M., Basagaña, X., Valentín, A., Martinez, D., Beelen, R., Cirach, M., Hoek, G., Jerrett, M., Brunekreef, B., & Nieuwenhuijsen, M. J. (2014). Residential proximity to major roads and term low birth weight: The roles of air pollution, heat, noise, and road-adjacent trees. *Epidemiology (Cambridge, Mass.), 25*(4), 518–525. https://doi.org/10.1097/EDE.0000000000000107

Data Driven Detroit. (2013). Southwest Detroit Neighborhood Profile. Retrieved from https://datadrivendetroit.org/files/SGN/SW_Detroit_Neighborhoods_Profile_2013_081913.pdf

Dillon, L., & Poston, B. (2021, November 11). The racist history of America's interstate highway boom. Los Angeles Times. Retrieved from https://www.latimes.com/homeless-housing/story/2021-11-11/the-racist-history-of-americas-interstate-highway-boom

Dudah, S., Percel, J., & Smith, G. (n.d.). *Displacement pressure in context: Examining recent housing market changes near the 606*. Institute for Housing Studies—DePaul University. Retrieved October 25, 2023, from https://www.housingstudies.org/releases/Displacement-Pressure-in-Context-606/

Epperly, B., Witko, C., Strickler, R., & White, P. (2020). Rule by violence, rule by law: Lynching, Jim crow, and the continuing evolution of voter suppression in the U.S. perspectives on. *Politics, 18*(3), 756–769. https://doi.org/10.1017/S1537592718003584

Esri. (2023). ArcGIS Pro (Version 3.1) [Computer software]. Environmental Systems Research Institute. https://www.esri.com/enus/arcgis/products/arcgis-pro/overview

Evans, F. (2023, September 21). *How interstate highways gutted communities—And reinforced segregation*. HISTORY. Retrieved from https://www.history.com/news/interstate-highway-system-infrastructure-construction-segregation

Federal Highway Administration. (n.d.-a). *Park east freeway removal project, Milwaukee, Wisconsin*. U.S. Department of Transportation. Retrieved December 3, 2023, from https://www.fhwa.dot.gov/ipd/project_profiles/wi_park_east_freeway.aspx

Federal Highway Administration. (n.d.-b). *Rochester inner loop east, New York, a freeway to boulevard*. U.S. Department of Transportation. Retrieved January 9, 2024, from https://www.fhwa.dot.gov/ipd/project_profiles/ny_freeway_to_boulvard_rochester.aspx

Federal-Aid Highway Act, 58 U.S.C.§ 838 (1944)

Freeway Park Association. (n.d.). *About the Park*. Freeway Park Association. Retrieved January 9, 2024, from https://www.freewayparkassociation.org/about-park/

Gershon, L. (2021, May 28). *The highway that sparked the demise of an iconic Black street in New Orleans | Smart News| Smithsonian Magazine*. Smithsonian

Magazine. Retrieved from https://www.smithsonianmag.com/smart-news/documenting-history-iconic-new-orleans-street-and-looking-its-future-180977854/

Graham, V. (Director). (2015, June 3). *Urban Renewal...Means Negro Removal. ~ James Baldwin* (1963). Retrieved from https://www.youtube.com/watch?v=T8Abhj17kYU.

Grant, D. L. (1993). *The way it was in the south: The black experience in Georgia.* Carol Publication Group.

Grimminger, A., & Kenny, S. (2023). *Divided by design.* Smart Growth America. Retrieved from https://search.issuelab.org/resource/divided-by-design.html

H. W. Lochner and Company, and DeLeuw, Cather and Company. (1946). *Highway and transportation plan for Atlanta, Georgia* (Prepared for the State Highway Department of Georgia and the Public Roads Administration, Federal Works Agency). Chicago, IL. Retrieved from Georgia Department of Transportation Office of Environment, via Georgia Department of Transportation Office of Environment. (2007). *Historic context of the Interstate Highway System in Georgia.* [Map] Retrieved from https://www.dot.ga.gov/AboutGeorgia/CentennialHome/Documents/Historical%20Documents/HistoricalContextof%20GeorgiaInterstates.pdf

Inwood, J. F. J. (2007). Sweet Auburn: Contesting the racial identity of Atlanta's most historically significant African American neighborhood.

Jo Black, K., & Richards, M. (2020). Eco-gentrification and who benefits from urban green amenities: NYC'S high line. *Landscape and Urban Planning, 204*, 103900. https://doi.org/10.1016/j.landurbplan.2020.103900

Keating, L. (2001). *Atlanta: Race, class, and urban expansion.* Temple University Press.

Kemp, B. H. (1965). *Social impact of a highway on an Urban Community.* National Capital Planning Commission and District of Columbia Health and Welfare Council of the National Capital Area.

Khalaj, F., Pojani, D., Sipe, N., & Corcoran, J. (2020). Why are cities removing their freeways? A systematic review of the literature. *Transport Reviews, 40*(5), 557–580. https://doi.org/10.1080/01441647.2020.1743919

Kruse, K. M. (2019). *How segregation caused your traffic jam—The New York times.* Retrieved from https://www.nytimes.com/interactive/2019/08/14/magazine/traffic-atlanta-segregation.html

Leijssen, J. B., Snijder, M. B., Timmermans, E. J., Generaal, E., Stronks, K., & Kunst, A. E. (2019). The association between road traffic noise and depressed mood among different ethnic and socioeconomic groups. The HELIUS study. *International Journal of Hygiene and Environmental Health, 222*(2), 221–229. https://doi.org/10.1016/j.ijheh.2018.10.002

Martenies, S. E., Milando, C. W., Williams, G. O., & Batterman, S. A. (2017). Disease and health inequalities attributable to air pollutant exposure in Detroit, Michigan. *International Journal of Environmental Research and Public Health, 14*(10), Article 10. https://doi.org/10.3390/ijerph14101243

Milwaukee County Board of Supervisors. (2004, December 16). Park East Redevelopment Contract.

Milwaukee County GIS & Land Information Office. (2000). 2000 Aerial Imagery [Map]. Retrieved from https://www.arcgis.com/home/item.html?id=3cc2843344404a6e8fe1ab5c3bab1982

Milwaukee County GIS & Land Information Office. (2020). 2020 Aerial Imagery [Map]. Retrieved from https://www.arcgis.com/home/item.html?id=2347970abf3f4d8593ad30bd11719af3

Mohl, R. A. (2002). The interstates and the cities: Highways, housing, and the freeway revolt. Poverty and Race Research Action Council. Retrieved from https://www.prrac.org/pdf/mohl.pdf

National Interregional Highway Committee (1944, January 12). Interregional highways. United States Government Printing Office.

Naylor, B. (2016, April 28). *After dividing for decades, highways are on the road to inclusion.* NPR. Retrieved from https://www.npr.org/2016/04/28/475985489/secretary-foxx-pushes-to-make-transportation-projects-more-inclusive

Office of Energy Justice and Equity. (n.d.). *Community benefit agreement (CBA) toolkit.* U.S. Department of Energy. Retrieved December 6, 2023, from https://www.energy.gov/justice/community-benefit-agreement-cba-toolkit

Office of Policy Development and Research. (n.d.). *Community benefits agreement guides development in Milwaukee's Park East Corridor.* U.S. Department of Housing and Urban Development. Retrieved January 12, 2024, from https://www.huduser.gov/portal/pdredge/pdr_edge_inpractice_072012.html

Oosterlee, A., Drijver, M., Lebret, E., & Brunekreef, B. (1996). Chronic respiratory symptoms in children and adults living along streets with high traffic density. *Occupational and Environmental Medicine, 53*(4), 241–247. https://doi.org/10.1136/oem.53.4.241

Patterson, R. F., & Harley, R. A. (2019). Effects of freeway rerouting and boulevard replacement on air pollution exposure and neighborhood attributes. *International Journal of Environmental Research and Public Health, 16*(21), 4072. https://doi.org/10.3390/ijerph16214072

Plessy v. Ferguson, 163 U.S. 537 (1896). *Retrieved from the Library of Congress.* Retrieved from https://www.loc.gov/item/usrep163537/

Popovich, N., Williams, J., & Lu, D. (2021, May 27). *Can removing highways fix America's cities?* The New York Times. Retrieved from https://www.nytimes.com/interactive/2021/05/27/climate/us-cities-highway-removal.html

Quintana, M. (2016, August 8). *Changing grid: Exploring the impact of the high line.* StreetEasy. Retrieved from https://streeteasy.com/blog/changing-grid-high-line/

Seligman, A. I. (2003). What is the second ghetto? *Journal of Urban History., 29*, 272–280. https://doi.org/10.1177/0096144202250377

Semuels, A. (2016, March 18). *The Role of Highways in American Poverty.* The Atlantic. Retrieved from https://www.theatlantic.com/business/archive/2016/03/role-of-highways-in-american-poverty/474282/

Servadio, J. L., Lawal, A. S., Davis, T., Bates, J., Russell, A. G., Ramaswami, A., Convertino, M., & Botchwey, N. (2019). Demographic inequities in health outcomes and air pollution exposure in the Atlanta area and its relationship to urban infrastructure. *Journal of Urban Health: Bulletin of the New York Academy of Medicine, 96*(2), 219–234. https://doi.org/10.1007/s11524-018-0318-7

Snyder, A. (2016). Freeway removal in Milwaukee: Three case studies (Doctoral dissertation, The University of Wisconsin-Milwaukee).

Sorensen, C., & Hess, J. (2022). Treatment and prevention of heat-related illness. *New England Journal of Medicine, 387*(15), 1404–1413. https://doi.org/10.1056/NEJMcp2210623

Spira-Cohen, A., Chen, L. C., Kendall, M., Lall, R., & Thurston, G. D. (2011). Personal exposures to traffic-related air pollution and acute respiratory health among bronx schoolchildren with asthma. *Environmental Health Perspectives, 119*(4), 559–565. https://doi.org/10.1289/ehp.1002653

Steuteville, R. (2020, February 13). *Park East Corridor: The freeway teardown that helped put Milwaukee on the national stage this summer.* Milwaukee Independent. Retrieved from https://www.milwaukeeindependent.com/syndicated/park-east-corridor-freeway-teardown-helped-put-milwaukee-national-stage-summer/

Sugiri, D., Ranft, U., Schikowski, T., & Krämer, U. (2006). The influence of large-scale airborne particle decline and traffic-related exposure on children's lung function. *Environmental Health Perspectives, 114*(2), 282–288. https://doi.org/10.1289/ehp.8180

Sugrue, T. J. (2005). *The origins of the urban crisis: Race and inequality in postwar Detroit: With a new preface by the author (1st Princeton classic ed).* Princeton University Press.

Sunyer, J., Esnaola, M., Alvarez-Pedrerol, M., Forns, J., Rivas, I., López-Vicente, M., Suades-González, E., Foraster, M., Garcia-Esteban, R., Basagaña, X., Viana, M., Cirach, M., Moreno, T., Alastuey, A., Sebastian-Galles, N., Nieuwenhuijsen, M., & Querol, X. (2015). Association between traffic-related air pollution in schools and cognitive development in primary school children: A prospective cohort study. *PLoS Medicine, 12*(3), e1001792. https://doi.org/10.1371/journal.pmed.1001792

Tajima, K. (2003). New estimates of the demand for urban green space: Implications for valuing the environmental benefits of Boston's big dig project. *Journal of Urban Affairs, 25*(5), 641–655. https://doi.org/10.1111/j.1467-9906.2003.00006.x

Testimony of U.S. Department of Transportation Secretary-Designate Pete Buttigieg. (2021). Before The Senate Committee on Commerce, Science, and Transportation. 117th Cong. 2 (testimony of Peter Buttigieg).

U.S. Agricultural Stabilization and Conservation Service Aerial Photography Division. (1978). *Fulton County, 1949: Aerial photography index [Map]*. Retrieved from https://dlg.usg.edu/record/gyca_gaphind_fulton-1978#item

U.S. Department of Commerce. (1955, September). *General location of national system of interstate highways*. Bureau of Public Roads. Retrieved from https://dlg.usg.edu/record/gyca_gaphind_fulton-1978#item

U.S. Department of Transportation. (n.d.). *Reconnecting communities and neighborhoods grant program*. U.S. Department of Transportation. Retrieved October 11, 2023, from https://www.transportation.gov/grants/rcnprogram

U.S. Environmental Protection Agency. (2014, June 17). *Learn About Heat Islands [Overviews and Factsheets]*. Retrieved from https://www.epa.gov/heatislands/learn-about-heat-islands

U.S. Geological Survey. (1952). Detroit Quadrangle Michigan—Ontario [map] 1:24,000 [Map].

U.S. Geological Survey. (1968). Detroit Quadrangle Michigan—Ontario [map] 1:24,000 [Map].

Walter P. Reuther Library, Archives of Labor and Urban Affairs, Wayne State University. (1959). *Souvenir pamphlet, Detroit Expressways, map, 1959, 12th Street Detroit [Map]*. Wayne State University. Retrieved from https://projects.lib.wayne.edu/12thstreetdetroit/items/show/347

Weingroff, R. (2017, June 27). *Designating the Urban Interstates*. U.S. Department of Transportation Federal Highway Administration. Retrieved from https://www.fhwa.dot.gov/infrastructure/fairbank.cfm

Weingroff, R. F. (n.d.). *Federal-aid highway act of 1956: Creating the interstate system*. U.S. Department of Transportation Federal Highway Administration. Retrieved October 25, 2023, from https://highways.dot.gov/public-roads/summer-1996/federal-aid-highway-act-1956-creating-interstate-system

Weyde, K. V., Krog, N. H., Oftedal, B., Evandt, J., Magnus, P., Øverland, S., Clark, C., Stansfeld, S., & Aasvang, G. M. (2017). Nocturnal road traffic noise exposure and children's sleep duration and sleep problems. *International Journal of Environmental Research and Public Health, 14*(5), 491. https://doi.org/10.3390/ijerph14050491

Wilhelm, M., Ghosh, J. K., Su, J., Cockburn, M., Jerrett, M., & Ritz, B. (2012). Traffic-related air toxics and term low birth weight in Los Angeles County. *California. Environmental Health Perspectives, 120*(1), 132–138. https://doi.org/10.1289/ehp.1103408

World Health Organization. (2011). *Burden of disease from environmental noise—Quantification of healthy life years lost in Europe*. Retrieved from https://www.who.int/publications-detail-redirect/9789289002295

Yusoff, S., & Ishak, A. (2005). Evaluation of urban highway environmental noise pollution. *Sains Malaysiana, 34*, 81–87.

Index

A
Acceptability and quality, 97, 99
Accessibility, 6, 20, 93, 96, 97, 99, 106, 139, 171–173, 181, 245, 249, 305
Accessibility of health services, 6, 171, 305
Access to healthcare, 6, 22, 54–56, 90–93, 97–99, 160, 197, 240, 287, 303, 305, 312
Activities of daily living, 137, 147, 150
Adverse effects on health, 5
Aggregation, 77, 203, 290, 292
Aging, 5–7, 40, 137, 138, 140, 141, 143, 150, 152, 153, 157, 158, 160, 161, 167, 168, 170, 173, 266
Aging in place, 6, 138, 144
Ambient assisted living, 157
Amenities, 35, 57, 63, 215–217, 219, 221, 222
Architecture, 143, 164
Assistive, 6, 142–144, 147, 149, 151–153, 157–173
Assistive robot, 159, 163–165, 173
Autocorrelation, 7, 58, 203, 252, 283, 287–289, 295
Availability, 6, 7, 74, 79, 92, 96–100, 106, 112, 115, 180, 182, 184, 199, 201, 245, 246, 249, 287, 292, 294, 295, 300, 303, 308

B
Behavioral factors, 140
Built environment, 6, 15, 17, 18, 25, 35–36, 40, 43, 53, 90, 109, 138–141, 152, 180, 186–189, 199, 214, 217, 220, 222, 245, 261, 262, 284, 294

C
Case studies, 5–8, 18, 51, 52, 58–66, 179–181, 184–187, 189, 190, 220, 222, 265, 317, 328
Characteristics of aging, 158–159
Child and family services, 122
Cohesion, 107, 122, 213, 215, 221, 226, 227, 239, 241
Colonialism, 42, 43, 45, 218
Community, 4, 13, 34, 53, 73, 90, 106, 119, 138, 160, 183, 197, 213, 237, 244, 266, 283, 299, 317
Community-based organizations, 8, 241
Community engagement, 53, 57, 107, 311
Community health, 8, 25, 80, 108, 111, 113, 200, 201, 205, 207, 304, 311, 327, 333, 334
Conceptual frameworks, 14, 18, 24, 25, 52, 91
Contextual determinant, 233
Course concepts, 180
Critical geography perspective, 5, 51
Critical praxis, 5, 33, 40–45

D
Design features, 144–146
Design strategies, 140, 306
Differential exposure, 248–250, 252–254, 301
Differential vulnerability, 248–250, 252–254
Disadvantage, 17, 35, 44, 55, 90, 95, 112, 119–133, 200, 205, 215, 245
Disease surveillance, 17, 287
Disparities, 4, 13, 51, 74, 90, 113, 173, 179, 197, 232, 248, 285, 301, 326